Inorganic Geochemistry
Applications to
Petroleum Geology

Inorganic Geochemistry
Applications to
Petroleum Geology

Dominic Emery & Andrew Robinson
BP Exploration, 4/5 Long Walk, Stockley Park
Uxbridge, Middlesex UB11 1BP, UK

WITH CONTRIBUTIONS FROM
Andrew Aplin *Newcastle University*
& Craig Smalley *BP Group Engineering & Research*

OXFORD
BLACKWELL SCIENTIFIC PUBLICATIONS
LONDON EDINBURGH BOSTON
MELBOURNE PARIS BERLIN VIENNA

This book is dedicated to our families:
Helen, Edward, David, Flo, Sandra, Bob, Avril and Dan

© 1993 by
Blackwell Scientific Publications
Editorial Offices:
Osney Mead, Oxford OX2 0EL
25 John Street, London WC1N 2BL
23 Ainslie Place, Edinburgh EH3 6AJ
238 Main Street, Cambridge
    Massachusetts 02142, USA
54 University Street, Carlton
    Victoria 3053, Australia

Other Editorial Offices:
Librairie Arnette SA
2, rue Casimir-Delavigne
75006 Paris
France

Blackwell Wissenschafts-Verlag GmbH
Meinekestrasse 4
D-1000 Berlin 15
Germany

Blackwell MZV
Feldgasse 13
A-1238 Wien
Austria

First published 1993

Set by Excel Typesetters Company, Hong Kong
Printed and bound in Great Britain
at the University Press, Cambridge

DISTRIBUTORS

Marston Book Services Ltd
PO Box 87
Oxford OX2 0DT
(*Orders*: Tel: 0865 791155
            Fax: 0865 791927
            Telex: 837515)

USA
Blackwell Scientific Publications, Inc.
238 Main Street
Cambridge, MA 02142
(*Orders*: Tel: 800 759-6102
            617 876-7000)

Canada
Oxford University Press
70 Wynford Drive
Don Mills
Ontario M3C 1J9
(*Orders*: Tel: 416 441-2941)

Australia
Blackwell Scientific Publications Pty Ltd
54 University Street
Carlton, Victoria 3053
(*Orders*: Tel: 03 347-5552)

A catalogue record for this title
is available from the British Library

ISBN 0-632-03433-5

Library of Congress
Cataloging-in-Publication Data

Emery, Dominic.
    Inorganic geochemistry:
applications to petroleum geology/
Dominic Emery & Andrew Robinson,
with contributions from Andrew Aplin
and Craig Smalley.
        p.   cm.
    Includes bibliographical references
and index.
    ISBN 0-632-03433-5
    1. Geochemistry. 2. Petroleum–Geology.
I. Robinson, Andrew. II. Title.
QE515.E44   1993
553.2'8 – dc20

Contents

Preface

During the spring of 1990 we were approached by a colleague, a geologist with a problem. He had read a paper in the latest issue of a major journal that presented fluid inclusion data from an area in which he was exploring. The paper had concluded that a Tertiary 'thermal event' had been responsible for generating petroleum; if this were the case, our friend assured us, it would have far reaching implications for the prospectivity of the basin. Could we read the paper and give an opinion as to the validity of the conclusion? We read the paper. Our view was that the data were probably fine but that they had been poorly and over-optimistically interpreted so that the evidence for a thermal event was very shaky indeed. Our friend was duly grateful.

It would be tempting to write that at this point we decided that we could make a lot of money by writing a book that would explain fluid inclusion and other geochemical techniques to petroleum geologists. In fact, we had been planning the project for a couple of months already but the incident did serve to confirm our view that there was room for a book that would enable non-specialists to make up their own minds about the large number of papers now appearing in print every month which include some facet of inorganic geochemistry as a major constituent. This book is the result. Its purpose is to bring together the most important inorganic geochemical methods in a single volume, to explain their potential and limitations in a form that is accessible to the non-specialist, and to demonstrate their application to a wide range of problems in petroleum geology, from exploration, through appraisal and development, to production. The book is therefore intended for geologists, geophysicists and production engineers in oil companies who wish to broaden their knowledge of the geochemical methods available for solving problems with which they are routinely faced. We hope that it will also be of interest to final year undergraduates, postgraduates with an interest in the inorganic geochemistry of sedimentary rocks and waters, and to those attending petroleum geology and related MSc courses.

Acknowledgements

Andrew Aplin (Newcastle University) contributed the majority of Chapter 4 and Craig Smalley (BP Research) wrote the sections on Rb–Sr, Sm–Nd and U–Th–Pb in Chapter 5. The book is far better for these contributions. Numerous friends and colleagues have also contributed to the book by providing prize specimen photomicrographs, and by reviewing the text at the many and varied stages of its development. Thanks in no particular order to Chris Rundle (British Geological Survey), Jim Marshall (Liverpool University), Ian Hutcheon (University of Calgary), Christine Knox, Jon Gluyas, Jonathan Henton, Ed Warren, Tim Primmer, Norman Oxtoby, Andy Brayshaw, Shona Grant, Andy Leonard, Mike Bowman, Joyce Neilson, Steve Rainey, Max Coleman and Keith Mills.

This book would have been particularly difficult to write had we not had the support of BP Exploration and BP Research which we thank for permission to publish. Organizations are the sum of individuals however and we would like to thank in particular Ian Vann and Jon Bellamy for their generous support. We are grateful to many other people throughout BP for allowing us to use case study material from the fields or areas in which they work. We would also like to acknowledge the co-operation of BP's partners in allowing us to publish information on many of the fields and licences mentioned throughout the text.

Dominic Emery & Andrew Robinson
BP Exploration, London

Chapter 1 Introduction

1.1 Background

Petroleum is not as easy to find as it used to be. Most of the accessible sedimentary basins in the world have been explored and a large proportion of the more obvious petroleum targets have been drilled. The more risky and costly exploration becomes, the more important it is to develop new discoveries as efficiently as possible and to extract a greater proportion of the petroleum in place from existing fields. The last 15 years have seen this imperative give birth to some important new topics within the earth sciences, most of which cross the boundaries between the traditional divisions of geology as taught to undergraduate students. For example, sequence stratigraphy − now a principal tool of the exploration geologist − has emerged from classical stratigraphy and seismic interpretation; and reservoir description has been born of a long overdue relationship between reservoir engineering, sophisticated geophysics, stratigraphy and sedimentology.

Inorganic geochemistry is one more relatively new weapon in the armoury of the petroleum geologist. In its broadest sense, the subject includes the study of all of the chemical constituents of rocks and subsurface fluids, excluding only organic components based on carbon. This book makes no attempt to cover this potentially vast field in its entirety. It may appear at first glance to contain rather a mixed bag of subject matter but the choice of what to include has not been arbitrary. The themes of the book are the characterization of fluids in sedimentary basins, understanding their interaction with each other and with rocks and the application of this information to finding, developing and producing oil and gas. This might include dating quartz cement growth in a sandstone in an attempt to predict porosity distribution, or determining the extent of seawater breakthrough into a reservoir during production. There is a considerable degree of overlap with the field of sediment diagenesis but this book covers important topics with a geochemical component which would never find their way into a diagenesis book, such as strontium isotope stratigraphy, correlation and production chemistry.

The subject of inorganic geochemistry is not of course in itself new, but its application to sedimentary basins has lagged behind igneous and metamorphic geochemistry. There are a number of reasons for this. One may be the relatively unglamorous nature of a sandstone when set beside a peridotite. Perhaps more important is the fact that sedimentary, particularly clastic rocks are inherently difficult to analyse in any meaningful way because they contain constituents with many different origins which for most purposes must be analysed separately. The study of sediment geochemistry has however been stimulated in recent years by the increased availability of core from deeply buried sediments taken by petroleum exploration companies and by their readiness to finance research into controls on porosity and permeability (we are ourselves beneficiaries of this largess). The last few years have also seen the development of new methods – the use of lasers, for example – which have helped to overcome sampling problems.

1.2 How is inorganic geochemistry applied to petroleum geology?

It is relatively easy to watch wave ripples form or to snorkel over a growing reef and study the deposition of sediments. It is a lot harder to study how sediments are modified during burial in a basin; processes cannot in the main be directly observed because they occur at substantial depths and in many cases may well have ended long ago. Only the products of fluid–rock interaction, diagenetic minerals and present day formation waters, are left to act as records. Inorganic geochemistry provides a means of interpreting these records. The main types of information it provides are as follows.

1 *Timing*. Relative and absolute ages of mineral growth and dissolution, and of the presence and migration of fluids (both water and petroleum).

2 *Temperature*. Temperatures at which minerals grew or dissolved and at which particular fluids were present in a rock's pores. Temperature and timing are correlated for a sediment in a subsiding sedimentary basin and can be related by modelling burial and thermal history.

3 *Chemical composition*. The bulk and isotope chemistries of minerals and water contain information about the history of fluids, especially their interaction with rocks.

It is one matter to obtain this information from minerals and fluids but another matter entirely to interpret the processes involved. Given a sample of quartz cemented sandstone, the geochemist will quite probably be able to find out what temperature it grew at and say something about the origin of the water involved, but will not be able to say *why* the rock is cemented or *what caused* the quartz to precipitate. This is the level of current understanding of most diagenetic phenomena: we can characterize them but not explain them. The trick to using inorganic geochemistry to solve problems in the field of petroleum geology is to accept these limitations and do the best possible with the information that can be obtained. In many cases, this is quite a lot. At the very least, geochemical methods provide a means of integrating diagenetic phenomena into the temporal framework that forms the basis of basin analysis. For example, the timing of mineral cementation and dissolution, and consequent changes in porosity and permeability can be related to phases in the development of a basin and to oil migration. The models that emerge for porosity and permeability prediction invariably involve a large empirical component, but they represent an improvement over empiricism alone.

1.3 What is in this book

This book has two parts. Chapters 2–5 describe the groups of techniques that we have found to be most useful in petroleum geology and Chapters 6–9 describe case histories – mostly from our own work or that of our colleagues – grouped according to the nature of the problem that inorganic geochemistry helped (or in some cases, failed) to solve.

The chapters on geochemical techniques emphasize applications to sedimentary rocks and the fluids in sedimentary basins. Particular attention is paid to precision and accuracy and to the questions of what information can be plausibly obtained and under what circumstances: what *do* data mean and what do they *not* mean? Particular difficulties and pitfalls are illustrated by the use of examples which are made up of complete sets of real data. The case histories that make up the second half of the book cover a wide range of applications but inevitably reflect our interests and biases. They include clastic and carbonate rocks from many parts of the world but a number are from the North Sea (Fig. 1.1). Many of

Fig. 1.1 (*Opposite page.*) Location of North Sea oil and gas fields mentioned in this book.

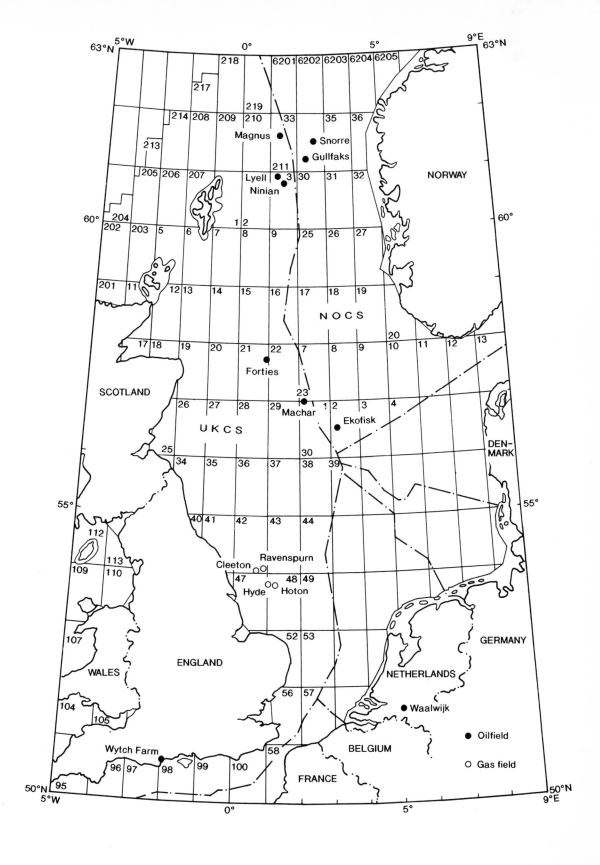

the case studies are from rift basins but there is usually no reason why analogous problems in other geological settings should not be tackled in similar ways. Coarse grained sediments figure more than mudstones because they are of importance as reservoir rocks and because, in the main, they are easier to study and more information can be obtained from them.

All of the case histories are taken from real life and we have not tinkered with the data or modified the conclusions. Some were successful in solving the problem they set out to solve, rather more solved it partially, some solved a completely different problem and others proved scientifically fascinating but of no use whatsoever. We feel that it is far more instructive, and interesting, to present relative failures alongside relative successes. Negative results rarely get published but are often as instructive as positive outcomes. In most cases, careful and circumspect science can take a problem only so far. When that point is reached − and we try to make it clear when it is − we do not feel embarrassed to speculate. Petroleum geology, particularly exploration, involves making the best of incomplete understanding and is an essentially optimistic enterprise.

1.4 Overview

Chapter 2 (Textural and Mineralogical Analysis) describes five basic and three more advanced techniques for mineral identification and quantification. Inclusion of this topic in a book on geochemistry stretches the definition of the term, but analysis of a sample's mineralogy is vital if any sense is to be made of geochemical information obtained from it. The basic techniques include thin-section petrography, cathodoluminescence microscopy, fluorescence microscopy, scanning electron microscopy and X-ray diffraction. Examples are given throughout the text of the application of the techniques to mineral identification, differentiation of detrital grains from diagenetic cements, mineral quantification and the construction of paragenetic histories for reservoir rocks. The three more advanced techniques covered are transmission electron microscopy, thermogravimetric-evolved water analysis and pore image analysis, which are playing an increasing role in chemical analysis, clay mineral quantification and porosity analysis respectively. The chapter stresses the application of techniques at the expense of a detailed description of apparatus and sample preparation (for which see Tucker, 1988). Emphasis is also placed on the value of combining the techniques with standard petrophysical and engineering methods for describing the reservoir quality of rock samples.

Chapter 3 (Fluid Inclusions) explains how the study of these minute samples of fluid trapped during mineral growth or fracture healing can provide information about the temperatures of diagenetic reactions, and about the composition of fluids passing through sedimentary rocks and at what temperatures they did so. The study of fluid inclusions in igneous rocks and particularly, metallic mineral deposits, goes back more than a century but only over the last 10 years has much effort been put into the study of inclusions in sediments. Much of the earlier work, and some current studies, suffer from overinterpretation and this chapter attempts to redress the balance by underlining some limitations of the technique. Particular attention is paid to the question of leakage because if fluid inclusions in diagenetic minerals routinely leak − and a case can be made for thinking that they do − then they will be of limited value for the study of sedimentary rocks.

The applications of stable isotopes, the subject of Chapter 4, to petroleum geological problems range from stratigraphic analysis to better understanding and predicting reservoir quality and reservoir fluid type. Stable isotopes are used principally as natural tracers for subsurface reactions and − in the case of oxygen isotopes − can provide information about reaction temperature. The chapter first describes the principles and nomenclature of stable isotope systems (a necessary evil but surprisingly painless), the basic analytical techniques, and the uncertainties associated with isotopic measurements. This is followed by a description of the most important stable isotope systems in the context of the fluids and minerals in which they are found: oxygen in water, silicate, sulphate and carbonate minerals; hydrogen in water and clays; carbon in carbonate minerals, CO_2 and CH_4; and sulphur in sulphate and sulphide minerals and H_2S.

Chapter 5 explains the use of radiogenic isotopes to unravel geological history. It begins with a refresher in the simple physics that forms the basis of radiometric dating, which serves to stress the common features of all of the dating methods. Four isotope systems are described: K−Ar, Rb−Sr, Sm−Nd and U−Th−Pb. K−Ar dating is almost routinely used to date K-bearing mineral cements, principally illite. The chapter explains why this must be done with great care even when the sample material is particularly suitable. Ar–Ar dating – a clever technique that also relies on the K–Ar decay series – is also covered. The other isotope systems can be used for dating sedimentary material only under rather restricted circumstances. However, radiogenic isotopes also have value as natural tracers. Strontium isotope ratios in particular can be used to trace the chemical evolution of natural waters. One consequence of this property is their use as a tool for dating marine carbonate and phosphate (strontium isotope stratigraphy).

Chapter 6 (Porosity and Permeability Prediction) is the first of four chapters that group case histories to illustrate a particular application of inorganic geochemistry. It is also the bulkiest because it has been up to now the most important application if the number of published studies is anything to go by. The introductory section explains how inorganic geochemistry contributes to porosity and permeability prediction by reducing risk (neither can be predicted using inorganic geochemistry alone). There follow six case histories which range from exploration in frontier areas where the amount of information available is very limited (Flemish Pass, Grand Banks) through progressively more mature exploration areas (offshore Louisiana and the Brent Province) to porosity prediction in appraisal, development and production in three North Sea fields (Magnus, Machar and Forties).

Chapter 7 (Fluid Migration) considers inorganic geochemical evidence for phases of water and petroleum migration. The value of knowing about petroleum migration need hardly be underlined but migration of waters is also of interest as these affect reservoir quality by interacting with rocks and can also alter oil through the processes of biodegradation and water washing. The four case histories cover the use of fluid inclusions for sorting out migration history in poorly and better known areas (Aquitaine and Weald Basins), the prediction of the occurrence of a rather odd diagenetic reservoir rock in the Central North Sea and the history of filling of a gas-condensate discovery (Waalwijk, Netherlands).

The next chapter (Correlation) describes how stable and radiogenic isotope systems can be used to correlate stratigraphic units, chiefly on a broad exploration scale, but also on the smaller scale of individual reservoir units within discrete oilfields. Four case studies are described. The first, from the Norwegian North Sea, shows how strontium isotope stratigraphy and radiometric dating can be used to refine the stratigraphic correlation of Tertiary clastic sediments, and assist in erecting a sequence stratigraphy. The second also uses isotope stratigraphy, but in this case oxygen isotopes are applied to provide a very high resolution stratigraphy of a few tens of thousands of years for Plio-Pleistocene sandstone reservoir targets in the Gulf of Mexico. The third and fourth case studies are both from oilfields in the Norwegian sector of the North Sea, the Ekofisk Chalk Field, and the clastic Gullfaks Field. In Ekofisk, strontium isotopes are used to correlate reservoir zones using data obtained from the rock matrix and from the formation waters. In Gullfaks, samarium and neodymium isotopic methods are used to predict the distribution of reservoir sand bodies by identifying changes in sand provenance.

Chapter 9 (Petroleum Recovery) outlines the basic principles of secondary and enhanced oil recovery, and the problems of corrosive fluid production (H_2S and CO_2). The first two case studies, from the Forties Field, UK North Sea, and from the Wytch Farm Field, onshore UK, demonstrate the value of oxygen and hydrogen isotopes as tracers for seawater breakthrough during oil production. The application of sulphur isotopes for understanding sour gas production in Wytch Farm is also outlined. The last two case studies are from the Cretaceous heavy oil sands of Alberta and Saskatchewan in western Canada. These demonstrate the importance of quantitative petrography linked to petrophysics for describing the effects of thermal recovery processes (steam injection and fireflooding) on shallow reservoir sandstones. The case study on steam injection also shows how carbon isotopes can be used to identify the source of CO_2 produced during thermal recovery.

1.5 What is not in this book

This is a book about the application of fluid–fluid and fluid–rock interaction to petroleum geology and as such contains only those methods which we have found to be useful in this field. We avoid describing methods of analysing whole rocks partly because we have found bulk chemistry to be of rather limited value in this respect and also because it is a subject in its own right that has been better covered than we could manage in other books. We have also avoided chemical modelling of water–rock interaction and modelling of fluid flow in sedimentary basins. These are used together by some (not by us) to predict porosity; we believe them to have explanatory power and to be useful for understanding diagenetic processes but feel that their predictive power is limited (see Section 6.1). Although geohistory analysis − especially burial and thermal modelling − is repeatedly referred to in the text, it also constitutes a separate subject and we have not as a consequence dealt with fission track analysis which is principally a means of calibrating burial history.

Chapter 2 Textural and Mineralogical Analysis

2.1 Introduction

Mineralogical and textural analysis of samples is the essential first step in any inorganic geochemical programme and provides the critical link between the quality of a reservoir – chiefly its porosity and permeability – and more advanced geochemical techniques such as fluid inclusion analysis (Fig. 2.1). Mineralogical and textural analysis can provide several types of information:

1 mineral identification;
2 differentiation of detrital from diagenetic phases;
3 quantitative analysis of mineral abundance;
4 mineral paragenesis (the sequence of mineral growth and dissolution);
5 mineral chemistry;
6 description and quantification of porosity; and
7 identification of the main factors influencing porosity and permeability.

The most appropriate technique for providing each of these is shown in Table 2.1. Note that, to be most effective, certain techniques need to be applied in conjunction with others. Clay mineral X-ray diffraction is most effective as a tool for quantifying clay mineral content in rock samples when applied with thermogravimetric-evolved water analysis. Similarly, cathodoluminescence microscopy of limestone samples is most effective when applied with observations made using transmitted light.

The objectives of this chapter are to introduce mineralogical and textural analytical techniques, and their applications to understanding the post-depositional evolution of siliciclastic and carbonate rocks in petroleum provinces. We will concentrate particularly on the *application* of the variety of techniques described here, rather than on details of instrumentation and sample preparation. References covering these aspects in more depth are given in the text.

2.2 Transmitted light microscopy

2.2.1 Introduction

Transmitted light microscopy is a basic tool of the trade for the description of sedimentary rocks and

7

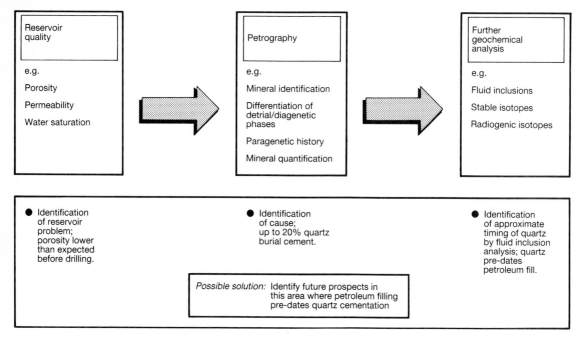

Fig. 2.1 The link between reservoir quality, petrography and further geochemical analysis.

there are many texts which cover all aspects of sedimentary petrology. Folk (1974), Pettijohn (1975) and Tucker (1981) are good general petrographic texts, whereas Moore (1989) and Pettijohn *et al.* (1973) concentrate on carbonate and siliciclastic petrography respectively. The *Atlas of Sedimentary Rocks* by Adams *et al.* (1984), and the AAPG colour guides to sandstones (Scholle, 1979) and carbonates (Scholle, 1978) contain superb colour thin-section photomicrographs and are well worth referring to.

The objective of this section is to explain how transmitted light petrography can be used to describe the diagenesis of sedimentary rocks (rather than the depositional fabric). Petrographic description can provide five main pieces of information:
1 mineral identification;
2 differentiation of detrital from diagenetic phases;
3 quantitative analysis of mineral abundance;
4 mineral paragenesis; and
5 description of porosity.
The problems associated with siliciclastic petrography are quite different from those associated with carbonates. Siliciclastic rocks tend to be polymineralic, with several different diagenetic phases such as clay minerals, quartz and carbonates and it is usually easy to distinguish these from the detrital

components of the sediment. In contrast, carbonate rocks are composed of fewer minerals which tend to be relatively unstable during diagenesis so that much of the primary depositional fabric may be obliterated. Recognition of subtle changes in carbonate fabric and chemistry is therefore essential to differentiate primary constituents from diagenetic phases, and to allow different diagenetic carbonates to be recognized.

2.2.2 Sample preparation

The thin-section is the basic sample requirement for microscope petrography. It consists of a rock wafer 30 μm thick, mounted on a glass slide. The thickness of the rock wafer is standard to ensure uniformity of birefringence colours (which are determined in part by sample thickness). Similarly, to ensure comparability of samples, the rock wafer is usually mounted on the slide in a medium of uniform refractive index. Basic thin-section preparation is described in more detail by Miller (1988).

Before the section is cut, samples are commonly impregnated with a dyed resin. This fills and colours porespace in the sample, allowing easier identification of porosity types in the sample, and also

Table 2.1 Summary matrix of technique versus application.

Technique		Application						
		Mineral identification	Differentiation of diagenetic[1] from detrital phases	Mineral quantification	Mineral[2] paragenesis	Mineral chemistry	High resolution mineralogical and textural analysis	Porosity description
Transmitted light microscopy		●	●	●[3]	●	●[4]		●[3]
CL	Cold	C	●		●[5]	●[6]		
CL	Hot	S	●		●			
CL	SEM	S	●		●			
UVF					●[5]			
SEM	SE	●	●		●	●[7]	●	●
SEM	BSEM	●		●[8]		●[9]		●[10]
TEM			●		●	●	●	
XRD	Whole rock	●[11]						
XRD	Fine fraction	●[12]		●				
TG-EWA				●				
PIA								●[10]

Key:
1 Includes sample screening
2 Includes cement stratigraphy and cement fabric analysis
3 With point-counting apparatus
4 Qualitative only for stained carbonates
5 Especially cement stratigraphy and cement fabric analysis in carbonates
6 Highly qualitative for Mn and Fe in carbonates
7 Qualitative with energy dispersive X-ray analysis
8 Software is available for BSEM quantification of very simple mineral mixtures
9 Semi-quantitative with energy dispersive X-ray analysis
10 BSEM-PIA provides quantitative porosity information only
11 Very rapid whole rock quantification
12 Especially for clay mineral identification

C, Chiefly carbonates
S, Chiefly siliciclastics
CL, cathodoluminescence
UVF, ultraviolet fluorescence microscopy
SEM, scanning electron microscopy
SE, emission mode
BSEM, backscatter mode
TEM, transmission electron microscopy
XRD, X-ray diffraction
TG-EWA, thermogravimetry-evolved water analysis
PIA, pore image analysis

prevents poorly consolidated rocks falling apart. Thin-sections are usually stained. Two stains are commonly used: a mixed stain for carbonates allowing the differentiation of ferroan and non-ferroan calcites and dolomites (Dickson, 1966); and a stain for feldspars allowing the differentiation of potassium feldspars, plagioclase and quartz (Houghton, 1980). Table 2.2 summarizes the effects of the staining procedures on carbonate and silicate mineralogies. Note that it is essential to apply the feldspar stain before the carbonate stain, otherwise the latter will be removed!

2.2.3 Mineral identification and differentiation of detrital grains from diagenetic cements

The main theme of this section is the differentiation of diagenetic phases from depositional grains, which can be a source of considerable ambiguity in thin-section description. The following provides a guide to the recognition of the most common diagenetic phases in siliciclastic and carbonate rocks, and outlines the pitfalls and problems of distinguishing detrital grains and matrix from diagenetic phases. A detailed description of sedimentary mineral iden-

Table 2.2 Typical stain colours for carbonates and feldspars.

Mineralogy	Stain colour
Non-ferroan calcite*	Very pale pink to red
Ferroan calcite*	Ranging from mauve to purple to royal blue with increasing Fe content
Dolomite*	No colour
Ferroan dolomite*	Ranging from pale to deep turquoise with increasing Fe content
Plagioclase[†]	Pink, intensity is proportional to Ca content
Pure Na-Albite[†]	No colour with rhodizonate stain
Alkali feldspars[†]	Yellow

* Carbonate stain colours using mixed stain of potassium ferricyanide and Alizarin red-S (Dickson, 1966).
[†] Feldspar stain colours using stain of sodium cobaltinitrate, followed by potassium rhodizonate (Houghton, 1980).

tification is beyond the scope of this book. The reader is referred to texts by Kerr (1959) and Deer *et al.* (1977). Table 2.3 summarizes the optical properties of common minerals in sedimentary rocks and their occurrence. If the nature of a mineral is in doubt from microscopy alone, microbeam techniques can be applied (see subsequent sections in this chapter).

Quartz cement
Quartz cement is the most common diagenetic silicate mineral in sandstones (McBride, 1989). It occurs chiefly in two forms: as microcrystalline cement and as syntaxial overgrowths on detrital quartz grains (Table 2.3). Microcrystalline quartz, also known as microquartz, chert or chalcedony, is relatively easy to differentiate from the depositional fabric of the rock (Fig. 2.2a). Syntaxial quartz overgrowths are the most common form of quartz cement but are often difficult to distinguish from detrital grains because of the optical continuity of quartz across the grain–cement boundary. Quartz overgrowths can be clearly differentiated using optical microscopy only if the grain–cement boundary is visible. Figure 2.2b shows a sandstone largely composed of quartz in which grain margins cannot be identified even though the euhedral crystal faces show that quartz cement is present. If quartz cementation had continued to fill porespaces completely, only sutured contacts between areas of quartz would be visible. One possible interpretation would be that the sutured contacts could represent compromise boundaries between growing quartz crystals. Alternatively, we could infer that there is no quartz cement, and that the sutured boundaries are pressure-solution contacts along which quartz cement has been *dissolved* rather than precipitated (Houseknecht, 1988). Distinction between these two quite different interpretations must involve recourse to further techniques such as cathodoluminescence microscopy (Section 2.3). Fortunately, many sandstones have a thin coating of depositional or diagenetic material, commonly haematite, between the grains and cement which allows grain outlines to be identified (Fig. 2.2c). Fluid inclusions are also often concentrated along grain–cement boundaries (see Fig. 3.3f).

Feldspars
Feldspar cements are relatively common in arkosic sandstones but usually form only a minor or trace component of the total rock volume (Waugh, 1978). Like quartz, feldspar cements chiefly occur as syntaxial overgrowths on detrital grains and so present the same problems of identification. However, there is a further complication in feldspar petrography: certain types of feldspar, whether grains or cement, cannot be distinguished from each other or, in some cases, from quartz unless the section is stained (see Section 2.2.2).

Three types of feldspar are usually distinguished in sedimentary petrography: plagioclase, orthoclase and microcline (Table 2.3). Plagioclase is easily identified by common multiple or lamellar twinning which gives the mineral its characteristic striped appearance under crossed polars (Fig. 2.2d; Deer *et al.*, 1977). Microcline, the low-temperature form of potassium feldspar, is also easy to identify owing to its cross-hatched 'tartan' twinning visible under crossed polars. Other feldspars are frequently untwinned. A common feature of feldspars, which also assists in their identification, is their diagenetic instability relative to quartz (Burley *et al.*, 1985). Feldspars dissolve more readily, leaving ragged, etched grain remnants in oversized secondary pores, or skeletal grains with microporosity (Schmidt & McDonald, 1979). As well as being leached, feldspars are commonly altered to fine-grained clay minerals.

Fig. 2.2 Silicate cements in sandstones. (a) Microcrystalline quartz cement between quartz and feldspar grains, Jurassic sandstone, Central North Sea. Plane polarized light. Courtesy A. Hogg. (b) Quartz cemented sandstone, Northern North Sea. The boundaries between quartz grains and cements are invisible. Plane polarized light. Courtesy A. Hogg. (c) Quartz cemented sandstone, Rotliegend Group, Southern North Sea. Quartz grains and cements can be distinguished by the presence of a dust rim. Plane polarized light. Courtesy A. Hogg. (d) Syntaxial overgrowth of plagioclase feldspar on a detrital grain. The grain–cement boundary is marked by a rim of haematite. Crossed polars. Width of photograph is 400 μm.

Clay minerals

Clay minerals are best observed in the scanning electron microscope because of their small size (usually a few to a few tens of micrometres). Nevertheless, optical microscopy can still provide useful general information on certain clay mineral types. There are two main problems to be aware of in clay mineral identification from thin-section petrography: the identification of clay mineral type and differentiation of clay mineral cements from matrix clay.

The common diagenetic clay minerals are summarized in Table 2.3. Kaolinite forms characteristic booklets (Fig. 2.3a, b) or vermiform aggregates with low birefringence (Fig. 2.3c). Chlorite, a ferromagnesian clay mineral, is also relatively easy to identify on account of its green hue in transmitted light and anomalous steely blue birefringence. It usually forms characteristic pseudohexagonal platelets or a complex meshwork. Illite, a potassium-bearing clay, commonly forms plates parallel to grain surfaces, or micrometre-thick fibres extend-

Table 2.3 Common minerals in sedimentary rocks and their optical properties. Compiled from Kerr (1959) and Tucker (1988).

Group/mineral	Crystal system	Colour	Cleavage	Relief	Birefringence	Other features	Cement	Form and occurrence
Quartz	Trigonal	Colourless	None	Low+	Grey		Overgrowth cement common	As detrital grains and cement/replacive phase
Cherts		Colourless	None	Low+	Grey		Chalcedony and microquartz	Chalcedony and microquartz are diagenetic unless as detrital grain
Feldspars								
Microcline	Triclinic	Colourless	Present	Low−	Grey	Crosshatch twins	Overgrowth cements	Present as detrital minerals, but often altered or leached. Generally minor diagenetic phase, except in some arkoses
Orthoclase	Monoclinic	Colourless	Present	Low−	Grey	Simple twins		
Plagioclase	Triclinic	Colourless	Present	Low−	Grey	Multiple twins		
Micas								
Muscovite	Monoclinic	Colourless	Prominent, planar	Mod+	Bright colours	Parallel extinction		Common detrital mineral
Biotite	Monoclinic	Brown−green	Prominent, planar	Mod+	Bright colours, masked by colour	Parallel extinction, pleochroic		Common detrital mineral
Clay minerals								
Chlorite	Monoclinic	Green/blue green	Planar	Mod+	Grey/blue		Common as cements	Present as detrital minerals, as alteration products of silicates and as cements
Kaolinite	Triclinic	Colourless	Planar	Low+	Grey			
Illite	Monoclinic	Colourless	Planar	Low+	Grey−bright	Fine grained		
Smectite	Monoclinic	Colourless	Planar	Low+	Grey			
Glauconite	Monoclinic	Green	Planar	Mod, masked by colour	Grey, masked by colour	Commonly replace pellets		Characteristic of low sedimentation rates, may infill foram tests etc.
Zeolites		Most colourless		Low, most−	Commonly grey		Common as cements in volcanics	Associated with volcanogenic sediments

	Crystal system	Colour	Cleavage	Relief	Birefringence	Distinguishing feature	Cement	Occurrence
Carbonates								
Aragonite	Orthorhombic	Colourless	Rectilinear	Mod-high	High			Commonly acicular
Calcite	Trigonal	Colourless	Rhombic	Low-high	Very high colours	Distinguished by staining (Table 2.2)	Carbonate cements display many cement morphologies (see Table 2.4)	Present as detrital grains, cements and replacive phases in carbonates and siliciclastics
Dolomite	Trigonal	Colourless	Rhombic	Low-high	Very high colours			
Siderite/ankerite	Trigonal	Colourless	Rhombic	Low-high	Very high colours			Common replacive phase in ironstones
Evaporites								
Gypsum	Monoclinic	Colourless	Planar	Low	Grey			Commonly crystalline, replacive after evaporites
Anhydrite	Orthorhombic	Colourless	Rectilinear	Mod	Bright colours		Common burial cement	
Celestite	Orthorhombic	Colourless	Planar	Low-mod	Grey		Burial cement	Commonly partially replacive
Barite	Orthorhombic	Colourless	Planar	Low-mod	Grey		Burial cement	Burial cement in sandstones, may be associated with sulphide mineralization
Halite	Cubic	Colourless	Rectilinear	Low	Isotropic	Only present if section prepared in oil	Burial cement	
Iron minerals								
Pyrite	Cubic	Opaque				Distinguished in reflected light — Yellow		Common diagenetic mineral
Magnetite	Cubic	Opaque				Grey-black		Common diagenetic mineral in siliciclastic red-beds
Haematite	Cubic	Opaque, brown tinge				Red-grey		
Chamosite/berthierine	Monoclinic	Green		Mod	Grey masked by colour			Occurs in ooids in ironstones and as partial pore-fills
Collophane	Non-crystalline	Browns		Mod	Isotropic			Replacive textures, commonly in carbonates
Bitumens	Non-crystalline	Opaque				Distinguished in fluorescence		Occurs in porespaces and in fluid inclusions

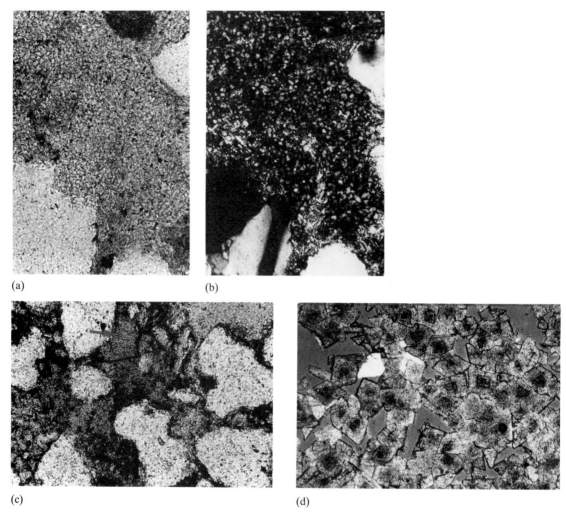

(a)　　　　　　　　　　　(b)

(c)　　　　　　　　　　　(d)

Fig. 2.3 Silicate cements and dolomitized limestone. (a) Authigenic kaolinite filling porespace between quartz grains, Brent Group, Northern North Sea. Plane polarized light. Width of photograph is 500 μm. (b) Field of view as above, crossed polars, showing individual kaolinite booklets. (c) Oil stained vermiform kaolinite (arrowed), Brent Group, Northern North Sea. Plane polarized light. Width of photograph is 700 μm. (d) Euhedral rhombs of replacive, sucrosic dolomite, Jurassic Arab Formation, Saudi Arabia. Plane polarized light. Width of photograph is 500 μm. Courtesy J. Dravis.

ing away from the grain surface, and can also be identified from its relatively high birefringence. Pure smectite is less common as a cement in sandstones, but interstratified illite/smectites may form fibrous cements. These can be distinguished from pure illites using X-ray diffraction (Section 2.7). Glauconite is relatively easy to identify owing to its green hue in thin-section and its common occurrence as pellets in marine sandstones. Differentiating detrital from diagenetic clay may be more difficult. Frequently, what is described as a clay matrix consists of a dense brown mess which may contain any combination of original detrital material, recrystallized detrital clay, or true diagenetic precipitate. Under such circumstances, clay mineral identification is better left to scanning electron microscopy (Section 2.5), transmission electron microscopy (Section 2.6) and X-ray diffraction.

Carbonates

Identification of carbonate cements in siliciclastic rocks is relatively straightforward and may be as-

Table 2.4 Descriptive terms for carbonate cement morphologies. Modified from Harwood (1988).

Cement terminology	Description and characteristic environment of precipitation (where appropriate)
Needle	Thin ($<10\,\mu$m) cements of single or *en-echelon* crystals
Pendant or microstalactitic	Cement forms in droplets beneath grains in vadose zone (zone where porosity is partially air- and partially water-filled)
Meniscus	Cement forms at or near grain–grain contacts, also characterizes vadose zone
Acicular	Thin, straight form (aspect ratios of 20–40, $\approx 10\,\mu$m width). Characterizes marine phreatic (pores wholly water-filled) environment
Peloidal	Dark, microcrystalline coating to grains and pores. Characterizes marine phreatic environment
'Micritic' or microcrystalline	Microcrystalline cement which may coat grains and form bridges between grains. Characterizes marine phreatic environment
Columnar	Broad cements ($\approx 20\,\mu$m+) commonly longer than broad
Circumgranular isopachous acicular	Equal thickness of acicular cements surrounding grains. Characterizes marine phreatic environments
Equant	Equidimensional crystals (commonly $\approx 100\,\mu$m+). Characterizes freshwater or burial phreatic environments
Circumgranular equant	Equidimensional cements surrounding grains. Characterizes meteoric phreatic environment
Overgrowth	Cement is in optical continuity with substrate
Sparry	Coarse ($\approx 300\,\mu$m+), commonly equidimensional crystals
Poikilotopic	Coarse cement crystals enclosing grains and pre-existing cement phases
Baroque (or 'saddle')	Coarse cement displaying undulose extinction. Characterizes burial environments

sisted by staining (Table 2.2; Dickson, 1966). Where samples are unstained, differentiation of carbonate cements from one another can be difficult and relies chiefly on cement fabric criteria which are much more subjective (Table 2.4; Harwood, 1988). Replacive and cementing dolomite commonly form euhedral rhombic crystals (Fig. 2.3d) or coarse crystals of baroque or saddle dolomite, showing undulose extinction. Calcite may display a wide variety of morphologies, from acicular (or less correctly, fibrous) through to poikilotopic, where coarse crystals of uniform extinction enclose detrital grains and earlier cement generations.

In carbonate rocks, the major problems are in distinguishing unaltered carbonate grains from those which have been diagenetically modified, and in differentiating genuine cements from neomorphic carbonate. Again, staining the thin-section helps. Under normal marine conditions, all organisms secreting a calcite test will precipitate non-ferron calcite with a low manganese content. Accordingly, the presence of any dolomite or ferroan calcite immediately indicates that the skeletal particle has undergone diagenetic alteration. However, the presence of non-ferroan calcite in itself is insufficient evidence to guarantee that no diagenetic alteration has taken place. Further screening for the presence of manganese in the calcite by cathodoluminescence (Section 2.3) should be carried out, as well as detailed observation of skeletal fabrics by scanning electron microscopy (Section 2.5).

Carbonate cements in carbonate rocks display

a wide variety of fabrics, many independent of mineralogy. It is often hard to distinguish pore-filling cement from neomorphic carbonate (usually calcite). Neomorphism refers to the transformation between one mineral and itself or a polymorph. In the case of calcite, this would be the transformation of aragonite to calcite and the transformation of calcite to calcite, believed to proceed via a thin solution film mechanism, dissolving a carbonate in front and precipitating calcite behind. For further discussion of carbonate neomorphism, the reader is referred to Tucker (1981) and Bathurst (1975). This process is quite different from direct precipitation of cement and, importantly, the isotopic and fluid inclusion compositions of neomorphic calcite are less likely to give any easily interpretable information about diagenetic environment. It is therefore necessary to distinguish neomorphic and cementing calcite prior to further geochemical analysis. Neomorphic calcites usually show irregular, curved or diffuse intercrystalline boundaries, irregular crystal size distributions and the presence of skeletal grains floating in coarse spar. Where skeletal grains have been neomorphosed, relics of internal structure of the shell may be preserved. The neomorphic spar may display a brownish hue and may be composed of an irregular mosaic of small and large calcite crystals with wavy, curved or straight intercrystalline boundaries. Cathodoluminescence is also a valuable tool in distinguishing neomorphic calcite from cement (Section 2.3; Dickson, 1983).

Evaporites
Several evaporite minerals, including halite and the sulphates gypsum, anhydrite, celestite and barite can occur as diagenetic phases in both siliciclastic and carbonate rocks. They are usually relatively easy to distinguish as a group from other minerals by their optical properties (Table 2.3). Halite, in particular, is the only common colourless cubic (and therefore isotropic) mineral found in sedimentary rocks. It is worth pointing out, however, that it is frequently missed by petrography because it dissolves during the standard preparation of thin-sections. If halite is expected, samples should be prepared in oil.

The most significant petrographic problem with the sulphate minerals is distinguishing one from another. Anhydrite can usually be recognized from its bright birefringence colours. Distinction between celestite and barite is often difficult. Where in doubt, X-ray diffraction (Section 2.7) or energy-dispersive

X-ray analysis in the SEM should be used (Section 2.5).

Iron oxides and pyrite
In sedimentary rocks, these opaque minerals are usually diagenetic in origin and may precipitate in both carbonates and clastics. Pyrite in particular is a common phase. Distinction between pyrite and the common iron oxides, haematite and magnetite is best achieved by reflected light microscopy (Table 2.3).

2.2.4 Mineralogical quantification

Point-counting is still the most widely used method for obtaining a quantitative analysis of sedimentary rock mineralogy (Galehouse, 1971a). The apparatus is very simple, consisting of a movable microscope stage on which the sample is mounted, connected to a multi-channel recorder. Each channel corresponds to a particular mineral or porosity type the operator has chosen to differentiate. Points correspond to a mineral or porosity type falling beneath the cross-hairs of the microscope. Each time the operator records a mineral or porosity type on a channel, the stage jumps forward a set distance which will normally correspond to the mean grain size of the sample. The operator can also specify the number of points to be recorded; clearly the greater the number of points, the more accurate the point-count analysis is likely to be. Table 2.5 illustrates a typical point-count record from a sandstone sample. Detrital and diagenetic minerals have been differentiated, and visible porosity has been recorded. The record comprises 400 points, recalculated as percentages.

Point-counting has numerous drawbacks and pitfalls. Its major drawback is that a lot of counts need to be obtained to get adequate precision (Fig. 2.4). Using our point-count analysis from Table 2.5, we can assign confidence limits to our mineral or porosity percentages, depending on the number of points counted and the estimated percentages. Quartz grain content (QG) which comprises 63% of the sample according to our analysis should really be expressed as $63 \pm 5\%$ (95% confidence limits); $\pm 5\%$ is acceptable for a major phase such as quartz. However, calcite cement (CC), comprising only 4% of the point-count has 95% confidence limits of $\pm 2\%$, an uncertainty of 50%. The problems of precision in point-counting can be overcome by very large numbers of counts, but this can be very

Table 2.5 Point-count analysis of mineral percentages.

Thin-section/sample: 010192   Date: 3.3.1992
Rock name: Brent Group − Etive Formation   Operator initals: MLC
Well: 211/18-12A
Depth (core): 3253.5 m
Sample preparation: stained for carbonates only

Mineral		%	Number of points	Comments
1	Quartz grains	63	252	Monocrystalline + polycrystalline quartz
2	Lithic grains	12	48	All lithic grains are mudclasts
3	Feldspar grains	6	24	Section not stained for feldspar
4	Undifferentiated matrix			
5	Quartz cement	6	24	Dust rim between quartz grains and cement
6	Non-ferroan calcite	4	16	Patchy distribution throughout section
7	Siderite			
8	Kaolinite	1	4	
9	Chlorite			
10	Porosity	8	32	Chiefly intergranular macroporosity
11	Opaque	Trace		
Total		100	400	

Accessory minerals (present but not at point-count sites): rutile, garnet, zircon.

tedious. To date, no single system has been developed which can provide an automated point-count, because none can provide all the information required for a detailed petrographic analysis, particularly mineral identification *and* differentiation of detrital from diagenetic mineralogies.

Identification of minerals during point-counting can be an additional problem depending on the preparation of the sample and experience of the operator. It might be difficult to subdivide different feldspar types on a point-count record if the thin-section had not been stained for feldspar. The point-count record should be designed according to the information that an operator believes he or she can reasonably get from a sample, depending on sample preparation and the experience of the operator.

2.2.5 Mineral paragenesis

Mineral paragenesis is the order in which diagenetic mineral phases grow. A full paragenetic description also includes information on mineral dissolution as well as precipitation and the timing of these events relative to phases of compaction, fracturing and, in petroleum reservoirs, petroleum filling. Unravelling a paragenetic history is an essential prerequisite to any further geochemical analysis as it provides a framework of the relative timing of mineral precipitation into which quantitative information can be inserted. The example shown in Fig. 2.5 was derived from several hundred thin-section analyses and was constructed by following a few simple guidelines. Firstly, the box or bar corresponding to a diagenetic episode is of uniform length because petrography alone provides no information about its duration. Although a greater volume of mineral type may intuitively suggest a greater length of time for precipitation, this need not necessarily be the case. Secondly, boxes representing diagenetic episodes overlap where there is ambiguity about the relative timing of diagenetic episodes. This may be due to the relatively low resolution of the petrographic microscope (in which case techniques capable of higher resolution may prove useful), may actually reflect the co-precipitation of different diagenetic phases, or may be due to an intrinsic ambiguity in the textural criterion employed. Only where there is unequivocal evidence of one phase consistently post-dating another should boxes not overlap.

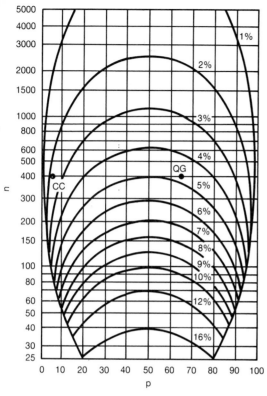

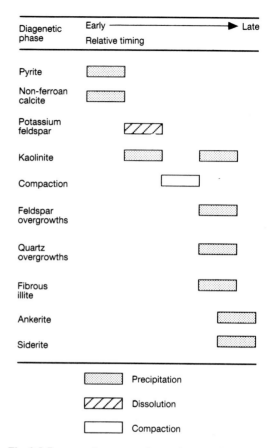

Fig. 2.4 Ninety-five per cent confidence limits for point-counted minerals (from Van der Plas & Tobi, 1965). p represents calculated percentage of minerals from point-count; n represents number of points counted. Percentages represent 95% confidence limits. See text for further details.

Fig. 2.5 Paragenetic sequence for sandstones of the Magnus Field (Northern North Sea).

Diagenetic phases are listed down the left-hand column of the paragenetic history approximately in order of occurrence. In the case of Magnus, some precipitated more than once. It may be necessary to construct separate paragenetic sequences to represent differences in diagenesis in different portions of a petroleum reservoir, for example in the zone of petroleum accumulation and in the underlying aquifer. Similarly, oilfields containing reservoirs deposited in different sedimentary environments should be represented on separate paragenetic history diagrams because the depositional environment may exert an influence on the nature of the diagenetic product (Burley *et al.*, 1985). An extreme example of this would be an oilfield with stacked sandstone and limestone reservoirs.

Several authors (Burley *et al.*, 1985; Harwood,

1988) have suggested that 'diagenetic environment' can be added to paragenetic sequences and have followed the scheme of Table 2.6 (Choquette & Pray, 1970). We feel that this classification can easily obscure more than it enlightens and would prefer to see it dropped in favour of simple textural terms to describe paragenesis, such as 'post-compaction'. Further information about diagenetic setting and absolute ages for diagenetic precipitates can be obtained by the application of geochemical techniques. Petrography is an excellent first-pass tool for paragenetic information, but its limitations should be fully appreciated.

2.2.6 Porosity description

Porosity is the proportion of a rock which is not composed of solid phases and is of great economic

Table 2.6 Summary of the different diagenetic environments of sedimentary rocks. Modified from Choquette and Pray (1970).

Diagenetic environment	Description
Penecontemporaneous (syndepositional)	Diagenetic processes which occur within the depositional environment
Eogenetic (near-surface)	Diagenetic processes which occur within the zone of action or surface-related process and surface-promoted fluid migration
Mesogenetic (burial)	Diagenetic processes that take place during burial, away from the zone of major influences of surface-related processes
Telogenetic (uplift or unconformity-related)	Diagenetic processes which are related to uplift and commonly result from surface-related fluid migration

significance as a reservoir parameter because it determines the *storage capacity** of a petroleum field, referred to as *oil-in-place*[†]. Throughout much of this book, we will be referring to geochemical methods of porosity prediction and we even devote a chapter to case studies (Chapter 6). Porosity is usually measured by downhole wireline log techniques or by core analysis. Pore image analysis (Section 2.9) is a sophisticated computer-based technique which can determine porosity from thin-sections, as well as statistically analyse pore shapes and pore size distributions. However, these techniques provide no information about the *origin* of porosity; this is the business of petrographic porosity description.

Porosity within a sediment may be *primary*, the voids remaining after deposition of the sand or carbonate has taken place, or *secondary*, resulting from dissolution of grains and/or cement, shrinkage of the sediment, or fracturing of lithified or partly lithified rock in the subsurface[‡]. Most primary porosity consists of the space remaining between the original components of the sediment (*intergranular*), although some grains, for example gastropods or foraminifera, may retain primary pores within them (*intragranular*). Secondary porosity may also be intergranular and intragranular. Intergranular secondary porosity is commonly observed in sand-

stones as *oversized* pores, so called because the remaining porespace is too large to be a primary pore (compare the mean primary intergranular pore sizes with the secondary pore in Fig. 2.6a). Oversized pores are interpreted to have originated from the dissolution of unstable grains such as lithic fragments or feldspars. In carbonates, secondary intra-particle porosity is very common, reflecting the mineralogical instability of carbonates, especially aragonite, under diagenetic conditions. A specific type of secondary intergranular porosity is the *mouldic* pore where clear evidence of whole or partial dissolution of skeletal (*biomouldic*) or other carbonate grains has taken place (Fig. 2.6b; Choquette & Pray, 1970). Intercrystalline porosity is commonly found in recrystallized carbonates, particularly dolomite. So-called 'oilman's dream' is actually sucrosic dolomite containing well-connected intercrystalline secondary porosity. Intercrystalline porosity may be found between authigenic clay fibres, but is usually too small to be resolved by conventional microscopy. Where individual pores cannot be resolved using standard petrographic methods, the term *microporosity* can be applied. This usually appears in thin-section as an area which has taken up some of the impregnating resin. Scanning electron microscopy is the most suitable tool for examining microporosity, which may be of primary or secondary origin. Other porosity types which cut across the original rock fabric and are all secondary in origin include fracture pores (which rarely contribute a large proportion of the total porosity, but may greatly assist the flow of petroleum), and vuggy and cavernous porosities which may be found in carbonate sediments which have experienced extensive leaching and karstification. Cavernous porosities (defined as man-sized pores) are for example common in karstified Tertiary limestone reservoirs of the Far East and present major drilling problems, such as dropped drill bits and massive drilling mud losses (Rudolph & Lehmann, 1989).

---

* The storage capacity of the rock is the amount of petroleum the rock can potentially hold per unit volume.
[†] Oil-in-place (OIP, or gas-in-place, GIP) is the amount of oil actually in the reservoir. OIP does not correspond to the reserves figure, which is the amount of *recoverable* oil in a field.
[‡] The terms used in petrographic porosity description are based largely on two papers: Choquette and Pray (1970) and Schmidt *et al.* (1977).

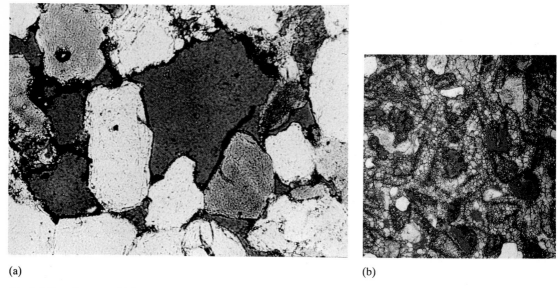

(a)

(b)

Fig. 2.6 Porosity types. (a) Secondary porosity in a Rotliegend Group sandstone, northeastern Germany. The secondary pores show remnant grain outlines in black. Plane polarized light. Width of photograph is 1 mm. (b) Mouldic pores in a limestone after leached mollusc fragments, Mesozoic of Texas. Plane polarized light. Width of photograph is 2 mm. Courtesy J. Dravis.

The reader will occasionally come across the term *minus-cement porosity*. This is (as its name suggests) not actually present-day porespace at all, but an estimate of porosity prior to cementation. Minus-cement porosity can be estimated for specific cement phases, such as minus-calcite porosity in a calcite-cemented sandstone nodule, or the quartz-cemented sandstone illustrated in Fig. 2.2c. The value of estimating minus-cement porosities is that they can provide general information on the burial depth at which cementation took place, assuming that the only change the sediment underwent before cementation was mechanical compaction. The minus-cement porosity can then be estimated from point-counting a cement phase, and the resulting porosity estimate can then be compared to known compaction curves which relate porosity to burial depth in normally pressured systems* (Robinson & Gluyas, 1991b). This approach is only really valid for moderately well sorted sandstones which show little or no evidence of diagenetic alteration prior to or during mechanical compaction. Carbonates, which display abundant diagenetic alteration, are wholly unsuitable for this type of analysis.

* A normally pressured system is one in which the down-hole increase in pressure is purely hydrostatic, with a continuous, connected water column through the pores of the buried sediment to the sea bed.

2.3 Cathodoluminescence microscopy

2.3.1 Introduction

Cathodoluminescence (CL) is the visible light emitted by the surface of a mineral when bombarded with electrons in a vacuum. The earliest geological use of CL was made by Long and Agrell (1965) and Sippel (1968). Since that time, commercial luminoscopes and luminescence detectors have become available and CL is now a standard technique in the petrographic description of a rock sample. The origin of CL is poorly understood. Recent detailed work on carbonates has demonstrated a link between the presence of certain cations substituting for Ca and Mg in the lattice of common carbonate minerals, and activation or quenching of luminescence[†] (Reeder, 1986; Mason, 1987; Walker *et al.*, 1989). Manganese is the best studied activator of luminescence, whereas iron is the most common quencher. The relative concentrations of these cations in calcite, for example, appear to control the intensity of the luminescence (Fairchild, 1983). Other cations such as lead and certain rare earth

[†] A more detailed discussion of the origin and history of CL is provided in the excellent book by Marshall (1988) and the review by Walker (1985).

elements may also control or influence luminescence intensity (Mason & Mariano, 1989).

CL in quartz has received increased attention in the last few years (Zinkernagel, 1978; Ramseyer *et al.*, 1988; Walker *et al.*, 1989). It is apparent that changes in the colour of quartz luminescence are controlled by trace element impurities in the crystal lattice and defects such as non-bonding Si-O (Ramseyer *et al.*, 1988). Depending on the nature of the defect and impurity, quartz luminescence may range in hue from brown to red to blue (Table 2.7). Diagenetic quartz, precipitated at relatively low temperatures, has a characteristic blue colour. Irrespective of our current level of understanding of the precise origin of CL, the technique is still of immense value as a petrographic tool for the differentiation of detrital from diagenetic phases and in understanding mineral paragenesis.

2.3.2 Analytical techniques

CL applications fall into three areas according to the apparatus used: *cold CL*, *hot CL* or *scanning electron microscopy-CL* (SEM-CL; Table 2.1). The

Table 2.7 CL colour and origin of quartz. From Zinkernagel (1978) and Mattern and Ramseyer (1985).

CL colour/zonation	Origin (corresponds to thermal history of quartz)
Violet	Typifies igneous quartz
Brown	Typifies certain metamorphic quartzes
Blue − non-luminescent may be zoned	Diagenetic quartz
Highly zoned	Hydrothermal quartz

most commonly used CL apparatus is the cold luminoscope. This device comprises an evacuated lead chamber, with lead glass windows permitting the passage of transmitted light across samples, usually polished thin-sections. An electron gun attached to the side of the chamber fires electrons at a low angle of incidence onto the sample surface, activating luminescence which can then be viewed through an ordinary optical microscope (Fig. 2.7).

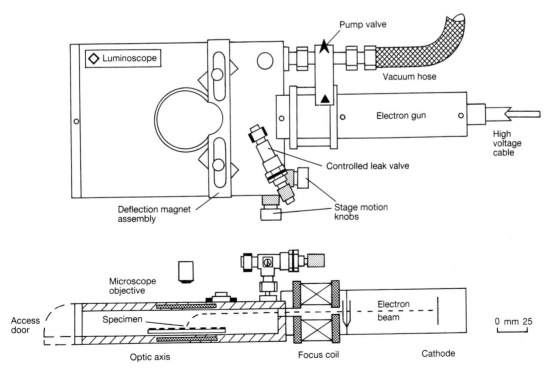

Fig. 2.7 The Nuclide ELM-2A cold cathode luminoscope specimen chamber in top view and cross-section. Courtesy D. Marshall, Nuclide Corporation.

The cold luminoscope is so-called because of its lower operating voltage of 15–20 kV, compared to the hot luminoscope. This lower voltage means that only minerals requiring low energy levels to activate CL, chiefly carbonates, and less commonly, quartz, can be studied.

Hot luminoscopes are identical in principle to cold devices, but require excitation of the electron gun to several tens of thousands of kilovolts. These instruments are most suitable for examination of sandstone samples because quartz luminesces less readily than carbonates. Quartz cements show up particularly well using hot CL but carbonates luminesce too brightly and may volatilize in the hot device (Burley *et al.*, 1989). Only a handful of hot luminoscopes exist, and there are no plans to mass produce them, probably because of the advent of high-quality SEM-CL detectors. This is perhaps a pity, because the information provided by hot CL colour images is informative and the pictures are stunning.

SEM-CL is currently the biggest growth area in luminescence technology. The SEM-CL device comprises a luminescence detector attached to a SEM. The basic SEM bombards samples with electrons in a vacuum as part of its normal operating procedure (Section 2.5) and visible (CL) light is emitted as part of this process. In the past, capturing the luminescence was achieved using an elliptical mirror focused onto an optical fibre. The mirror design led to problems with inefficient light capture which have since been overcome by use of a parabolic mirror (Kearsley & Wright, 1988). The SEM-CL detector is used principally for the examination of sandstones because the slow decay of carbonate luminescence leads to bright smears across the image as the electron beam scans across the sample (Kearsley & Wright, 1988). The advantages of the SEM-CL detector over the hot device include its relatively simple operation and the availability of other SEM-based analytical tools which can be used on the same sample area.

2.3.3 Sample preparation

CL is a surface phenomenon and luminescence can be obtained from any highly polished surface, whether it be a rock chip or a thin-section. Usually, uncovered polished thin-sections are used for CL so that the sample can also be used for conventional petrography. Doubly polished fluid inclusion wafers (Section 3.2) are also suitable. The main require-

ment for CL is that the surface be extremely smooth. Conventional thin-sections are not highly polished enough, and poor luminescence imaging and shadowing may arise from points of high relief on the specimen. For best results, samples should be finished with a 0.25 µm diamond paste. Stained samples can be used for CL, but the stain tends to be discoloured by the electron beam and the reaction of the staining solution with the rock surface will also reduce the quality of the surface finish. In general, staining techniques should be applied to samples after CL, or alternatively the stain should be removed prior to CL by polishing.

2.3.4 Applications of CL

Despite the range of CL technology, the three devices supply essentially the same information for carbonates and siliciclastics (Table 2.1):
1 differentiation of detrital from diagenetic phases; and
2 information on mineral paragenesis.
In carbonates, qualitative information on the oxidation state of the diagenetic environment may also be obtained because of our better understanding of the origin of their luminescence.

CL has been used with spectacular success to differentiate detrital from diagenetic quartz (the problems associated with conventional microscopic differentiation have been discussed in Section 2.2.1). Figure 2.8a is a SEM-CL image of sandstone taken from the same sample as Fig. 2.2b. The grey luminescent quartz cement can be readily distinguished from the brightly luminescent detrital grain. CL has also been used to distinguish detrital from diagenetic feldspars. In the cold CL, detrital feldspars may luminesce very brightly, as brilliant blues, reds and greens, whereas diagenetic overgrowths tend to be non-luminescent or very dully luminescent (Kastner, 1971). Feldspar luminescence has been used in sediment provenance studies, although at this stage there is no evidence to suggest any link between feldspar luminescence hue or intensity and specific igneous or metamorphic origins (Stow & Miller, 1984). CL has also been used in sandstones to provide quantitative information about the diagenetic behaviour of cemented and uncemented sandstones from a variety of maximum burial depths. In a classic paper, Houseknecht (1988) used cold CL (which proved to be surprisingly effective in this case) on samples of Palaeozoic and Mesozoic sandstones to determine whether sutured

(a)

(b)

Fig. 2.8 CL petrography.
(a) Quartz-cemented sandstone, Northern North Sea, from the same sample as Fig. 2.2b. Quartz grains are brightly luminescent (white), quartz cement rimming grains is grey and slightly mottled. Feldspar grains are slightly darker grey than the cement, but are generally difficult to distinguish from CL grey level alone. Porosity is black. SEM-CL. Width of photograph is 1 mm.
(b) Zoned calcite cements, Carboniferous Limestone, Wales. Non-luminescent calcite cement contains no manganese or iron, and was probably precipitated in an oxidizing environment from fresh waters. The brightly luminescent calcite contains manganese, and was precipitated under reducing conditions. The dully luminescent calcite contains both iron and manganese, and was also precipitated under reducing conditions (see Fig. 2.9). Cold CL. Width of photograph is 500 μm. Courtesy A. Dickson.

boundaries were pressure-solution seams or compromise boundaries between diagenetic quartz cements, or a combination of the two. By estimating the amount of quartz cement that had been precipitated or removed by pressure solution, Houseknecht was able to classify sandstones at certain depth ranges as silica importers or exporters.

In carbonates, CL has been used chiefly to distinguish original depositional material from altered depositional carbonate and cement. CL is commonly used as a screening tool to determine whether calcite tests of certain organisms are original precipitates, or whether they have been diagenetically altered. This is particularly important when samples are to be used for isotopic analysis in isotope stratigraphy studies. Samples of the Carboniferous brachiopod *Gigantoproductus* were screened in some detail by Popp *et al.* (1986b) using the cold CL.

Areas of brightly luminescent calcite containing manganese (which is not precipitated by organisms under normal marine conditions; Frank *et al.*, 1982; Barnaby & Rimstidt, 1989) were rejected. Areas of non-luminescent calcite were believed to be diagenetically unaltered and were sampled for stable and strontium isotopic analysis. Recently, however, Rush and Chafetz (1990) have questioned this luminescence screening procedure by demonstrating that luminescent, obviously altered brachiopods and non-luminescent brachiopods have identical isotopic compositions, suggesting that all have undergone significant diagenetic alteration. A lack of luminescence is a requirement for demonstrating a lack of diagenetic alteration, but does not necessarily prove it.

CL has been used successfully in the study of mineral paragenesis, particularly in *cement stratigraphy* (the study of the spatial relationships of similar cement generations) to subdivide one or two generations of calcite (or other carbonates) established using conventional light microscopy, into many further 'zones' or 'stages' (Miller, 1988). Cement stratigraphy in carbonates involves building up a paragenetic sequence of different cement generations using CL (and other complementary petrographic disciplines), and correlating the distributions of individual cement zones (Fig. 2.8b). There are no rules as to how the different generations should be subdivided, although Miller (1988) has suggested a unifying terminology for description of carbonate luminescence. The champion of carbonate cement stratigraphy is Meyers, who in 1974 and 1978 published classic papers on the cement stratigraphy of Mississippian carbonates across many sample locations in New Mexico. The distribution of these cement zones suggested that a regional freshwater aquifer system controlled diagenesis for much of the post-depositional history of the carbonates. The different luminescence intensities were shown by microprobe analyses to be related to the relative concentrations of manganese and ferrous iron in the lattice of the calcite cements (Section 2.3.1), and hence to the oxidation state of the diagenetic environment (Fig. 2.9; Barnaby & Rimstidt, 1989). Input of fresh, oxidizing waters into the carbonates resulted in precipitation of non-luminescent, iron- and manganese-free calcite. As the fresh waters became more reducing, brightly luminescent calcite with manganese was precipitated, followed by dully luminescent calcite containing both ferrous iron and manganese. The characteristic luminescence

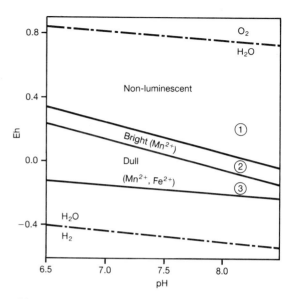

Fig. 2.9 Conditions of precipitation of carbonate cements with particular luminescence characteristics (from Barnaby & Rimstidt, 1989). Non-luminescent, iron- and manganese-free carbonate will be precipitated in area 1 of the diagram, brightly luminescent carbonate will be precipitated in area 2, and dully luminescent carbonate with both iron and manganese will precipitate in area 3.

sequence in these cements was black–bright–dull. Similar reasoning was used by Horbury and Adams (1989) to identify the effects of freshwater diagenesis on Carboniferous platform carbonates in northern England, by Emery and Dickson (1989) to identify a much more restricted generation of zoned freshwater lens cements in Jurassic limestones of eastern England and by Read and his students (e.g. Grover & Read, 1983; Dorobek, 1987) to identify the effects of freshwater diagenesis in Palaeozoic carbonates of Virginia. On a cautionary note, the production of complex zonation patterns in carbonates need not be related solely to iron and manganese concentrations, as other elements are known to activate or quench luminescence (Mason, 1987). Furthermore, Reeder *et al.* (1990) have produced synthetic zoned calcites in the laboratory from a single starting fluid with constant Eh and pH, demonstrating that factors unrelated to the bulk fluid composition or chemical environment may be responsible for causing zonation in carbonate cements (possibly kinetic effects related to crystal growth).

Quartz cement stratigraphy is a less well established technique, partly because the appropriate technology has only been widely used in recent years, but mainly because the relationship between CL colour, chemistry and prevailing diagenetic conditions is less well understood than it is for carbonates. Probably the best work to date on quartz cement paragenesis and stratigraphy is by Burley *et al.* (1989) on quartz (and many other) cements in the Jurassic Piper Formation sandstones of the Tartan Oilfield, UKCS.

The final application of CL, particularly in carbonates, is in cement fabric description. Dickson (1983) has shown how neomorphic calcite can be distinguished from calcite cement based on CL zonation patterns. CL can also help reveal ghosts of precursor grains or cements (Lee & Harwood, 1989), and complex zonation patterns, such as sector zoning (Reeder & Grams, 1987) which may not be visible by any other technique. For further reading on the types and origin of complex intracrystalline zonation patterns, the reader is referred to Sellwood (1989).

As a concluding point, the application of CL technology to understanding siliciclastic and carbonate diagenesis can no longer be regarded as an optional extra, and should form a standard petrographic tool because its value in differentiating detrital from diagenetic phases, unravelling cement paragenesis and stratigraphy, and in cement fabric analysis is unrivalled. Future lines of CL research lie in improving our understanding of its origin, especially in silicate minerals, and ultimately in using CL colour and intensity to provide quantitative information about elemental concentration and lattice defects.

2.4 Ultraviolet fluorescence microscopy

2.4.1 Introduction

Ultraviolet fluorescence (UVF) is the visible light response of materials exposed to ultraviolet light (in the wavelength range 200–300 nm). UVF is used routinely in petroleum geology in the detection of oil on rock samples, commonly drilling cuttings or cores. These are known as hydrocarbon *shows* and are indicated by fluorescence of the oil on the samples. This simple principle has been extended to microscope work, with UV light sources available for conventional microscopes. Various wavelengths of UV can be selected by means of filters which can be interposed during UV emission. Two different types of information are available from UVF

microscopy. Firstly, petroleum fluid inclusions, often present in diagenetic minerals, can be detected relatively easily by their bright fluorescence (Section 3.5.4). The second application of UVF is in mineral paragenesis studies, particularly in cement fabric analysis. There are two main shortcomings with UVF microscopy. The first is that inorganic materials such as calcite generally display only weak UV fluorescence, and only certain samples appear to provide any paragenetically useful information. UVF is certainly not a substitute for CL microscopy in this respect. The other problem is our limited understanding at present of the origin of UVF. It is clear that included organic matter causes fluorescence but there may be other causes, such as trace elements and/or rare earth elements. The technique could therefore potentially provide information about the chemistry of fluids which have precipitated fluorescing phases.

2.4.2 Applications of UVF

Sample preparation for UVF microscopy is essentially the same as for CL; a highly polished thin-section is ideal. Doubly polished fluid inclusion wafers are also suitable. UVF has been most successful with carbonates. Dravis and Yurewicz (1985) were able to resolve the timing of burial dolomitization relative to porosity evolution using blue-light (slightly longer wavelength than UV) fluorescence to see relict grains which were invisible with standard microscopy. UVF can also be used to show cement fabrics which may be invisible using both standard and CL microscopy (Fig. 2.10). Emery and Marshall (1989) demonstrated the value of UVF in conjunction with CL in emphasizing sector zonation in calcite cements and Miller (1988) has discussed how organic-rich fossils in reefs show up very well in UV light, but cannot be seen in the visible part of the spectrum. At present, UVF is a useful tool for obtaining qualitative cement fabric information in carbonates, but its limitations will remain until we better understand the origin of fluorescence.

2.5 Scanning electron microscopy

2.5.1 Introduction

The high magnifications ($\times 20$ to $\times 100\,000$) attainable, and extreme depth of field make SEM an excellent tool for examining the details of mineral morphology, grain–cement relationships, and

Fig. 2.10 UVF petrography. Zoned, pore-filling calcite cement, Smackover Formation, Mississippi, USA. These concentric zones are only visible using blue light fluorescence; in cathodo-luminescence the sample is totally black. This example demonstrates that fluorescence and CL do not always respond to the same activators. Width of photograph is 1 mm. Courtesy J. Dravis.

porosity, especially microporosity. Since the first commercial SEM was produced in 1965, a whole range of additional techniques and applications have become available to complement standard high magnification microscopy. These include backscatter electron imaging, SEM-CL (discussed in Section 2.3) and energy dispersive X-ray analysis.

The basic operation of the SEM is illustrated in Fig. 2.11. An electron gun produces a stream of electrons in a vacuum, across which an accelerating voltage is applied. Electromagnetic lenses produce a small demagnified image of the electron source which is scanned or 'rastered' across the surface of the specimen. On striking the specimen, two types of electron are emitted: high energy *backscattered* electrons and lower energy *secondary* electrons, some of which may be completely absorbed at certain depths within the specimen. Cathodoluminescence and X-rays will also be generated during electron scanning (see Sections 2.3 and 2.7 respectively). To create an image, the electron beam scan is synchronized with that of a cathode ray tube and a picture built of the scanned area. SEM operation can be carried out in two modes, both of which yield different types of information. The microscope is generally operated in *emission* mode (SE), in which secondary and backscattered electrons are collected, providing a high magnification image. The contrast in the image produced during normal high magnification operation is a function of the topography and orientation of the specimen, its chemistry, and

differences in electrical potential on its surface. The second mode of operation is *backscattered* mode (BSE) in which only the backscattered electrons are captured. The efficiency of electron backscatter from the surface of a mineral depends on its chemistry,

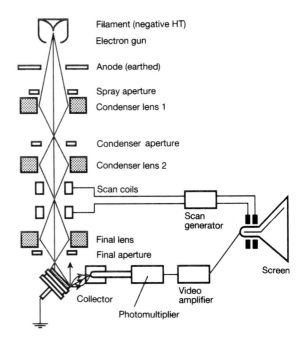

Fig. 2.11 The basic features of a scanning electron microscope (from Trewin, 1988).

specifically its mean atomic number. Backscatter mode operation requires a backscattered electron collector, such as the Robinson device (no relation; Pye & Krinsley, 1984).

2.5.2 Sample preparation

Basic sample preparation for observation of the surfaces of rock chips in emission mode is relatively simple, but several important rules need to be followed. Firstly, rocks which are oil saturated or contain oil residues such as bitumen need to be cleaned in solvents, otherwise oil will be vaporized by the electron bean and make the SEM column dirty, reducing image quality. Samples which previously contained water and have dried out in air do not need to be treated in solvent. Secondly, samples need to be mounted flat against the mounting stub using a glue which is stable at high temperatures, and any loose material on the sample surface should be gently removed. The specimen is then ready for coating with a conductive material (usually gold or carbon) to take away the electrical charge which builds up on the specimen surface during SEM operation.

One of the major problems with basic sample preparation is that the air drying process can lead to damage of delicate crystals by surface tension forces during the passage of the air–water interface across them. Fibrous illite in sandstones is particularly susceptible to damage, causing the delicate fibres to appear matted on the pore surface, rather than protruding into the porespace (Fig. 2.12a). A technique known as critical point drying can eliminate the surface tension forces present during air drying (McHardy et al., 1982). This process is time consuming and requires special apparatus, but the contribution it makes to an improved understanding of the sample texture can be considerable (McHardy & Birney, 1987; Trewin, 1988).

Samples for backscattered electron imaging must have a highly polished surface. Polished thin-sections are ideal, and are of course suitable for other techniques such as CL and UVF. Prior to BSEM analysis, however, the sample must be carbon coated and mounted on a stub.

2.5.3 Applications of emission mode SEM

The main applications of emission mode SEM are (Table 2.1):
1 mineral identification;
2 differentiation of detrital from diagenetic phases;
3 mineral textural observations; and
4 porosity description.
These applications parallel closely those of transmitted light microscopy. However, SEM comes into its own when higher magnifications are required to resolve ambiguous paragenetic relationships, or details of the morphology of fine grained minerals, such as the fibrous illite illustrated in Fig. 2.12a.

Mineral identification using the SEM is helped by the energy dispersive X-ray analysis system (EDS), which permits elemental distributions to be obtained for single spots of $1\,\mu m$ diameter, or for the entire screen area (Trewin, 1988). This information is commonly used to supplement morphological and textural observations of mineral types. The SEM is also a useful tool for distinguishing detrital from diagenetic minerals, especially clays. Figure 2.12b shows a sandstone sample with platy or 'leafy' illite and fibrous illite. The platy illite is interpreted to represent an earlier, probably detrital phase, introduced by muddy water flowing through the sand and trapped against the grain surface, whereas the fibrous illite is interpreted to be a later diagenetic phase. The interpretation of two illite generations with quite different origins would have implications for any further geochemical analysis, such as K–Ar dating. There would be little point in dating all the illite in this sample as the date would simply be a weighted average of the age of the detrital phase (of uncertain origin, but commonly older than the depositional age of the sandstone itself) and that of the authigenic fibrous illite. To determine the age of the fibrous illite only, the two would have to be physically separated. SEM observations on the size ranges of the different illites can be used to design the separation process.

Textural information obtained from emission mode SEM can be used to construct paragenetic histories (Burley, 1984, 1986; Huggett, 1984, 1986; Kantorowicz, 1985). Figure 2.12c shows quartz cement enclosing kaolinite, suggesting that the kaolinite was precipitated first. In Fig. 2.12d, from the same sample, the kaolinite appears to rest on the surface of the quartz grain, suggesting that the quartz came first. We could conclude from these observations that quartz and kaolinite precipitation were broadly contemporaneous, or that there are two generations of kaolinite, one earlier than and one later than the quartz.

The SEM in emission mode is also useful for observing porosity, particularly microporosity and

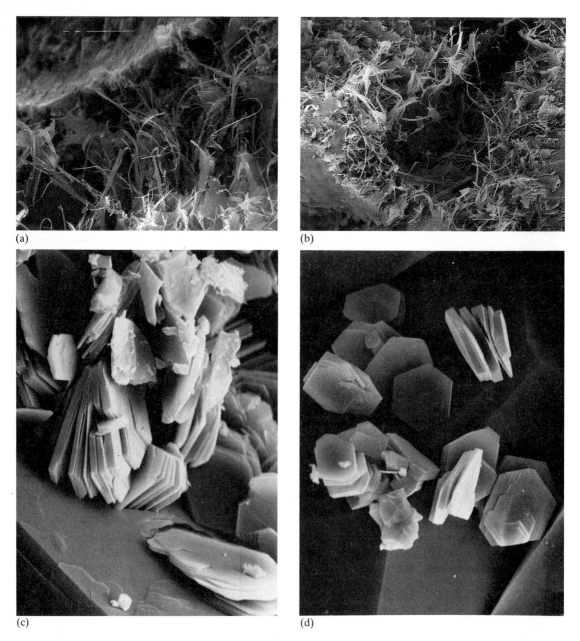

(a)

(b)

(c)

(d)

Fig. 2.12 SEM petrography. (a) Fibrous or 'hairy' illite from a Rotliegend Group sandstone, Southern North Sea. Although discrete illite fibres can be seen protruding into the porespace, some matting of fibres is also evident. Width of photograph is 100 μm. (b) Two types of illite? A leafy illite appears to be the earliest form from and on which later authigenic fibres have grown. See text for details. Rotliegend Group Sandstone, Southern North Sea. Width of photograph is 200 μm. Courtesy T. Primmer. (c) Quartz cement enclosing kaolinite, Magnus Sandstone Member, Northern North Sea. Width of photograph is 20 μm. (d) Kaolinite resting on the surface of quartz cement, Magnus Sandstone Member, Northern North Sea. Width of photograph is 30 μm. See text for details of this and Fig. 2.12c. (e) Dissolved feldspar grain with etch pits (microporosity) extending parallel to cleavage, Jurassic, Central North Sea. Width of photograph is 100 μm. Courtesy T. Primmer. (f) Replacive dolomite with well-connected porosity, Carboniferous of Texas. Width of photograph is 1.2 mm. Courtesy A. Dickson. (g) Microporosity between coccolith plates, Cretaceous Chalk, Central North Sea, K = kaolinite. Width of photograph is 20 μm. Courtesy R. Maliva.

(e)

(f)

(g)

Fig. 2.12 *Continued*

connectivity between pores. Figure 2.12e shows a partially dissolved feldspar grain with microporosity. The orientation of the etch pits is parallel to the cleavage planes, but their size suggests that they would probably not contribute to the effective porosity (connected porosity within a rock through which fluids will flow) of the rock. The texture and packing of dolomite crystals in a dolomitized limestone show that effective, connected porosity is present (Fig. 2.12f). The SEM is commonly used to examine the microporosity that is characteristic of chalks, which is very difficult to see using a petrographic microscope (Fig. 2.12g).

2.5.4 Applications of backscatter mode SEM

The main applications of backscatter mode SEM (BSEM) are (Table 2.1):
1 mineral identification;
2 mineral chemistry; and
3 porosity description.

Mineral identification by BSEM relies on the increase in the efficiency of electron scattering as the mean atomic number of the mineral increases. Quartz, with a lower mean atomic number than calcite, will reflect electrons less efficiently and give a duller grey image. Pyrite, with a very high mean atomic number, will reflect electrons very efficiently and produce an intense white image. Grey levels can therefore be used as a qualitative guide to mineralogy (although minerals with entirely different chemical compositions, but similar mean atomic numbers, will give similar grey levels). Figure 2.13a shows a BSEM image of a sandstone: quartz grains are medium grey and form the bulk of the sample, which is cemented by light grey carbonate and very dull grey booklets of kaolinite. The bright grain right of centre is rutile; other bright specks are probably pyrite. A few light grey feldspars, of similar intensity to the carbonate can also be distinguished. Porosity is shown in black because the mean atomic number of the polymer resin filling the porespaces is very low.

BSEM can be used to detect chemical zonation in minerals which may not be visible in stained thin-sections or by cathodoluminescence. Figure 2.13b shows zoning in ankerite that would not be visible in CL because of the quenching effect of the iron (Section 2.3). The change in the mean atomic number of the ankerite from zone to zone is clearly picked out by BSEM.

BSEM is a particularly powerful tool for porosity description because of the great mean atomic number contrast between the colour of the impregnating polymer resin (appearing as black) and the rock matrix (variable mean atomic number, ranging from dark grey to white). This contrast can be detected by image analysis software and has given rise to a new branch of petrographic science referred to as pore image analysis (PIA; Section 2.9).

2.6 Transmission electron microscopy

2.6.1 Introduction

Transmission electron microscopy (TEM) involves the transmission of an electron beam through a thinned sample in a vacuum, analogous to optical microscopy using transmitted light. The main requirement for specimen observation is that specimens should be thin enough to allow electrons to pass through. Individual clay fibres may be suitable but other minerals have to be thinned by an ion beam to a thickness of a few nanometres. The advantage of extreme sample thinness is that accurate quantitative chemical analyses can be obtained from areas as

(a)

(b)

Fig. 2.13 BSEM petrography. (a) Carbonate-cemented sandstone, Jurassic, Northern North Sea. Width of photograph is 1.5 mm. (b) Zoned ankerite cement, Carboniferous sandstone, Bothamsall Oilfield, East Midlands, UK. The change in backscatter intensity from the centre to the margins of the ankerite indicates an increase in the iron content of the ankerite. Width of photograph is 500 μm. Courtesy E. Warren.

small as 5–10 nm in diameter as there is negligible spreading of the beam and no absorption or fluorescence effects to correct for.

The performance of today's TE microscopes is so advanced that high resolution images showing sub-unit cell detail can be obtained, with magnifications up to × 500 000 commonly used to obtain lattice fringe images. Highly detailed textural information can be obtained at magnifications of a few thousand to about × 30 000. This range is particularly suitable for examination of clay minerals (Sudo *et al.*, 1981). However, the quality of the information obtained in

the TE microscope is only as good as the sample preparation and orientation.

2.6.2 Sample preparation

TEM analysis requires preparation of samples so that they are transparent to electrons. Rock samples are normally prepared as polished thin-sections which are then thinned by ion-beam milling until holes begin to appear (Phakey *et al.*, 1972). Areas of the sample adjacent to the holes may then be thin enough to be transparent to the electron beam. This

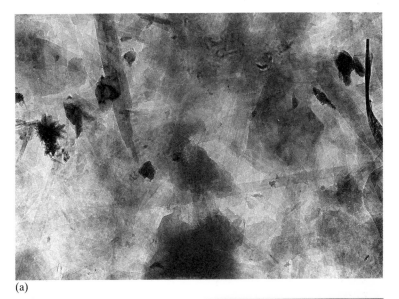
(a)

(b)

Fig. 2.14 TEM petrography. (a) Fine fraction mount under the TEM. Laths of fibrous illite are visible, as well as unidentifiable opaque matter. Width of photograph is 3 μm. Courtesy T. Primmer. (b) Mineral paragenesis, showing quartz cement including kaolinite platelets, suggesting that some of the quartz cement post-dated kaolinite precipitation. A later phase of illite is also visible on the left of the photograph. Width of photograph is 30 μm. Courtesy E. Warren.

tends to be a bit of a hit and miss process. Clay mineral separates can also be analysed using TEM. Sample separation is essentially the same as that described for clay fraction X-ray diffraction (Section 2.7), following which the clay particles are dispersed in very dilute suspensions and mounted on a metal microgrid (Nadeau & Tait, 1987).

2.6.3 Applications of TEM

Applications of TEM fall into the area of high resolution mineralogical and textural analysis, particularly of clay minerals, and provide the following information:
1 differentiation of detrital from diagenetic phases;
2 mineral textural relationships; and
3 mineral chemistry.

TEM is particularly useful for differentiation of detrital from diagenetic clays and other fine particulate matter. Figure 2.14a shows a fine fraction mount under the TEM. Laths of fibrous illite − almost certainly of diagenetic origin − are visible and there is a smaller quantity of unidentifiable opaque matter. This sample was intended for further K−Ar dating and oxygen and hydrogen stable isotope analyses but the presence of opaque matter suggested that further separation would be required before the sample became sufficiently pure. The very high resolution of TEM is potentially excellent for clarifying ambiguities in mineral paragenetic sequences. In Fig. 2.14b, quartz cement includes kaolinite platelets, suggesting that some of the quartz cement post-dated kaolinite precipitation. A later phase of illite is also visible. In practice, many samples need to be prepared to provide textural information because of the rather random nature of the ion-beam milling process which cannot be guaranteed to thin areas of interest.

Chemical analysis by TEM can also be performed on thinned samples, or on dispersed clay fractions and can be a useful supplement to textural information, particularly on clay mineral fractions prior to isotopic analysis. Illite and illite−smectite for example tend to look rather similar. Chemical analysis using TEM confirms that those in Fig. 2.14a are true illites (Table 2.8; see also Ireland et al., 1983; Warren & Curtis, 1989).

2.7 X-ray diffraction

2.7.1 Introduction

X-rays are electromagnetic radiation produced when an electron beam hits a substance and causes rapid deceleration of the electrons. In conventional diffraction, the X-ray tube is arranged so that electrons strike a target which produces X-rays of known wavelengths, which can be further filtered to produce radiation of a single wavelength which can be directed at the sample in the diffractometer. The wavelength of X-rays used ranges from 0.05 to 0.25 nm. The sample is rotated in the beam so that crystallographic planes diffract X-rays as they reach the appropriate angle. The relationship of X-ray wavelength to the angle of diffraction and the characteristic lattice (or d) spacing of the mineral under examination forms the basis of all X-ray diffraction (XRD) analysis and interpretation, and is known as Bragg's law (Fig. 2.15). Further details of the theory of X-ray diffractometry and diffraction equipment are available in Klug and Alexander (1974), Brindley and Brown (1980) and Eslinger and Pevear (1988).

The introduction of XRD into the geological realm allowed the structural identification of minerals previously identified by their thin-section

Table 2.8 Transmission electron microscope analysis of illite illustrated in Fig. 2.14a.

Analysis Number	Si	Al (IV)	Al (VI)	Fe (III)	Mg	Oct	Ca	K	Na	Al (Total)
1	6.845	1.155	3.404	0.273	0.282	3.959	0.008	1.509	0.034	4.559
2	6.629	1.371	3.591	0.170	0.274	4.035	0.042	1.561	0.005	4.962
3	6.541	1.459	3.520	0.202	0.266	3.988	0.012	1.581	0.158	4.979
4	6.547	1.453	3.550	0.237	0.197	3.984	0.014	1.613	0.057	5.003
5	6.683	1.317	3.635	0.256	0.012	3.879	0.053	1.519	0.046	4.952

IV, Al in tetrahedral sites
VI, Al in octahedral sites
Oct, total octahedral occupancy

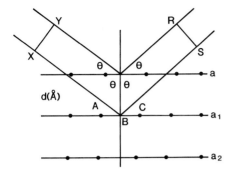

Fig. 2.15 Illustration of Bragg's law (from Hardy & Tucker, 1988). a, a_1 and a_2 are lattice arrays of atoms that can be regarded as an infinite stack of parallel, equally spaced planes. If a wavefront X–Y is incident on a–a_1 the reflection path from the lower plane (a_1) is longer, i.e. AB + BC = Δ = difference in paths of wavefronts

$$d \sin \theta + d \sin \theta = 2d \sin \theta = \Delta$$

For diffraction to occur Δ must equal a whole number of wavelengths: $2d \sin \theta = n\lambda$.

properties alone (Brindley & Brown, 1980). In petroleum geology, most XRD work is performed on powdered samples, providing qualitative information on minerals present in whole-rock samples, and semi-quantitative information on separated fine fraction (usually clay size) samples (Table 2.1). Fine fraction analysis is by far the most valuable application in diagenetic studies (Burley *et al.*, 1985), particularly when used in conjunction with other techniques such as thermogravimetry-evolved water analysis (Section 2.8), when near-quantitative analyses can be obtained. XRD preparation is destructive of fabric, so that no textural or paragenetic information can be obtained, and detrital and diagenetic phases cannot be differentiated.

2.7.2 Sample preparation

For whole rock analysis, samples must be reduced to 5–10 μm powder for satisfactory results. Grinding must be performed carefully to reduce the risk of damaging crystals by strain, as this can lead to diffraction line broadening (Klug & Alexander, 1974). Once the sample is ground to the appropriate size, it can be mounted in two ways for exposure to the X-ray beam: cavity mounting and smear mounting. Cavity mounting tends to produce better the random particle orientation required for a good semi-quantitative analysis: powder is packed into an

aluminium cavity mount holder from the rear (with limited pressure to avoid orienting the crystals; Klug & Alexander, 1974). Smear mounting involves smearing the powder onto a glass slide with an organic solvent such as acetone. This process can lead to some particle orientation.

Fine fraction analysis is carried out to determine the mineralogy of particles <2 μm in size, the bulk of which are clay minerals. Sample preparation is more involved as the fine fraction needs to be separated from coarser material. This may not be a problem with mudrocks but for sandstones a thorough separation procedure needs to be followed (Hardy & Tucker, 1988). Sample pre-treatment requires removal of carbonate, usually by weak organic acids to prevent damage to the clay particles. If large quantities of organic matter are present, this can be removed by hydrogen peroxide (Jackson, 1979). The sample is then ready for disaggregation prior to separation, usually achieved by ultrasonic means. The disaggregated sample is transferred to a settling column (or a centrifuge to speed the settling process) and allowed to stand until the required size fraction, calculated according to Stokes' law, can be removed (Galehouse, 1971b; Hardy & Tucker, 1988). A major difficulty with clay mineral separation is the tendency for particles to flocculate, forming larger aggregates which may settle faster and fall outside the desired fine fraction size range. Flocculation can be minimized by adding a deflocculant such as sodium hexametaphosphate (Calgon). The separated fine fraction can be mounted as an oriented or unoriented mount for XRD analysis. The objective of orienting the sample is to ensure that the clay plates lie parallel to the slide or tile surface and to avoid differential settling of the clays. In practice, this is often difficult to achieve (Towe, 1974). Oriented samples can be prepared by smearing the clay slurry onto a glass slide, by evaporating the clay suspension onto a slide in a beaker, or by sedimenting the suspension onto a porous substrate such as a ceramic tile under vacuum. Unoriented samples are prepared by cavity mounting, although the pressure applied to fill the cavity usually results in some orientation of the sample (this can be minimized by embedding the samples in a resin such as Araldite, or by regrinding the particles to approximately equal lengths).

Further treatments may be applied to help distinguish different clay types, particularly where diffraction peaks overlap. In addition to normal air drying, these treatments usually comprise glycol

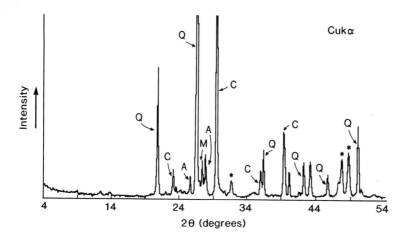

Fig. 2.16 Typical whole-rock XRD trace from a reservoir sandstone. See text for details.

solvation, heating to 375°C, and further heating to 550°C. Glycol treatment will cause expandable clays such as smectites to swell, increasing their *d* spacing. Heat treatment will cause certain clays partially or entirely to collapse, or may increase the intensity of diffraction peaks (chlorite; Starkey *et al.*, 1984).

2.7.3 Applications of XRD

The two main applications of XRD to petroleum geology are (Table 2.1):
1 rapid mineralogical analysis of whole-rock samples; and
2 clay mineral identification and quantification from mineral separates.

Whole-rock XRD provides information about the mineralogy of a sample without differentiating its detrital or diagenetic origin. It is particularly useful in rapid identification of minerals that may cause problems during drilling, well testing*, petroleum production or reservoir stimulation and can be applied to samples which cannot be prepared for thin-section analysis, such as poorly consolidated core samples, or drilling cuttings. The speed of XRD whole-rock analysis is a major advantage in petroleum geology where the mineralogical composition of a reservoir interval needs to be determined very rapidly, often while a well is still drilling.

---

*Testing (or drill-stem testing) a well involves dropping the pressure in the well bore and allowing petroleum to flow from the reservoir, through the drill-stem to the surface. Testing can provide information about the permeability and potential rate of flow of petroleum from the reservoir, and its dimensions and continuity.

Some clay minerals affect the performance of a reservoir during testing and their presence needs to be rapidly looked for by XRD. These include loosely bound kaolinite, which may physically migrate though the reservoir (Khilar & Fogler, 1984; Amaefule *et al.*, 1987), and swelling clays such as smectites which may expand and block pores (Hewitt, 1963). In addition, certain reservoir intervals may produce petroleum more readily if they are acidized[†] before testing. Some clay minerals may react with acids, particularly HF, causing the precipitation of silica (Crowe, 1986). Figure 2.16 is a typical whole-rock XRD trace from a reservoir sandstone. From a quick inspection of the peaks and relative peak heights, it is evident that the sample contains significant amounts of quartz and calcite, minor feldspars, and possibly small quantities of clay minerals. Calcite might be susceptible to dissolution and reprecipitation during acidization.

The main application of XRD to sediment diagenesis is in fine fraction analysis. This technique is slower than whole-rock XRD because sample preparation is more complicated, but it is far better for identifying clay mineral types and can provide semi-quantitative information on the relative proportions of clay minerals in samples. When used in conjunction with thermogravimetry-evolved water analysis (Section 2.8), fine fraction analysis can provide near-quantitative data on very small proportions of clay minerals in reservoir rocks. Figure 2.17 is from a fine fraction separated from a reservoir

---

[†]Acidization of a well is the introduction of acid into the reservoir to leach cement phases which may be occluding pores.

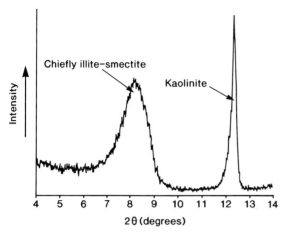

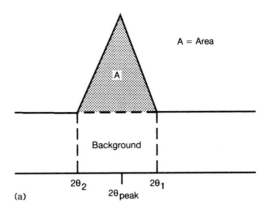

Fig. 2.17 Fine fraction XRD trace showing kaolinite and interstratified illite–smectite present in the sample. See text for details.

sandstone. In this case, an oriented mount has been used which has been subjected to various treatments (air drying, glycol solvation, heating to 375°C and heating to 550°C) to help identify different clay mineral components. The diffraction intensity (I) for each clay mineral can be determined from the air dried trace (Fig. 2.17) by integration of peak area above background (Fig. 2.18), and can be calculated for single peaks and overlapping peaks. To calculate clay mineral percentages, Weir et al.'s (1975) equation is commonly used:

$$\frac{I(\text{Kaolinite})}{2.5} + I(\text{Illite}) + I(\text{Smectite})$$

$$+ \frac{I(\text{Chlorite})}{2.0} = 100\%$$

The divisors beneath kaolinite and chlorite correct for the relatively greater X-ray responses of these minerals (this passes any error in one mineral percentage on to the others). For the reservoir sandstone in Fig. 2.17, a known percentage of an internal standard, boehmite (γ-AIOOH with a characteristic diffraction pattern; Griffin, 1954), was added to the fine fraction. The following clay mineral percentages were obtained: kaolinite, $84 \pm 2\%$; illite (probably interstratified illite–smectite), $16 \pm 2\%$; no pure smectite or chlorite.

However, both methods may be affected by two factors which will limit the accuracy of quantitative analysis.

1 Errors associated with estimating the peak areas and hence the intensities, particularly in complex

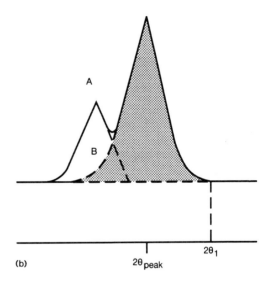

Fig. 2.18 Measurement of intensity of a diffraction peak by integration of area for (a) single peak, area A and (b) overlapping peaks, areas A and B (from Hardy & Tucker, 1988).

mixtures. An estimate of this error has been incorporated in the results for Fig. 2.17.

2 Probably the greatest disadvantage is that quantitative clay mineral information relates solely to that fraction being analysed and does not give any information about the proportions of clay minerals in the whole rock. This problem may be overcome by the simultaneous application of clay mineral fine fraction quantification and thermogravimetry-evolved water analysis (Section 2.8).

It is essential to bear in mind the shortcomings of XRD data when interpreting whole rock and fine fraction traces. Nevertheless, it remains an essential tool for rapid scrutiny of whole rock samples and differentiation of clay mineral types in fine fractions.

2.8 Thermogravimetry-evolved water analysis

2.8.1 Introduction

Thermogravimetry (TG)-evolved water analysis (EWA) is the quantitative analysis of the water evolved from a sample on heating and is suitable for analysis of rocks containing a proportion of hydrous phases such as clay minerals. Although TG-EWA will not easily distinguish between different clay minerals in a sample, it can provide very accurate information about its *total* clay mineral content because the amount of water vapour driven off from the clay lattice is directly proportional to the total clay content (Primmer & Thornley, 1991). Such information cannot be obtained with any accuracy by whole-rock XRD. Accurate determinations of clay minerals amounts in reservoir sandstones are important, as even small quantities of certain clays (particularly fibrous illite) can seriously impair reservoir quality.

2.8.2 Analytical techniques

The sample requirement for TG-EWA is the bulk rock *plus* the fine fraction separated for XRD analysis. The fine fraction XRD analysis must be done before thermogravimetry, as the heating process destroys the clay minerals. TG-EWA instrumentation comprises a high precision thermobalance which is used to record continuously the weight loss of a sample as it is heated at a controlled rate to temperatures of 1000°C (Fig. 2.19). Dry nitrogen gas is passed around the sample which is located within a high temperature furnace. Water evolved from the sample is transferred with the nitrogen carrier gas to an infrared water vapour analyser which monitors the water vapour concentration in the carrier gas and expresses the dehydroxylation information in terms of structural water content as a weight fraction of the clay mineral. Most hydroxyl groups in clay minerals are driven from the lattice at temperatures between 300 and 900°C. A typical TG-EWA profile for the dehydroxylation of sandstone with a variety of clay mineral types is shown in Fig. 2.20. TG-EWA alone cannot distinguish between these clay mineral types: clay minerals do

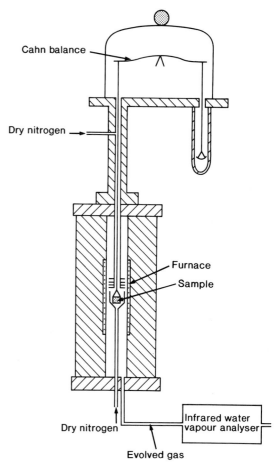

Fig. 2.19 Thermogravimetry-evolved water analysis instrumentation. See text for details.

dehydroxylate over different temperature intervals, but their peaks are broad and cannot be resolved.

2.8.3 Applications of TG-EWA

The main application of TG-EWA is in the quantification of clay mineral quantities in sandstones, often to very high degrees of accuracy. To demonstrate this application, we will return to the reservoir sandstone fine fraction analysis discussed in Section 2.7.3. Knowing the clay mineral percentages as a proportion of the fine fraction, we will use TG-EWA to determine what percentage of the sandstone the clay fine fraction makes up. For this purpose, the fine fraction and sample from which it was disaggregated must be recombined before placing in the thermobalance. TG analysis reveals

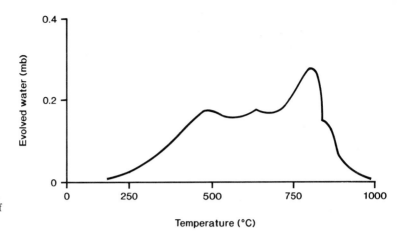

Fig. 2.20 Example TG trace from a reservoir sandstone with a variety of clay mineral types.

a weight loss from the sandstone of 0.54%. Comparison with the EWA trace indicates that this weight loss is entirely due to dehydration of hydrous phases. From this information, the calculated fine fraction percentage in the whole rock is $3.7 \pm 0.3\%$. The fine fraction XRD results (Section 2.7.3) give us the relative proportions of the clay minerals, so their absolute abundances can now be calculated: kaolinite $3.57 \pm 0.16\%$ and illite $0.68 \pm 0.12\%$.

Most of the uncertainties of the TG-EWA XRD method are introduced during XRD analysis. However, if there are significant quantities of hydrous phases which are too coarse to form part of the fine fraction, this can introduce an additional error. For example, detrital mica flakes (particularly muscovite) will lose water, but may not constitute part of the fine fraction. A correction can be applied by point-counting thin-sections to estimate the contribution of mica to the TG-EWA response, but this will lead to further error, as well as adding to the length of time required for sample processing.

2.9 Pore image analysis

2.9.1 Introduction

Pore image analysis (PIA) is a technique which quantifies and analyses the size and shape of pores seen in two-dimensional rock thin-sections (Ehrlich *et al.*, 1984; Ehrlich & Davies, 1989; McCreesh *et al.*, 1991). This is done by digitizing the images of resin-filled pores in polished thin-sections using backscatter SEM or conventional microscopy. It does not provide any information about the origin of the pores, but it does give an immense amount of information about pore geometries which can be empirically related to petrophysical properties, such as permeability and *water saturation**. If a reliable relationship can be established between features of pore size distributions and permeability and water saturation, PIA will have two main advantages over conventional methods of measuring permeability and saturation:

1 speed of analysis, which equates to cost; and
2 measurements can be made on samples as small as drilling cuttings.

2.9.2 Analytical techniques

Pore image analysis can be carried out using conventional microscope images or using backscatter electron microscopy. BSEM has two advantages: its higher resolution capable of measuring features of pores as small as $1\,\mu m$ in diameter and its ability to differentiate between minerals so that mineralogical image analysis may be performed on the same sample. Both methods require image analysis software, with a significant computer memory to store and analyse thin-section images. A monochrome BSEM image is divided up into a series of pixels and an appropriate number of grey levels chosen by the operator. The greater the number of grey levels specified, the larger the memory requirement for the computer. The resulting digital image can then be manipulated using a range of software routines.

For image analysis of pores alone, a single grey

---

*Water saturation (S_w) is the total saturation of the reservoir with water, expressed as a fraction. $1 - S_w$ is the oil saturation, S_o.

level detection threshold must be specified so that pores can be distinguished from minerals. Pores are identified under BSEM by their blackness, and in conventional microscopy by their intensity in transmitted light, making them relatively easy to separate from the rest of the image. Once this is done, the image processing routines can be used to measure the dimensions and certain morphological parameters of the pores. Pore parameters that are commonly measured include the area (A), perimeter (P), length (longest diameter) and breadth (shortest diameter). From these measurements, three other parameters are commonly derived: the equivalent circular diameter (ECD = $4\pi A^{1/2}$), the roundness (R = $A4\pi/P^2$) and a shape factor (shape = length/breadth). These can be rapidly obtained for thousands of pores per sample, at a magnification predetermined by the operator. PIA will often be carried out at several different magnifications on the

same sample, as pore sizes may range from 1 µm up to hundreds of micrometres.

2.9.3 Applications of PIA

Once a PIA dataset is obtained, the task of comparing the parameters with permeability or petrophysical measurements begins. This is usually achieved by cross-plotting permeabilities against PIA parameters which are likely to exert a control on permeability. Figure 2.21 shows a cross-plot of the 95th percentile of the ECD (equivalent circular diameter of the largest pores) versus log permeability for two reservoir sandstones from the North Sea. In both cases the correlation is good; the slopes of the best-fit lines are however quite different demonstrating that there is no unique relationship for all sandstones. The relationship suggests that permeability is controlled predominantly by the largest pores,

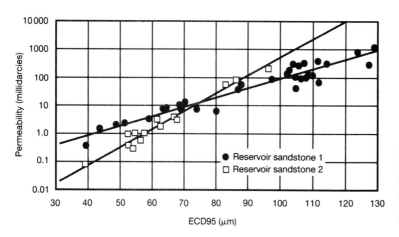

Fig. 2.21 Cross-plot of the 95th percentile of the ECD (equivalent circular diameter of the largest pores) versus log permeability for two reservoir sandstones from the North Sea. See text for details.

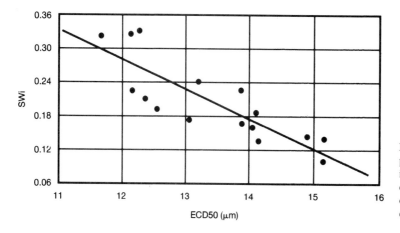

Fig. 2.22 Cross plot of the 50th percentile of the ECD versus irreducible water saturation for one of the reservoir sandstones examined above. See text for details.

because higher proportions of larger pores provide better connectivity through which fluids can flow. The relationships can be used to predict permeability from ECD (95th percentile) alone, quickly and from small samples unsuitable for permeability measurements by core analysis.

There is also a relationship between ECD (50th percentile) and irreducible water saturation (S_{wi})* for one of the reservoir sandstones examined above (Fig. 2.22). Although the correlation is not as good as in Fig. 2.21, there is a clear fall in S_{wi} with increasing pore size. This relationship can also be interpreted geologically: smaller pores are more likely to retain formation water than larger ones.

PIA does have significant limitations. It will be of little use when there is a poor correlation of an essential reservoir property with a PIA parameter, and when no clear geological explanation exists for an empirically derived correlation. Under such

---

*Irreducible water saturation (S_{wi}) is water that cannot be removed from the reservoir, such as water held in microporous clays. It is a subset of water saturation (S_w).

circumstances, PIA parameters should not be used to measure reservoir quality. Additional limitations concern the analytical technique itself, chiefly whether two-dimensional porosity information can really be used to give higher-dimensional information such as permeability. Proponents of PIA argue that the sample coverage of thousands of pores is statistically significant enough to provide a three-dimensional view of pore parameters. More recent work has involved taking parallel serial thin-sections every few tens of micrometres from a sample and comparing PIA parameters from sample to sample to give a three-dimensional PIA view. This is very time consuming and rather defeats the object of quick and cheap analyses from small samples. To sum up, PIA is still in its infancy and further research is required before we fully understand the meaning of many of the PIA–petrophysical correlations. For the present, if a good PIA–petrophysical correlation exists for a sound geological reason, there is no reason why it should not be used to measure petrophysical parameters.

Chapter 3 Fluid Inclusions

3.1 Introduction

Fluid inclusions are microscopic pockets of liquid and/or gas trapped within crystals (Fig. 3.1). They can provide three basic types of information. Firstly and most obviously, they are samples of fluids – waters, petroleum and gases – once present in the subsurface, which are available for analysis. Secondly, provided that there is no leakage of material into or out of the inclusions after trapping, they can provide an estimate of the temperature at which the fluid became included in the mineral host. Indeed, together with oxygen isotope analysis which we will discuss in Chapter 4, fluid inclusions constitute the most important type of geothermometer. Thirdly, they can provide estimates of fluid density. We are most interested in temperature and fluid composition because these can be translated into three kinds of information of real value to petroleum exploration and development (Pagel *et al.*, 1986; McLimans, 1987; Burruss, 1989):

1 temperature, age and duration of mineral cement growth;

2 temperature and timing of episodes of fluid migration (including water, oil and gas); and

3 temperature of fracture healing.

The temperature at which a mineral cement grows can be transformed into an approximate age by means of a modelled burial history. In this way, absolute ages can be ascribed to a relative sequence of mineral growth established by the petrographic methods described in Chapter 2. Mineral growth and dissolution are of course major influences on porosity and permeability, so fluid inclusion geothermometry provides a means of timing modification of reservoir quality relative to oil or gas migration, or indeed to any other feature of geological history*. Fluid inclusions can also tell us about the petroleum itself. Petroleum fluid inclusions are common in sedimentary rocks and can be used to determine what kinds of oil or gas were passing

---

*The timing of oil and gas generation and migration is generally estimated by modelling the thermal history of the source rock (Allen & Allen, 1990).

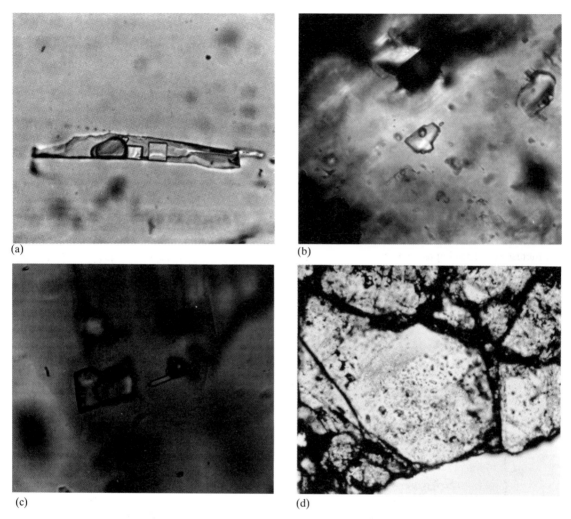

Fig. 3.1 Some fluid inclusions. (a) This spectacular fluid inclusion (about 150 μm long) contains three fluids − two immiscible liquids and a vapour bubble − and three solid phases. The cubic minerals are probably halite and sylvite. This example comes from an emerald from Muzo, Colombia. Fluid inclusions in diagenetic minerals are at best an order of magnitude smaller and tend to have relatively simple chemistry. (b) Simple two phase inclusion about 20 μm long in diagenetic calcite cement. It contains liquid water (with dissolved ions and CH_4) and a spherical vapour bubble. (c) Two phase liquid + vapour fluid inclusion about 20 μm long in diagenetic calcite. In this case, both liquid and vapour are mixtures of CH_4 and H_2S. The inclusion contains no detectable water. (d) Crystals of diagenetic celestite ($SrSO_4$) with abundant fluid inclusions up to 25 μm across containing two petroleum phases (liquid and vapour). Again, these inclusions contain no water.

through the rock sample and approximately when they did so. If a particular group of fluid inclusions became trapped in a mineral *after* it grew, during healing of a fracture set, any temperatures obtained from them are not related to mineral growth, but to palaeo-stress regimes. Stress history is an important influence on the way in which a reservoir will behave during a 'Frac' − hydrofracturing de-

signed to enhance flow of petroleum from a reservoir rock into the well bore or flow of injection water from well bore to rock.

If this kind of information is to be reliable, it cannot be gathered without exercising a considerable amount of care. This ought to extend from sample selection and preparation through data collection to, in particular, interpretation. Unfortunately, fluid inclusion data from sedimentary rocks are frequently gathered and reported perfectly competently and then interpreted with a lack of circumspection that represents a triumph of hope over expectation. It is difficult for non-specialists to understand where optimistic or unjustified assumptions or jumps in logic have been made, or even to evaluate them when they are spelled out. In this chapter, we are going to try to provide readers with the knowledge to interpret data for themselves and make their own judgements. We hope to do this in a digestible way by concentrating on the practicalities of extracting a *geologically meaningful* set of conclusions from fluid inclusion measurements. We will explain theory and data collection only to an extent that will allow readers to understand how to interpret for themselves. The chapter begins with a section on the relationship of fluid inclusions to the mineral host. It is vital to sort this out at an early stage. After all, there's not much point in acquiring an extensive set of measurements if it's not possible to work out what they refer to. The bulk of the chapter is then devoted to microthermometry – the measurement of the temperatures of phase changes in inclusions as they are heated and cooled on a microscope stage.

This is a non-destructive technique that provides information about both temperature and fluid composition and is by far the most important method for acquiring useful data from fluid inclusions. The chapter concludes with two sections on the chemical analysis of the fluids in the inclusions. Fluid inclusions are notoriously difficult to analyse, mainly because they are extremely small, particularly so in mineral cements. As a consequence, almost every analytical technique devised has at one time or another been thrown at them (see review by Roedder, 1990). We will cover only those that have been proved, or are likely in the near future to prove of some use for analysing those types of fluid inclusion that are commonly found in sedimentary rocks.

3.2 Relationship to host mineral

Any fluid inclusion study should begin with a petrographic study. Its purpose is to describe the setting of the inclusions in the rock fabric and so ascribe them to groups related to features of the history of the sample (Fig. 3.2). Mineral growth and fracturing are the most obvious such features. If this step is not followed or is done badly, any data

Fig. 3.2 Relating fluid inclusions to features of a sample's history: camera lucida drawing from a thin-section of a cemented sandstone. Inclusions A are isolated within the clastic quartz grains and were probably trapped during crystallization of the granitic rock from which they were eroded; inclusions B lie along a healed fracture but do not pass either through diagenetic cements or two adjacent grains, so were probably trapped in the source terrane; inclusions C are isolated in a quartz overgrowth and will provide information about the conditions of quartz cement precipitation; inclusion D is secondary, trapped in the sediment after quartz cementation; inclusion E is related to calcite cement growth; and inclusions F are secondary and later than both quartz and calcite cement.

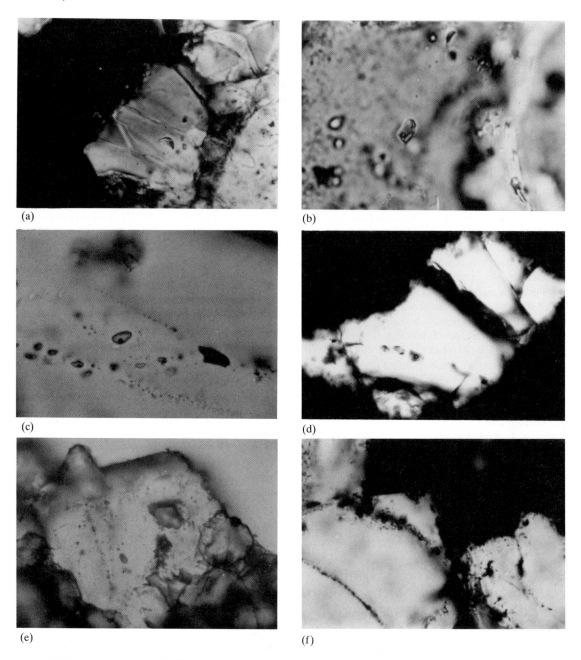

Fig. 3.3 Primary and secondary fluid inclusions. (a) Isolated inclusion 15 μm long in a calcite cement crystal. This would be interpreted as primary. (b) Isolated and therefore primary inclusion 10 μm long in quartz cement. The curved boundary between grain and overgrowth can be seen to the right of the photo. (c) Secondary aqueous inclusions along a healed fracture that cuts a calcite cement crystal. (d) Secondary inclusions cutting across an anhydrite fracture fill in a micrite. Both liquid and vapour are petroleum. (e) A euhedral diagenetic celestite crystal. The specks lying along planes parallel to the crystal faces are tiny petroleum fluid inclusions up to about 10 μm long. The inclusions lie along growth zones and must have been trapped as the crystal grew. Such inclusions are sometimes described as pseudosecondary because, like secondary inclusions, they lie along planes. (f) The boundary between quartz grain and overgrowth is marked by numerous minute fluid inclusions up to 5 μm long. These must have been trapped as the overgrowth began to grow.

subsequently obtained from the inclusions will be at best useless and at worst positively misleading.

Most work on fluid inclusions, including petrography, is carried out on specially prepared thin-sections. The sections are polished on both sides and prepared without unnecessary heating that might damage the inclusions, so that the same sample material can be used on the heating–cooling microscope stage for microthermometry (Shepherd *et al.*, 1985). Though sometimes prepared with thicknesses up to 200 μm so as to contain as many inclusions as possible, sections much thicker than the 30 μm of standard thin-sections are not of much use for sandstones because of the small crystal size of the mineral cements. Inclusion petrography *can* be carried out on normal thin-sections, but this is best avoided since the process by which the sections are made may destroy or damage some inclusions.

Populations of fluid inclusions are denoted *primary* or *secondary* (Fig. 3.3; Roedder, 1984). Primary inclusions are those trapped during mineral growth while secondary inclusions are those trapped at some time afterwards. These distinctions are extremely important. Data obtained from primary inclusions can tell us something about mineral growth which is often (though not quite always) what we want to know. Data obtained from secondary inclusions will not tell us anything at all about mineral growth but can provide information about a phase of fracture healing that post-dated growth of the host mineral. Secondary inclusions can be positively identified by their relationship to a healed fracture: if disposed along a plane that cross-cuts the mineral host, they are likely to be secondary. Unfortunately, it is usual to have to resort to negative evidence in order to identify primary inclusions: if they are isolated within the mineral host with no apparent relationship to planar features, then they are usually considered to be primary. One further term is often used to classify fluid inclusions. *Pseudosecondary* inclusions are those that look secondary, that is, disposed along planes, but which were in fact trapped during mineral growth. Two examples are shown in Figs. 3.3e and f. Criteria like these are not often found but when they are they constitute the best kind of textural evidence for relating a set of fluid inclusions to mineral growth.

Fluid inclusions in diagenetic minerals tend to be both rare and small. This is especially true of inclusions in diagenetic quartz. It is not unknown for a trained operator to scan thin-sections of quartz-cemented sandstone for an entire day without finding a single inclusion in the cement. When they do turn up, they are invariably only a few micrometres long (when less than about 4 μm long, no analysis of any kind can be made). Petroleum inclusions of this size appear quite colourless and cannot be distinguished from aqueous inclusions in transmitted light. It is therefore vital to scan sections under an incident ultraviolet light microscope. Oil inclusions

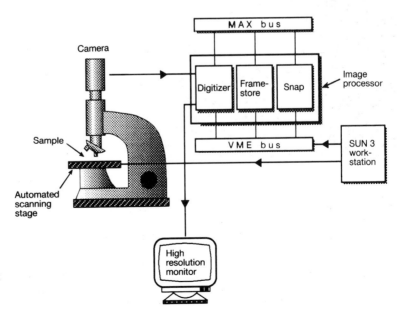

Fig. 3.4 Image analysis equipment for automatic location of fluid inclusions.

will fluoresce strongly under UV light (though gas inclusions will generally not). Cathodoluminescence is another useful tool for linking fluid inclusion populations to mineral petrography (Section 2.3). The standard cold cathode luminoscope can be used for carbonates while either a hot cathode device or a SEM with a visible light collector will normally be used to detect luminescence in quartz. The heating involved in inducing luminescence may damage inclusions and render them unsuitable for subsequent microthermometry, so CL should be done afterwards to locate inclusions already analysed.

Because finding fluid inclusions in diagenetic minerals can take a very long time, we have experimented with automatic inclusion location by image analysis. The equipment is illustrated in Fig. 3.4 (p. 45). A computer is used to move the microscope stage in three dimensions so that an entire doubly polished thin-section can be scanned in a series of fields of view. Each is digitized and scanned for inclusions by the image analyser (Fig. 3.5). The locations of all inclusions are then recorded so that the operator can later examine each in turn.

3.3 Microthermometry I – principles

3.3.1 Introduction

Imagine a sandstone buried to a depth of several kilometres in a sedimentary basin and undergoing cementation by quartz. As it grows, the quartz will trap some samples of the pore fluid as fluid inclusions. Usually, the pore fluid is a solution of a few major salts, and for the purposes of this example

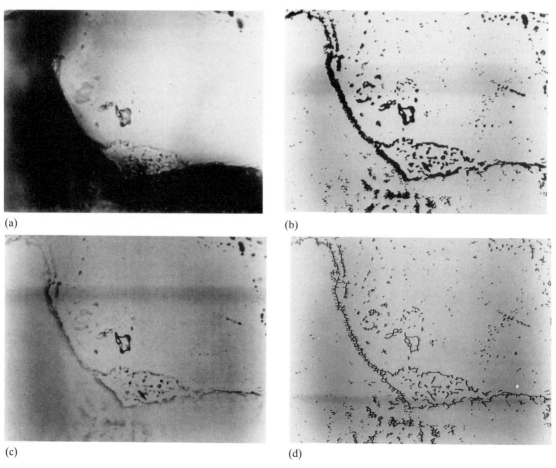

(a)　　　　　　　　　　　　　　　(b)

(c)　　　　　　　　　　　　　　　(d)

Fig. 3.5 Automatic detection of fluid inclusions. (a) Photograph of a thin-section showing a two phase fluid inclusion in quartz. (b) The image is digitized. (c) A filter removes noise. (d) Another filter enhances edges and the inclusion is identified by the occurrence of one closed contour within another. Its location in the section is then stored.

we will assume that it is a simple NaCl solution. Let us imagine that the sample is then cored and brought to the surface. At the reservoir temperature and pressure, the brine was a single, liquid phase but when we observe the inclusion under our microscope at room temperature it will contain the characteristic assemblage, liquid plus vapour bubble. On a special microscope stage, we can cool the inclusion still further and freeze the brine. Suppose we now begin to reverse the process and heat the inclusion. Salt and other solutes depress the melting point of ice (which is why $CaCl_2$ is spread on roads in winter) so the temperature at which the ice in the inclusion melts will reflect the salinity of the fluid. If we keep heating, the vapour bubble will eventually disappear and the inclusion will once again contain a single, liquid phase.

Microthermometry is the observation of the temperatures of phase changes like these that are induced within fluid inclusions as they are cooled and heated on a special microscope stage. The example above shows the two basic kinds of information that the method can provide: the temperature of ice melting tells us something about fluid chemistry and the temperature of bubble disappearance is a minimum value for the temperature at which the inclusion was trapped. The more complicated the chemistry of an inclusion fluid, the greater the number of possible phase changes. Fortunately in practice, it is usually only necessary to consider a fairly limited number. These involve:

1 melting of solids like ice, in the presence of a vapour bubble (solid + liquid + vapour equilibrium);

2 disappearance of a vapour bubble into the liquid phase − the *homogenization temperature** (liquid + vapour equilibrium); and

3 dissolution of a crystal (*daughter mineral*) in very concentrated solutions (solid + liquid equilibrium). In the following sections, we will describe the first two but not the third, because it is extremely unusual to find mineral crystals in fluid inclusions in diagenetic minerals and even more unusual for these to dissolve on heating (perhaps suggesting that they fell into the inclusion cavity before it was sealed rather than exsolving from the fluid on cooling).

3.3.2 Melting temperatures of solid phases

Salt solutions
Microthermometry involves interpreting phase changes so the best way of understanding how it works is to look at some simple phase diagrams. Figure 3.6 is part of the phase diagram for the system NaCl–water. A frozen inclusion with bulk composition I that consists only of Na^+, Cl^- and water will contain a mixture of ice and the intermediate salt, hydrohalite ($NaCl.2H_2O$). On heating, the first liquid will appear at the eutectic temperature, E ($-20.8°C$) as hydrohalite melts and eventually disappears. As heating continues and ice

* The homogenization temperature is strictly defined as that at which the inclusion becomes single-phase on heating. At room temperature, highly saline fluids can contain a 'daughter' crystal of halite (or another mineral) which may dissolve at a temperature higher than that at which the vapour bubble disappears. The temperature of halite dissolution is then the homogenization temperature. However, in sedimentary rocks, inclusions where homogenization does not involve the disappearance of a vapour bubble into a liquid phase are very unusual.

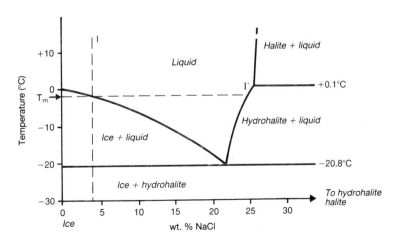

Fig. 3.6 Part of the vapour-saturated phase diagram for NaCl–water. See text for explanation.

melts, the proportion of liquid increases until at T_m, the last ice crystal disappears. Three pieces of information would be recorded:

1 the temperature at which liquid first appears;
2 the temperature of final solid melting; and
3 the last solid to melt (ice or hydrohalite?).

The first − the eutectic temperature − confirms that the inclusion fluid is indeed a NaCl solution and the second two − melting of ice at $T_m°C$ − together fix the bulk composition at I. Note that if hydrohalite rather than ice had been the last solid to melt at $T_m°C$, a very different bulk salinity would be indicated (I′). This idealized example shows that the composition of an inclusion fluid can be determined provided that:

1 the composition of the fluid is known; and
2 the depression of melting point has been experimentally calibrated; that is, that the phase relationships in the system are known.

Most natural waters are not pure NaCl solutions and, in practice, we rarely have even a qualitative ionic analysis of inclusion fluids. The composition of inclusion fluids can however sometimes be guessed from the temperature at which liquid first appears because the eutectic is an invariant property of a particular system. For example, it is quite common to observe first melting at temperatures close to −52°C, the eutectic in the system $CaCl_2$−NaCl−water. The main difficulty is that a eutectic is impossible to measure accurately, simply because a small amount of liquid is almost impossible to see. This problem becomes acute with the minute inclusions that are so characteristic of most diagenetic minerals. The approach taken is therefore a pragmatic one. The temperature at which liquid is first seen is recorded and often indicates at very least that the solution is not a simple NaCl solution. Provided that the last solid to melt is ice, the temperature of final melting is then interpreted as though the solution were a simple NaCl brine, producing a number known as *equivalent weight per cent NaCl* (eq. wt% NaCl). This is only a semi-quantitative estimate of salinity but is still useful for distinguishing between fluid inclusions, and fluids, of different origins.

Gas hydrates in aqueous inclusions

Many fluids in sedimentary basins contain substantial amounts of gas in solution. The most common are CH_4 and CO_2, but H_2S and N may locally be abundant and many others can be present in small quantities. Most of these gases can form hydrates known as *clathrates* that are stable at temperatures close to 0°C. Petroleum geologists will no doubt be familiar with CH_4 clathrates because they form in apparently vast quantities in sediments beneath deep water and in permafrost regions (their stability requires a combination of low temperature and moderate pressure). Clathrates also form when aqueous fluid inclusions that contain substantial amounts of dissolved gas are frozen. They can be detected by their low refractive indices (though this makes them difficult to see), or more usually by melting temperatures above 0°C.

Clathrates exclude cations from their structures when they form, causing the salinity of any remaining water to increase. Ice melting then becomes a misleading indicator of bulk fluid salinity. The melting temperature of the clathrate is however itself depressed by the presence of ions in solution and can be used to estimate salinity provided that the particular clathrate can be identified and that the depression of melting point has been experimentally calibrated. Individual gas clathrates can be identified by laser Raman spectroscopy (see Section 3.5.2) but the effort involved is considerable and in any case the melting point curve is known only for CO_2-clathrate (Collins, 1979). Clathrate identification may nonetheless be important to petroleum geologists if, for example, they are interested in the presence of CH_4.

Gas-rich inclusions

Most fluid inclusions in diagenetic minerals are either aqueous, with or without minor dissolved gas, or petroleum. Occasionally, however, we have found inclusions that contain liquid + vapour at room temperature that are essentially water free, comprising mixtures of CH_4 and H_2S − natural 'sour' gas (Fig. 3.1c). A solid phase does not form in these inclusions unless they are cooled below about −150°C and melts on heating at around −90°C. This is below the triple point of pure H_2S and reflects a depression of the solid + liquid + vapour equilibrium by the presence of an extra component, CH_4. These identifications were subsequently confirmed by laser Raman spectroscopy (see Fig. 3.23).

3.3.3 Homogenization temperatures

Pure water and salt solutions

Figure 3.7 is another phase diagram, a pressure−temperature plot of the phase relations for pure water. Consider a fluid inclusion trapped in the liquid part of the phase diagram at point T_t, P_t

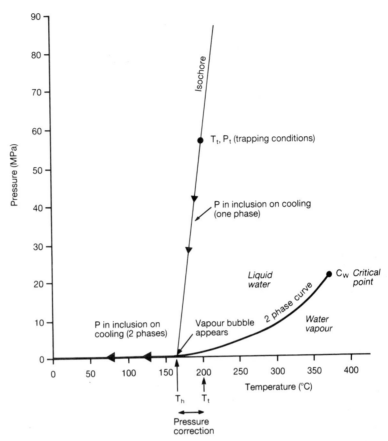

Fig. 3.7 Pressure–temperature plot of the phase diagram for pure water. See text for explanation.

representing the temperature and pressure in an oil reservoir. When the sample is brought to the surface, it cools. We make the assumption that over the temperature interval of interest, the fluid in the inclusion has constant density. This requires that:

1 the inclusion volume remain constant; and

2 there is no leakage of material in or out of the inclusion*.

As the inclusion cools, its internal pressure is determined by the *PVT* properties of water at constant density and follows a path known as an *isochore*, a relationship between the pressure and temperature of water at constant density (equivalent to constant molar volume, *V*). When the internal pressure has dropped to the point where the isochore intersects

---

* These assumptions are generally justified. It is possible to calculate the error involved in assuming that the inclusion volume is constant and this turns out to be small. Equally, though leakage of inclusions sometimes does happen during analysis, it can generally be recognized.

the two-phase curve (or 'boiling point' curve) at T_h, a vapour bubble nucleates and continues to grow on further cooling. At room temperature, the inclusion then looks like everybody's idea of what a fluid inclusion should be: it contains two phases – liquid and a vapour bubble.

If the natural cooling process is reversed by heating the inclusion on the microscope stage, the bubble will decrease in size and disappear at T_h, the homogenization temperature. T_h is a minimum value for the trapping temperature T_t and defines the fluid density provided that the composition of the fluid and its *PVT* properties are known. Note that T_h is *not* equal to T_t (which cannot in fact be measured at all!). The homogenization temperature is indeed always less than the trapping temperature, by an amount that is confusingly known as the *pressure correction*. Determining the pressure correction is one of the greatest problems in the interpretation of fluid inclusion data and one to which we will return in Section 3.4.3.

The behaviour of fluid inclusions containing NaCl solutions is qualitatively similar to those that contain pure water. The two-phase curve gets an extra degree of freedom with the addition of one extra component, expanding into an area and changing its position in $P-T$ space. Homogenization takes place on a *bubble point curve*. The slopes of the isochores are also somewhat different. More complex salt solutions are usually assumed to behave in a similar manner (though their *PVT* properties are mainly poorly known).

Regardless of the composition of the fluid, the homogenization temperature can always be measured and is always a minimum value for the trapping temperature as long as two assumptions are true:

1 the fluid was trapped as a single phase; and
2 fluid density has not changed since trapping.

The effect of dissolved gases
Figure 3.8 shows the phase behaviour of a fluid inclusion trapped at T_t, P_t that contains dissolved CH_4. Nucleation of a bubble on cooling, and bubble disappearance (homogenization) on heating, takes place on the bubble point curve. This lies at much

higher pressures than two-phase curves in salt–water systems. The difference between T_t and T_h, the pressure correction, is therefore small and the measured homogenization temperature is a good estimate of the trapping conditions (Hanor, 1980). Figure 3.8 is constructed for a CH_4 solubility of 3200 ppm (0.2 molal), a reasonable value for a formation water in a sedimentary basin. At room temperature, the partial pressure of CH_4 in the vapour bubble is about 9 MPa. Laser Raman spectroscopy is capable of identifying CH_4 in fluid inclusions at partial pressures as low as about 0.1 MPa, so it ought to be easily detectable in our hypothetical inclusion (see Section 3.5.2).

Petroleum fluid inclusions
Petroleums have two phase envelopes in $P-T$ space, bounded by bubble point and dew point curves (Fig. 3.9). Homogenization in petroleum fluid inclusions usually occurs by vapour bubble disappearance along the bubble point curve. Isochores for petroleums are not as steep as those of aqueous fluids and, as a consequence, the pressure correction tends to be quite substantial, frequently many tens of degrees Celsius. The homogenization temperatures of petro-

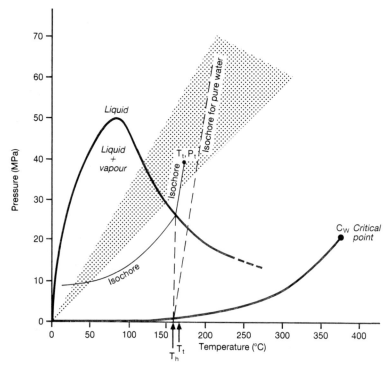

Fig. 3.8 The effect of dissolved CH_4 (modified from Hanor, 1980). Phase diagram for water containing 3200 ppm methane. The stippled area represents reasonable *P-T* conditions in sedimentary basins. The boiling point curve for pure water is shown for reference only. See text for explanation.

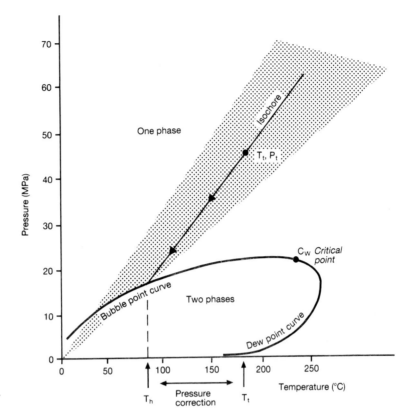

Fig. 3.9 A phase diagram for a crude oil. The stippled area represents reasonable *P–T* conditions in sedimentary basins. See text for explanation.

leum fluid inclusions therefore tend to be poor estimates of the temperatures at which the petroleum became trapped so that petroleum inclusions are far less useful than aqueous inclusions when it comes to geothermometry.

3.3.4 Data collection — precision and accuracy

Microthermometry is carried out on a heating–cooling microscope stage. Several designs are commercially available. All permit cycling between temperatures close to those of liquid N_2 (which is used in most as a coolant) and an upper limit of several hundreds of degrees Celsius, at the same time permitting microscopic observation in plane polarized light. All stages need to be carefully calibrated using ultra-pure compounds of known melting point (MacDonald & Spooner, 1981)* or synthetic fluid inclusions grown at known temperatures from fluids of known composition.

* Details of how the measurements are made can be found in Shepherd *et al.* (1985).

Measurements of each phase change are usually presented as histograms in which the frequency of measurement is plotted against a temperature interval, usually 5 or 10°C for homogenization temperature and 2 or 5°C for ice final melting temperature. Histograms can be plotted either for a single or several samples, but these should come from about the same depth[†]. Figure 3.10 shows a typical sample of homogenization temperature measurements from quartz cement. How accurate and precise are these data? The precision of the temperature measurements is equivalent to their repeatability. Usually, this is around ± 0.1°C for an ice final melting point and better than ± 0.5°C for a homogenization temperature. The accuracy depends both on precision and on the calibration of the stage. For our stage, we have estimated the accuracy of the respective measurements to be around ± 0.2 and ± 1°C. So microthermometric measurements are both precise

[†] Data from samples from a wide range of depths should not be lumped together as depth and T_h are often correlated (Robinson *et al.*, 1992b).

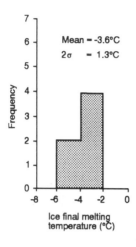
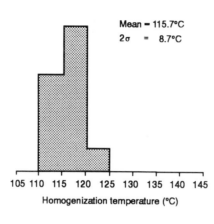

Fig. 3.10 Homogenization temperatures of fluid inclusions in quartz cement. Histograms of ice final melting (left) and homogenization (right) temperatures measured from primary aqueous fluid inclusions in quartz cement, Tarbert Formation, Brent Group, UKCS well 3/8b-10; sample depths 4054 and 4062 mBRT (below rotary table).

and accurate. But, what are they precise and accurate measurements of? Not usually of the number that we want to know — trapping temperature — but of the temperature of a phase change which has to be interpreted to provide trapping temperature. The message is that effectively all the uncertainty surrounding fluid inclusion data is contained in the assumptions made in their interpretation. The accuracy and precision of measurements like those presented in Fig. 3.10 are so good that they can be ignored. In other words, the data in Fig. 3.10 are highly accurate measurements of a minimum value for trapping of the inclusion fluids but their accuracy as estimates of the *actual* trapping temperature is unknown.

When dealing with fluid inclusions in diagenetic minerals, it is extremely time consuming or even impossible to collect a large number of measurements simply because suitable inclusions are so scarce. Figure 3.10 shows 11 measurements of T_h obtained from two samples. The mean is 115.7°C and 2σ is 8.7°C. Do we need more data? The answer depends on whether we feel that the data already collected represent to an acceptable degree the population of all fluid inclusions that share this common origin. What is acceptable depends on what the data are to be used for. The main applications in petroleum geology would be to estimate an age for mineral growth from the mean and perhaps its duration from the variance (as described by the standard deviation). The uncertainties involved in both procedures are likely to be large relative to any changes in the mean and standard deviation of our sample that might arise if more data were collected. We would call it a day and go for a beer!

3.4 Microthermometry II — interpretation

3.4.1 Introduction

As with all analytical methods, there are complications and difficulties involved in fluid inclusion work. Some are really procedural and can be overcome with a mixture of application, good fortune and experience. In this category we would include, for example, the small size and scarcity of fluid inclusions in diagenetic minerals and the consequent difficulty in generating adequate datasets. There are however two uncertainties associated with interpretation that are less easy to deal with. The first is the extent to which inclusions have undergone *natural* stretching or leakage *before* the sample is collected. If they have, measured homogenization temperatures will not reflect the original conditions of trapping and may be almost impossible to interpret in any geologically meaningful way. This is because if an inclusion stretches or leaks, one of the key assumptions of fluid inclusion interpretation — constant density since trapping — will not hold. We will consider this question in some detail because it is fundamental to the use of fluid inclusion studies in sedimentary geology. It is also a question that has been dodged by many who use the technique. The conclusion that we reach is that only fluid inclusions in quartz can be relied upon to provide information about mineral growth temperature (though those in other minerals may under certain circumstances). The reliability of fluid inclusions in quartz is widely but not universally accepted: Osborne and Haszeldine (1993), for example, believe that they routinely leak and break the constant density assumption. The second question concerns the calculation of trapping temperature from homogenization temperature – making the pressure correction. We will see that this process is fraught with difficulty, raising the question as to whether it should even be attempted.

3.4.2 Stretching and leakage – a terminal problem?

Why should inclusions stretch or leak?

In order to interpret information about trapping temperature from fluid inclusions, it is necessary to assume that the density of the fluid in each inclusion cavity has remained constant since trapping. If this assumption is not valid, homogenization temperatures measured in the laboratory will bear no systematic relationship to fluid trapping temperatures and will be at very least extremely difficult to interpret geologically. Under certain circumstances, fluid inclusions can undergo natural non-elastic

deformation: they may stretch or even burst open ('leak'). In the first case, the volume of the cavity and therefore fluid density would change; in the second, gain or loss of fluid to or from the inclusion would again alter the density of that remaining in the cavity. Either way, the critical assumption is violated.

The diagenetic environment is unhappily one of those in which the constant-volume assumption tends not to hold (Robinson *et al.*, 1992b). Indeed, the increase in temperature followed by a sediment in a subsiding sedimentary basin promotes stretching/leakage of fluid inclusions. Figure 3.11 helps to explain why. Consider an inclusion of density $0.90\,\mathrm{g.cm^{-3}}$ trapped in a mineral cement in a porous sediment at a pressure of 57 MPa (point T). The P–T path followed by the fluid in the sediment pores during burial is calculated assuming hydrostatic pressure and a geothermal gradient of $30°\mathrm{C.km^{-1}}$. After 0.5 km of further subsidence, the pore pressure

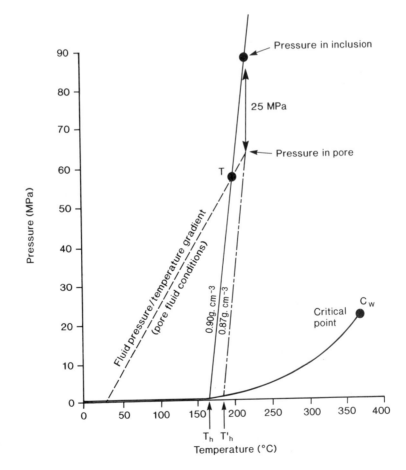

Fig. 3.11 Why fluid inclusions in diagenetic minerals tend to stretch or leak. See text for explanation.

in the sediment will have increased to about 63 MPa. The pressure in the inclusion, however, is constrained to follow a constant density isochore as temperature increases. For aqueous fluids (though not for petroleum), these isochores are steep relative to the $P-T$ path followed by the fluid in the pores. The result of even a small amount of heating beyond the trapping temperature is therefore a large differential pressure between inclusion and pore. In the example in Fig. 3.11, just 15°C of overheating (half a kilometre of further burial) leads to a differential pressure of about 25 MPa. If this exceeds the strength of the mineral, non-elastic deformation of some kind will result. The inclusion fluid may even re-equilibrate with that in the pores, reducing its density (from 0.90 to 0.87 g.cm^{-3} in the example). The homogenization temperature that we would measure for this inclusion (T_h') would bear no systematic relationship to the trapping temperature, T.

It has now been demonstrated beyond reasonable doubt that aqueous fluid inclusions in many minerals generally do leak or stretch during diagenesis and consequently *do not* provide reliable information about trapping temperatures (Burruss, 1987; Prezbindowski & Tapp, 1991). This is true of calcite (Comings & Cercone, 1986; Goldstein, 1986; Prezbindowski & Larese, 1987; Barker & Halley, 1988) and probably of all carbonates. It is also likely to be true of other minerals with cleavages and/or low mechanical strengths, such as sulphates and feldspars. Below we will briefly consider the evidence for and against stretching of fluid inclusions in diagenetic minerals, paying particular attention to quartz. We begin by describing patterns shown by fluid inclusion data in mineral cements and explain why they may be interpreted as indicating wholesale stretching or leakage. We then review the experimental work on stretching, noting its possible implications and limitations. The next section is devoted to describing evidence from the rocks themselves. The comforting conclusion is that inclusions in quartz probably do not generally stretch or leak.

Patterns of inclusion data – evidence for stretching?
Homogenization temperatures of primary fluid inclusions in diagenetic minerals are very commonly close to maximum burial temperatures (e.g. Jourdan *et al.*, 1987; Hogg, 1989; Walderhaug, 1990). In cases where data have been gathered from the same formation over a range of depths, there is a clear correlation between homogenization temperature and depth (and, of course, maximum burial tem-

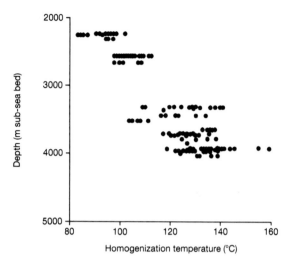

Fig. 3.12 Homogenization temperatures of primary fluid inclusions in quartz cement correlate with depth. Data from quartz cement in the Garn Formation, Haltenbanken, offshore Norway (from Grant & Oxtoby, 1992). Plotted against depth sub-sea floor.

perature since depth and temperature are themselves correlated; Fig. 3.12). These correlations are open to two interpretations. If the inclusions retain information about mineral growth temperatures, they must indicate geologically rapid (less than about 10 Ma ago), late-diagenetic periods of cementation that affect much of the formation at about the same *time* (Robinson & Gluyas, 1992a). The alternative explanation is that the correlation is due to stretching or leakage and that homogenization temperatures are responding to increasing degrees of overheating as a result of progressive burial (Prezbindowski & Tapp, 1991).

Experimental stretching of fluid inclusions
One obvious way of trying to resolve the question of whether stretching/leakage is a serious problem is to try to simulate the process in the laboratory. Experiments have been carried out on fluorite and sphalerite (Bodnar & Bethke, 1984), barite (Ulrich & Bodnar, 1988), calcite (Prezbindowski & Larese, 1987) and on quartz itself (Leroy, 1979; Binns & Bodnar, 1986; Bodnar *et al.*, 1989; Sterner & Bodnar, 1989). The studies involve determining inclusion homogenization temperatures and then repeating the measurements after heating by a particular increment above the initial value. The results show that the amount of overheating required to

induce stretching/leakage depends firstly on the strength of the mineral; secondly on inclusion size; and thirdly, on inclusion shape.

The most important factor is mineral strength. Of the inclusions in barite, 42% stretched or burst open (*decrepitated*) with <10°C of overheating. Inclusions in fluorite began to stretch with 20–30°C of overheating and after 80°C 90% had done so. Stretching became significant for inclusions in sphalerite only with overheating of 30–50°C and none recorded temperatures greater than 25°C above the original homogenization temperature. Stretching was more common than leakage (decrepitation) in these minerals and was difficult or impossible to recognize petrographically. The corresponding experiments performed at 1 atmosphere confining pressure on quartz are however more encouraging. They suggest that <10% of fluid inclusions in quartz will be modified provided that the internal pressure is not greater than about 200 MPa*. For an inclusion containing pure water, this figure corresponds to about 200°C of overheating past the homogenization temperature. For a salt solution, the amount of overheating needed would be even higher.

One of the most important conclusions to be drawn from the experimental studies is that the amount of differential pressure required to induce non-elastic deformation depends on inclusion size (and to a lesser extent on shape). In general, small inclusions are more resistant to stretching than large ones and equant inclusions more resistant than irregular ones. Relationships of this kind have been found for fluorite, barite, and also for quartz. There are two alternative equations for quartz:

$$P_{(int)} \text{ bars} = 2525 - 162\log_{10}V - 58.5\log_{10}^2V + 36Y \tag{1}$$

(Binns & Bodnar, 1986)

$$P_{(int)} \text{ bars} = 3890V^{-0.141} \tag{2}$$

(Bodnar *et al.*, 1989; see Fig. 3.13). *V* is the inclusion volume in cubic micrometres and *Y* is a shape factor that varies from 0 for highly irregular inclusions to 10 for spherical inclusions or those with a negative crystal shape. For a spherical inclusion with a diameter of 5 μm, equation 1 predicts that the pressure required to induce decrepitation would be

---

* The experiment was carried out at a confining pressure of 1 bar. The critical factor controlling stretching is probably *differential pressure* between inclusion and pore rather than absolute inclusion pressure.

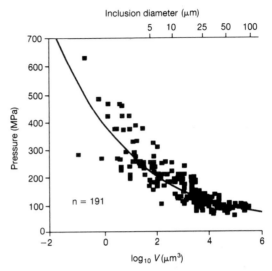

Fig. 3.13 Experimentally determined relationship between inclusion size and the internal pressure required to stretch or decrepitate inclusions in quartz (from Bodnar *et al.*, 1989).

240 MPa; if the diameter were 3 μm, the pressure would be 262 MPa. Equation 2 predicts corresponding pressures of 216 and 268 MPa.

The experimental work provides then some confidence that differential pressures of the order of 200 MPa (equivalent to overheating by more than 200°C) are needed to cause wholesale natural deformation of a population of fluid inclusions in quartz cement. However, there is some doubt as to whether these results can be extended to pressure–temperature and time regimes characteristic of diagenesis; whether for example, over geological time, even relatively small differential pressures might lead to inclusion volume changes. Also, deformation behaviour may differ under natural conditions. In the experiments carried out at atmospheric confining pressure, inclusions in quartz tended to respond to overheating by decrepitating. In those performed at elevated confining pressures, inclusions tended to stretch rather than decrepitate. This is important because inclusions that have decrepitated can probably be recognized and excluded from a dataset. Stretching cannot be detected petrographically.

Evidence for absence of stretching
Experimental studies show relationships between inclusion size and the amount of overheating

required to induce deformation. Small inclusions in diagenetic minerals might therefore be expected to retain low (unmodified) homogenization temperatures better than large ones and the result should be a correlation between T_h and inclusion size in a population of deformed inclusions. The homogenization temperatures of fluid inclusions in quartz cement are *not* in our experience correlated with inclusion size (Fig. 3.14). However, in some cases, neither is the expected relationship displayed by fluid inclusion data from calcites and dolomites when there is independent evidence that stretching or leakage has in fact occurred (Burruss, 1987). It seems then that the lack of a clear correlation between homogenization temperature and size does not demonstrate that the inclusions have not stretched.

Perhaps a more convincing line of evidence comes from the internal consistency of fluid inclusion data. The homogenization temperatures of primary fluid inclusions in quartz cement in a single sample (or a number of samples from a restricted range of depths in a well) generally cluster tightly about a well-defined mean (Robinson & Gluyas, 1992a). It seems unlikely that a sample of inclusions that had been stretched to an extent that should depend on size, fluid composition and, probably, location within the mineral host (proximity to dislocations, fractures, etc.) would show such consistency. This pattern, so characteristic of inclusions in quartz, certainly contrasts with the large ranges of T_h often shown by

inclusions in diagenetic carbonate cements (Fig. 3.15; Visser, 1982; Goldstein, 1986). It is nonetheless conceivable that an internally consistent dataset may not constitute good evidence against deformation. Experimental stretching of inclusions in calcite by Prezbindowski and Larese (1987) did lead to broadening of an originally tighter distribution about a much higher mean temperature, but not by amounts that could be easily identified in a natural example. The histograms of Fig. 3.15a−c might then represent samples of stretched inclusions that had originally slightly tighter T_h distributions, but quite different (lower) means.

If homogenization temperatures obtained from fluid inclusions are to be interpreted in terms of mineral growth temperatures, they must be compatible with information from independent sources. Illite cements, often found in sandstones cemented also by quartz, can be dated radiometrically by the K−Ar method. Their textural relationship to quartz cement can frequently be established from SEM and thin-section petrography. K−Ar dating of illite cements has its own uncertainties (see Chapter 5), but we have found that relative ages of quartz and illite cement determined petrographically generally are in agreement with estimated absolute ages derived, in the case of illite, from careful K−Ar dating, and in the case of quartz from fluid inclusion data referred to a modelled burial history. This suggests that the fluid inclusions have not stretched.

The last and possibly most persuasive line of evidence is the preservation of temperature differences between related inclusion populations. While it is conceivable that different inclusion populations in a cement might all undergo stretching and still retain homogenization temperature distributions suggestive of their separate origins, it seems likely that stretching would smooth out or obliterate original differences. In the absence of stretching/leakage, different generations of inclusions in a sample ought then to show different homogenization temperature distributions. In a sedimentary rock, those which can be shown by textural relationships to have been trapped earliest would usually be expected to exhibit the lowest temperatures. It is often possible to recognize a sequence of growth zones in quartz cements using cathodoluminescence. In those cases where it has proved possible to relate primary fluid inclusions to a cement zone stratigraphy, the homogenization temperature distributions are consistently significantly different for inclusions from different

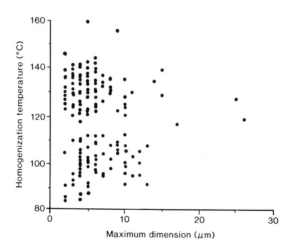

Fig. 3.14 Homogenization temperatures of fluid inclusions in quartz cement are not correlated with inclusion size. Data from quartz cement in the Garn Formation, Haltenbanken.

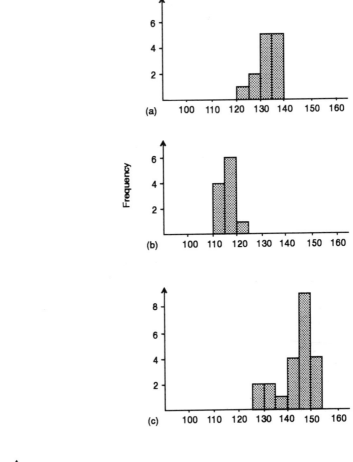

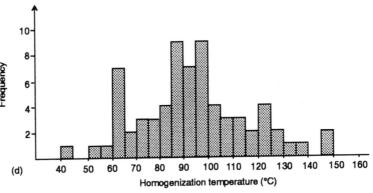

Fig. 3.15 Distributions of homogenization temperatures of inclusions in quartz cement tend to be narrower than those in carbonates (from Robinson *et al.*, 1992b). (a) Quartz cement, Garn Formation, Haltenbanken well 6506/12-6, samples from 4223.9–4330.7 mBRT. (b) Quartz cement, Tarbert Formation, UKCS well 3/8b-10, samples at 4054 and 4062 mBRT. (c) Quartz cement, Gyda Formation, NOCS well 2/1-4, samples from 4044–4053.75 mBRT. (d) Phreatic zone calcite cement, Mississippi Valley Lake Formation, New Mexico (outcrop samples; from Goldstein, 1986).

zones (e.g. Burley *et al.*, 1989). Perhaps an even more compelling example is illustrated in Fig. 3.16. This shows histograms of homogenization and ice final melting temperatures of two types of primary inclusion in quartz cement from some samples of

Carboniferous sandstone from the Southern North Sea. Inclusions at grain–cement boundaries would have been the first to be trapped as quartz cement began to grow. They homogenize at lower temperatures (105–113°C) than those which are isolated

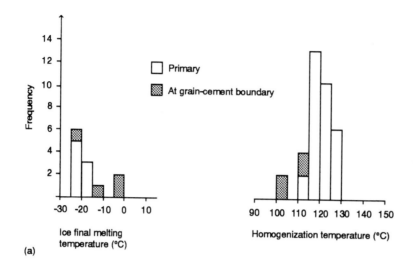

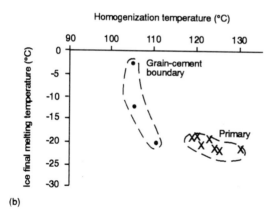

Fig. 3.16 Fluid inclusions in quartz cement retain a record of progressive burial and changes in pore fluid composition. (a) Histograms of ice final melting (left) and homogenization (right) temperatures for inclusions at grain–cement boundaries and those isolated within quartz overgrowths (primary). (b) Plot of ice final melting against homogenization temperature for the same inclusions.

within the overgrowths (113–129°C) and are in general less saline. It seems unlikely that this detailed record of changes in pore fluid composition and progressive burial would have survived had the inclusions undergone significant natural deformation.

Do petroleum inclusions stretch?
Petroleum isochores are much less steep on a pressure–temperature diagram than are those of water and salt solutions. This is the reason why the homogenization temperatures of petroleum inclusions tend to be much lower than their trapping temperatures. It is also the reason why petroleum inclusions are far less prone to leakage/stretching.

Because petroleum isochores are almost parallel to the pressure–temperature path followed by the pore fluid of a subsiding sediment (Fig. 3.9), a substantial differential pressure between pore and inclusion is not likely to develop, regardless of how much the inclusions are heated beyond their homogenization temperatures.

Summary
Although no single line of evidence may be regarded as totally conclusive, we believe that taken as a whole, the body of evidence, experimental and geological, suggests that aqueous fluid inclusions in diagenetic quartz cement do not generally stretch

and that they provide reliable information about the conditions under which they were trapped. This is probably true of other diagenetic minerals only in exceptional circumstances. Aqueous inclusions in calcite, barite, sphalerite and fluorite undoubtedly stretch naturally with small amounts of overheating and cannot be relied on to provide information about mineral growth temperature. The same is probably true of authigenic feldspars, other carbonate minerals and sulphates. Petroleum inclusions are far less susceptible to stretching in any mineral host, but for the same reason are poor geothermometers.

3.4.3 Pressure corrections: can we and should we?

The pressure correction is the amount (in degrees Celsius) that must be added to the measured homogenization temperature to obtain the trapping temperature. In order to calculate it, we must be able to determine both:
1 the slope of the isochore; and
2 the trapping pressure, P_t, or an independent $P–T$ relationship (in order to fix the point on the isochore).

There are two widely used approaches to calculating the slopes of isochores. The first involves measuring the *decrepitation temperature* (T_d) at which the fluid inclusion bursts open. The point T_d is plotted on a $P–T$ diagram at a decrepitation pressure either (as is more general) assumed to be a constant for the particular mineral host, or calculated from an equation such as (1) or (2) above. This point must lie on the isochore for the inclusion fluid, so the final step is to join this point to the homogenization temperature and pressure with a straight line. The procedure, illustrated in Fig. 3.17, defines the slope of the isochore. This method is quite widely employed, particularly by French and Russian scientists. However many, including ourselves, are highly suspicious of it. The main reason is the uncertainty surrounding the decrepitation pressure. It seems clear that there is no single decrepitation pressure applicable to inclusions of all sizes and shapes in a given mineral. Equations like (1) and (2) predict quite different values and are in any case regression lines fitted to data that show considerable scatter (see Fig. 3.13).

The second method of determining isochore slopes is however just as fraught because it requires two items of information that, certainly in the case of fluid inclusions in sedimentary samples, are extremely difficult or impossible to determine.

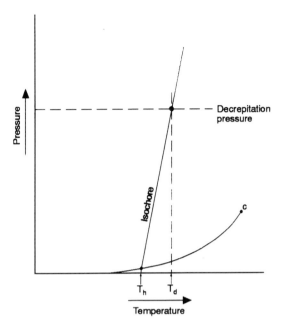

Fig. 3.17 How to calculate an isochore from decrepitation temperature. See text for explanation.

These are:
1 a quantitative chemical analysis of the inclusion fluid; and
2 the *PVT* properties of that chemical system.
Virtually the only case when *PVT* properties *may* be accurately known is if the inclusion fluids are petroleum. *PVT* properties are regularly determined by petroleum engineers by recombining oil and gas collected from the separator tank at the well head. Experimental *PVT* determination is usually preferred although equations of state are also used. In order to co-opt these *PVT* properties for use with a particular set of fluid inclusions, it is necessary to make an assumption: that the oil in the reservoir and that in the inclusions are chemically identical. This can be sometimes be checked by certain analytical methods (described below) but is unlikely in general to be strictly true. Fortunately many of the so-called 'black oils' have isochores with rather similar slopes, so the error involved in making the assumption may not be large.

Even if we were able to muster enough courage to calculate an isochore, we would still be faced with the problem of locating the trapping temperature along that line, either by obtaining an independent estimate of trapping pressure, or a pressure–temperature equation. The first is something of a no-hoper as there are no reliable geobarometers

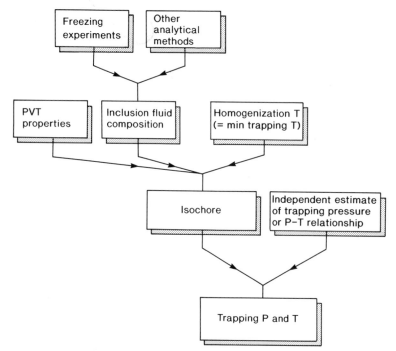

Fig. 3.18 The information required to calculate trapping temperature.

that can be applied to sedimentary rocks. There are however at least two ways of obtaining an independent pressure–temperature relationship. One involves a simple calculation for a normally (hydrostatically) pressured pore fluid in a sedimentary basin. Though easy to do, this involves making assumptions about palaeo-thermal gradients and pressure regimes that may well be incorrect. The second method is known as that of *crossed isochores*. If two immiscible fluids are trapped simultaneously, the trapping conditions of both must lie where the isochores for the two fluids cross. This phenomenon is not uncommon: oil and water are immiscible and often seem to be trapped together in growing diagenetic cements. The two examples that we will consider in a moment both use the method of crossed isochores.

This discussion ought to convince the reader that the process of determining trapping temperatures from homogenization temperatures (summarized in Fig. 3.18) involves numerous assumptions that are difficult to evaluate. Our approach is not even to try unless circumstances are unusually favourable and to accept homogenization temperatures for what they are: *highly accurate minimum values* for trapping temperatures.

3.4.4 Example 1: calcite filled fractures, Little Knife Field, North Dakota

This example involves the interpretation of a set of fluid inclusion data from fracture filling calcite in the Little Knife Field, North Dakota (Narr & Burruss, 1984). It provides a good illustration of the kinds of problem that we now know must be faced when interpreting fluid inclusion data in diagenetic minerals, problems which are still routinely glossed over in published case histories.

Two main groups of inclusions were identified: one containing liquid water plus vapour and one with liquid petroleum plus vapour. A few inclusions contained all three phases. Aqueous inclusions homogenized at temperatures between 102 (one at 76) and 141°C, though the authors believed (for reasons that were not specified) that those numbers in excess of 126°C were spurious (Fig. 3.19). All but one of the petroleum inclusions homogenized between 90 and 106°C. Most solid melting temperatures in the aqueous inclusions ranged from −3 to −5°C. The authors thought that the solid was probably hydrohalite rather than ice, so estimated the salinity of the inclusion fluid to be around 25 eq. wt% NaCl. The bottom hole temperature was 121°C

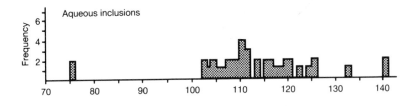

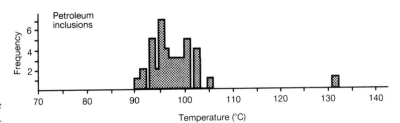

Fig. 3.19 Homogenization temperatures of fluid inclusions in fracture filling calcites, Little Knife Field (from Narr & Burruss, 1984).

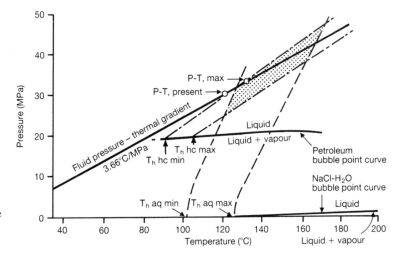

Fig. 3.20 Interpretation of the Little Knife inclusion data. The inferred trapping conditions are indicated by the stippled parallelogram. Petroleum and brine isochores are indicated by broken lines. See text for explanation.

and the maximum likely temperature was estimated at about 132°C (about 300 m of sediments had been eroded).

To the diligent reader who has struggled through to this point of the chapter, two features of the aqueous inclusion data should set alarm bells ringing. Firstly, the homogenization temperatures show a considerable scatter with some values well in excess of the maximum temperature to which the rocks had probably been subjected. Secondly, the inclusions are in calcite. These two together suggest to us that the inclusions have probably stretched or leaked. If this has happened, they are useless for estimating the temperature of calcite growth in the fractures. Leaving aside the question of stretching,

let us look at how the data were interpreted. The authors believed that petroleum and aqueous inclusions were trapped together and so used the method of crossed isochores to estimate trapping P and T from homogenization temperatures (Fig. 3.20). Isochores were obtained for petroleum inclusions by assuming that they had the same PVT properties of Little Knife crude (which were known). The aqueous inclusions were assumed to be NaCl brines so that isochores could be calculated from PVT data available in the literature. The parallelogram formed by the intersection of the two sets of isochores is the range of pressures and temperatures within which all inclusions should have been trapped. The range of trapping P and T suggested by this interpretation is mainly well in excess of the likely maximum P and T experienced by the reservoir and the authors realized that this posed problems. They believed that the most likely explanation was that the aqueous inclusions contained small amounts of CH_4 (and possibly H_2S) that were difficult to detect, pointing to one measurement of solid melting above 0°C which might have been a CH_4 clathrate. If the aqueous inclusions did contain CH_4, their homogenization temperatures would be very close to the true trapping conditions and the embarrassing need to suggest that the reservoir had experienced temperatures as high as about 170°C would disappear.

We would make two points about this interpretation. Firstly, though it seems and probably is quite reasonable, there is another. It is also possible to take the crossing isochores at face value, as indicating a period during which hot water flowed through the reservoir, raising temperatures as high as 170°C. The question is not without significance: petroleum geologists would certainly want to know if temperatures had been so high, as this would have important implications for the maturity of petroleum source rocks. This brings us on to the second point. In a real world where there is *always* considerable uncertainty about the composition of fluid inclusions, calculating an accurate pressure correction and choosing between competing interpretations such as these will not usually be possible.

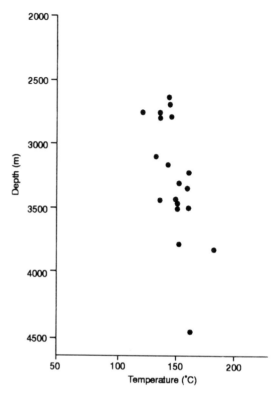

Fig. 3.21 Homogenization temperature of primary fluid inclusions in several minerals plotted against depth, offshore Angola (after Walgenwitz *et al.*, 1990).

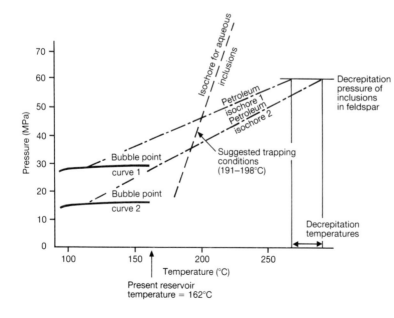

Fig. 3.22 Interpretation of the offshore Angola inclusion data. Petroleum and brine isochores are indicated by broken lines. See text for explanation.

3.4.5 Example 2: mineral cementation, offshore Angola

Walgenwitz *et al.* (1990) describe fluid inclusion data from diagenetic feldspar and anhydrite (with a couple of measurements from dolomite and one from quartz) from several formations from offshore Angola. The data come from a wide range of depths and are sensibly presented not as histograms, but as plots of T_h against depth (Fig. 3.21). Both aqueous and petroleum inclusions were found; the homogenization temperatures of the aqueous inclusions are close to current temperatures at the respective sample depths. Seismic sections suggest that the area has not been recently uplifted so that current burial is a maximum.

In this study, the approach the authors adopted to interpreting their data was somewhat different from that of Narr and Burruss (Fig. 3.22). They determined isochore slopes for petroleum inclusions using the decrepitation temperature (assuming for example that this occurred at a constant 60 MPa in feldspar). Well aware of the influence of dissolved CH_4 on water *PVT* properties, they searched for it using laser Raman spectroscopy. Unhappily, it was not possible to prove or disprove its presence because fluorescence of the host minerals masked any Raman signal from CH_4 that might have been present. They then resorted to a highly indirect method based on decrepitation temperatures to infer that the inclusions contained no CH_4. Finally, having inferred the presence of $CaCl_2$ brines in the aqueous inclusions, the authors assumed that these would have the *PVT* properties of NaCl solutions and calculated isochore slopes from the published *PVT* data. The final step was to find where petroleum and aqueous inclusion isochores intersected – the interpreted trapping temperatures and pressures.

Walgenwitz *et al.* ended up with the same problem as Narr and Burruss. Interpreted trapping temperatures were as much as 40°C higher than present-day temperatures (which had been thought to be maximum values for the area). Their approach to this difficulty was, however, quite different: they believed their interpretation and attempted to find a geological explanation for the high temperatures. They ended up by concluding that thermal gradients were substantially greater in the Miocene, although such a 'thermal event' would not be predicted by any of the models for the thermal evolution of a passive continental margin.

The preferred interpretation suffers from the usual two difficulties. Firstly, all the data come from minerals with low mechanical strengths, in which inclusions would be expected to stretch. Homogenization temperatures of primary inclusions would not then provide any information about mineral growth. Secondly, even if stretching were not a concern, so many assumptions are involved in calculating a pressure correction that the ultimate conclusion – that there has been an important Miocene thermal event – must, we feel, be regarded with scepticism. In our view it would be preferable to accept that homogenization temperatures of fluid inclusions that have not stretched (if these can be identified) are minimum estimates of trapping temperatures and that further interpretation is not possible.

3.5 Non-destructive analysis of individual inclusions

3.5.1 Introduction

Look back to the sandstone sample illustrated in Fig. 3.2. Supposing by crushing a few grams of rock we obtained a bulk sample of the inclusion fluids and analysed it. What would it represent and how would we be able to interpret it? The sample would contain fluids from inclusions trapped in (now clastic) quartz crystals during crystallization of perhaps several igneous intrusions of unknown age or location in the sedimentary provenance area; possibly some secondary fluid inclusions trapped as the intrusions cooled; and small numbers of inclusions, possibly of several distinct origins, trapped during sediment diagenesis. An analysis of a mixture of these would plainly be meaningless. This problem – the presence of multiple inclusion generations – is common to most sedimentary (especially clastic) rocks and makes any analytical method that can be applied to selected individual inclusions particularly attractive. At present, all such methods are non-destructive. Microthermometry may itself be considered as one; indeed, the most important of these. Three spectroscopic methods have also been applied with varying success to fluid inclusions in diagenetic minerals. The application of laser Raman spectroscopy (LRS) to fluid inclusion analysis has a successful history dating back to the early 1970s, although there have been relatively few studies of sedimentary samples. Fourier transform infrared (FTIR) spectroscopy has only been tried on fluid inclusions relatively recently and shows some promise. Though both spectroscopic methods are based on molecular vibrations, LRS and FTIR spectroscopy are complementary because they differ in mechanism. LRS depends on *scattering*: an incident photon perturbs a molecule causing it to undergo a particular transition or rotation. FTIR spectroscopy relies on *absorption* of the incident radiation. The use of UV fluorescence to provide quantitative information about oil composition is in its infancy but the method is extremely useful in a qualitative sense. In this part of the chapter, we will look at the capabilities and potential of each of these methods to see what information we may reasonably expect to obtain about the chemistry of individual fluid inclusions in sedimentary rocks.

3.5.2 Laser Raman spectroscopy

When any substance is excited by a source of monochromatic electromagnetic radiation, the radiation may be scattered in three ways:

1 *Rayleigh scattering.* Elastic scattering of the incident radiation without a change in energy (or, therefore, in wavelength*).

2 *Fluorescence.* Complete absorption and re-emission of radiation at lower energy (with an increase in wavelength).

3 *Raman scattering.* Inelastic scattering (not absorption) by covalent bonds with a shift in wavelength that is characteristic of the bond. The effect is several orders of magnitude weaker than Rayleigh scattering or fluorescence, even for the strongest Raman scatterers.

Raman shifts, the basis of Raman spectroscopy, are usually sharp spectral features because they correspond to the energies of particular vibrational or rotational transitions of the scattering molecule. In the microscope-based version of the technique, a monochromatic visible light laser is introduced into a modified optical microscope and focused by a high magnification glass objective lens onto a sample which sits on the microscope stage. The Raman signal is collected simultaneously through the same objective and analysed by a spectrometer. Spectra are traditionally plotted as intensity against the shift from the excitation *wavenumber* (Δv)[†].

The same doubly polished thin-sections used for microthermometry and petrographic study are also

* Scattering is proportional to the fourth power of frequency which is why the sky is blue.
[†] Wavenumber is the reciprocal of wavelength, measured in cm^{-1}.

ideal for LRS. The only requirements of the host mineral are that it should be transparent to the laser and to the Raman signal, and that it should not fluoresce (both quartz and calcite sometimes do). The laser can be focused down to a spot size of about 3 μm, close to the theoretical maximum resolution of about 0.7 μm (limited by the wavelength of the laser light). The signal, and thus analytical information, comes from an irradiated volume whose exact size cannot be determined exactly for fluid inclusions, but which is certainly of the order of only a few cubic micrometres. This means that analyses can be obtained from individual phases — liquid, vapour and solid — within even very small fluid inclusions. In fact, LRS has such extraordinary spatial resolution that it is not at all good for analysing bulk inclusion compositions, because although it is possible to defocus the beam to irradiate a larger volume, this reduces the power of the excitation and thus the Raman signal strength. The technique is generally non-destructive, though we have known inclusions to introduce an element of excitement by exploding unpredictably under the laser.

What then can be detected using LRS? The Raman effect relies on scattering by covalent bonds so there is no response from ionic species. Equally, some covalently bonded compounds are such poor Raman scatterers that they are very difficult to detect. For example, water is a poor Raman scatterer with an unusually broad Raman spectrum which can make it difficult to detect above a noisy background. Nonetheless, a large number of species of geological importance contain bonds that are reasonable or strong Raman scatterers. These include virtually all S-bearing compounds (H_2S, SO_4^{2-}, SO_2, S^0, etc.), CH_4, CO_2, CO, N_2 and O_2. C–H, C–C, C=C and C≡C bonds can all be detected and distinguished, but analysis of petroleum by LRS is usually impossible because the laser tends to cause strong fluorescence which totally obscures the much weaker Raman signal.

Figure 3.23 shows Raman spectra from one of the fluid inclusions that we discussed in Section 3.3.2 (illustrated in Fig. 3.1c). On cooling, the liquid in these inclusions froze to a solid that melted at around −90°C, slightly lower than the triple point of H_2S. LRS confirmed that the inclusions contain a mixture of H_2S and CH_4. Small dark specks in some of the inclusions were identified as elemental sulphur. The main use for LRS in fluid inclusion analysis has, just as in this example, been as a tool for the *qualitative* identification of volatiles. In

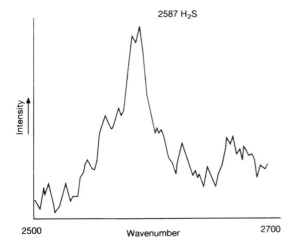

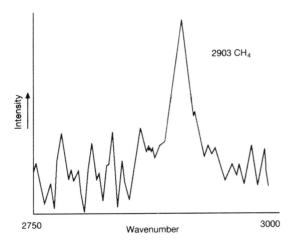

Fig. 3.23 Laser Raman analyses of a liquefied natural gas inclusion.

fact, LRS cannot produce a quantitative chemical analysis from this type of sample. This is because the strength of the Raman signal reaching the spectrometer is dependent not only on the number of moles of sample being irradiated, but also on several factors related to sample geometry and instrumentation that are impossible to determine. However, it is possible to ratio out these indeterminate factors and calculate the *relative amounts* of chemical components in the phase being analysed (Wopenka & Pasteris, 1986; Pasteris *et al.*, 1988). The ratio of two components, expressed as mole fraction or partial pressure, can be calculated from their *Raman scattering cross sections* (which describe

their strengths as Raman scatterers) and the areas of their Raman peaks. However, even this process is not without problems. Raman scattering cross sections are known only for low density fluids; are not appropriate to the spectrum collection geometry of the equipment used in microanalysis; and are in any case disputed. Also, if peak intensities are low, there may be a significant error involved in determining their areas. The precision of an analysis of this type is equivalent to the ability of the equipment to reproduce relative peak areas. This depends on the counting statistics and the background and can often be very high, say a few relative per cent. Accuracy however is determined both by precision and by uncertainty surrounding the true value of the scattering cross sections, and is far from easy to quantify. Wopenka and Pasteris (1986) estimated the accuracy of a LRS analysis of a $CO_2 + CH_4$ inclusion by assuming that the tabulated values for Raman cross sections (with a quoted uncertainty of $\pm 10\%$) were applicable. The results of the analysis were 14.8 ± 2.0 mol% CH_4 (14 relative %) and 85.2 ± 2.0 mol% CO_2 (2.3 relative %). Even making this (incorrect) assumption, the estimated accuracy of ± 2 mol% was dominated by the uncertainty in the values of Raman cross sections. One way of trying to overcome the problem of uncertain Raman cross sections is to produce standards (although these are applicable to only one machine). Pasteris *et al.* (1988) used gas-filled capillaries at pressures of 1.0–1.5 MPa to estimate the minimum concentrations of several volatiles of geological interest that their machine was capable of detecting. Detection limits, determined as partial pressures, varied from 0.05 to 0.15 MPa.

By far the most important application of LRS to fluid inclusion analysis has been this type of qualitative to semi-quantitative analysis of covalently bonded volatiles. The method has however been used in some subtly different ways. Though ionic species themselves produce no Raman signal, they can sometimes be detected in a fluid inclusion by means of a clever trick. If a heating–cooling stage is fitted to the LRS microscope, the inclusion can be frozen. Many of the solids formed in concentrated salt solutions – hydrohalite for example – will produce a characteristic Raman spectrum (Dubessy *et al.*, 1982). Clathrates can be identified in the same way. The position of Raman peaks is also sensitive to pressure. This has been calibrated for CH_4 so that the method can be used to estimate the pressure in inclusions (Fabre & Couty, 1986). It is even

possible to estimate salinity independently of microthermometry from the shape of the water Raman
peak (Mernagh & Wilde, 1989)!

3.5.3 Fourier transform infrared spectroscopy

The bonds of molecules are continually vibrating
due to interactions with molecules in the immediate
environment. If the bonds are excited by infrared
radiation, the amplitudes of the vibrations will
increase, subsequently decreasing again as energy
is lost to surrounding molecules. Only certain
(quantized) increases in vibrational energy are
possible and these constitute a characteristic absorption spectrum. The more bonds that are excited,
the more energy is absorbed, so the strength of
the spectrum depends on concentration. For most
molecules, there are several possible types of vibra-
tion (*vibrational modes*). These are based either on
stretching, which may be asymmetric or symmetric,
and bending, which may take place in several poss-

ible planes and directions. As a consequence, infrared absorption spectra of even simple substances
can be complicated, with many peaks attributable
to different vibrational modes (Brown *et al.*, 1988).
Only asymmetric vibrations which produce a fluctuating electrical dipole will absorb infrared and
produce characteristic spectra.

In FTIR spectroscopy, a polychromatic infrared
source (λ typically from 1.1 to 200 µm) is focused
onto a sample using cassegrainian (mirror) objectives. These can be exchanged for glass objectives
so that samples can be located using visible light.
The infrared beam emerging from the objectives is
still several hundreds of micrometres across and so
has to be stopped down by mechanical apertures.
The beam passes through the sample to a detector
located beneath it which scans the range 1.8–14 µm
(Wopenka *et al.*, 1990). FTIR spectroscopy produces an analysis of everything along the beam path,
including the atmosphere (CO_2 and H_2O absorb
infrared!), as well as the mineral host. The problem

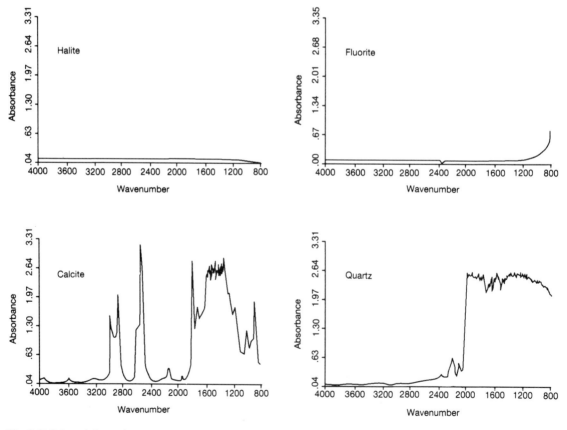

Fig. 3.24 Infrared absorption spectra of common diagenetic minerals (from O'Grady *et al.*, 1989).

must be solved by collecting a background spectrum from a part of the mineral with no inclusions, using identical operating conditions.

Though doubly polished thin-sections are good preparations for FTIR spectroscopy, the nature of the sample is more critical than is usually the case with LRS. The mineral host clearly should be as far as possible transparent over the entire infrared range, but of the common diagenetic minerals, only halite is (Fig. 3.24). Quartz has strong absorption bands at wavenumbers less than about $2000 \, cm^{-1}$ but is quite transparent at higher wavenumbers and calcite has absorption bands that overlap those of some petroleum components. Furthermore, as it passes through the polished section, the infrared beam must pass only through the fluid inclusion being analysed. Samples with a high inclusion density will be unsuitable (unless these are all of one generation so that spectral contamination is acceptable). Spatial resolution − controlled by closing the aperture − is much poorer than in LRS because of the longer wavelength of the infrared excitation source. If the aperture is narrower than about $20 \times 20 \, \mu m$, the beam is diffracted so that some photons hit the surroundings and contaminate the spectrum. This means that the smallest inclusion that can be analysed is about $10 \times 20 \, \mu m$, larger than most fluid inclusions in sedimentary rocks. Because of the relatively poor spatial resolution and because the beam passes through the sample, FTIR spectroscopy is better at analysing entire inclusions than individual phases.

Like LRS, FTIR spectroscopy does not respond to ionic substances or even all those with covalent bonds. Simple diatomic molecules like H_2, O_2, N_2 vibrate symmetrically and so cannot be detected by the method. Even some molecules that absorb infrared do so very weakly and are difficult to detect (for example, H_2S). As with LRS, however, bonds in many species of geological interest − including water, CO_2 and CH_4 − are readily detected by FTIR spectroscopy. In suitable (large) inclusions, FTIR spectroscopy is better than LRS for detecting CO_2 but not as good for detecting CH_4. It is also better for detecting water because of the low intensity of the water Raman peak (Vry *et al.*, 1987). The method has in particular considerable potential for the analysis of petroleum in fluid inclusions (Barres *et al.*, 1987; O'Grady *et al.*, 1989; Guilhaumou *et al.*, 1990a; Moser *et al.*, 1990; Rankin *et al.*, 1990). Oil usually fluoresces under a visible light laser so that any Raman signal is swamped, but the various *func-*

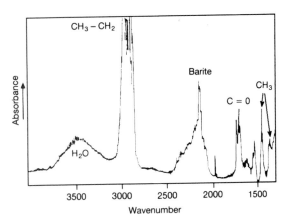

Fig. 3.25 Infrared spectrum of an oil in a fluid inclusion (from Guilhaumou *et al.*, 1990b).

tional groups (groups of compounds characterized by bond types such as C–H, C=H, C=O, etc.) can be identified from their infrared spectra (Fig. 3.25). Any quantitative (or perhaps more plausibly, semiquantitative) analysis would therefore take the form of ratios of functional groups and would not be directly comparable to an oil analysis made by a standard technique such as gas chromatography. FTIR spectroscopy is not capable of detecting individual compounds in oil.

Like LRS, FTIR spectroscopy is of value principally as a method for the *identification* of species in fluid inclusions. Quantitative analysis depends on knowing relative peak areas and *infrared absorptivities* (analogous to Raman scattering cross sections) and does not appear to have been much attempted on fluid inclusions. One problem is that the absorptivities are quite uncertain and known to vary with density. It is hard to quantify precision, accuracy and detection limits. Precision (reproducibility) is usually excellent provided that the sample does not present difficulties, but accuracy is determined largely by the uncertainty surrounding the appropriate absorptivity. Detection limits for any compound will vary with machine, analytical conditions and the nature of the sample.

3.5.4 Ultraviolet fluorescence

Many organic compounds, and most oils, fluoresce under UV light. The incident energy is absorbed by the π orbital of C=C bonds and is then re-emitted, causing fluorescence. Aromatic compounds − present in small quantities − are therefore respons-

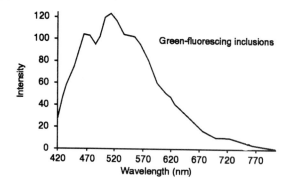

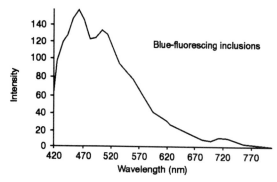

Fig. 3.26 UVF spectra of petroleum fluid inclusions in diagenetic celestite, Machar Field, Central North Sea.

ible for the fluorescence of oils. UVF has been used mainly for simple identification and classification of petroleum fluid inclusions in doubly polished thin-sections and most studies report inclusion groups simply as 'blue' or 'green' or even 'white' fluorescing. Nonetheless, there is more information to be obtained. If a spectrometer is attached to the microscope (commercial equipment is available), fluorescence spectra can be easily collected from even very small individual fluid inclusions in a matter of seconds. Figure 3.26 shows UVF spectra for petroleum inclusions identified as blue- and green-fluorescing. The spectra are quite different, though the main peaks occur at the same wavelengths, probably because fluorescence is dominated by a relatively small number of polycyclic aromatic hydrocarbons. It proved possible to analyse these petroleums by gas chromatography–mass spectrometry (see below) which showed that the blue-fluorescing petroleum was more mature (generated from a source rock at a higher temperature). A shift

towards the blue end of the spectrum with increasing maturity has been noticed in other studies (Hagemann & Hollerbach, 1986; McLimans, 1987). Figure 3.27 shows fluorescence spectra of oils from the UK North Sea (all sourced from the Kimmeridge Clay Formation), arranged in order of API gravity*. The spectra of the lighter and more mature oils (higher API gravity) are clearly weighted towards the blue end of the spectrum. The difficulty in using relationships of this kind is partly one of identifying parameters which can be used to measure the shape changes of the spectra, and can then be correlated with maturity (for an attempt, see Guilhaumou *et al.*, 1990b).

3.6 Bulk analysis of petroleum inclusions

3.6.1 Introduction

We have just spent a fair amount of effort thinking about microspectroscopic analytical methods for a good reason: the difficulty of extracting enough fluid from one generation of fluid inclusions to make a meaningful bulk chemical analysis possible. For some problems, for example the bulk analysis of aqueous inclusions in quartz cements, we believe the position to be hopeless: bulk analysis will never be possible and microspectroscopy of some kind offers the only way forward. There are nonetheless some cases where extraction of sufficient fluid for analysis is possible and one of these is of great interest – the analysis of petroleum trapped in fluid inclusions. There are two reasons for this. One is that petroleum fluid inclusions are often surprisingly abundant; and the second is that a mineralogically and texturally complex rock usually contains several generations of aqueous inclusions, but often only one set containing petroleum. Crushing a bulk rock sample will therefore liberate petroleum from only the one inclusion generation.

In this section, we will look first at how to get the petroleum out of the inclusions, and then at how the sample of fluid can be analysed. The information available includes the following.

1 Nature of source rock: is its kerogen derived from land plants or marine algae?

2 Maturity: at what temperature was the petroleum generated from the source rock?

3 *PVT* relationships (for calculation of isochores):

---

* API gravity = $(141\,500/\text{density at }60°F\text{ in kg.m}^{-3})$ – 131.5; denser oils have lower API gravity.

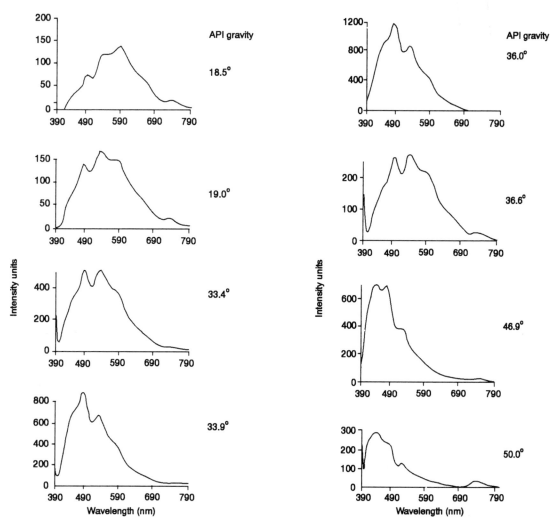

Fig. 3.27 UVF spectra of oils from the UK North Sea, arranged in order of API gravity. Note the shift in the dominant wavelength towards the blue end of the spectrum with increasing API gravity (decreasing density). API gravity is a crude indication of maturity, so UVF has potential for estimating the maturity of oils in individual fluid inclusions.

obtained by analysis of the major component hydrocarbons.

3.6.2 Isolation of a fluid sample

The key step in analysis of a bulk sample of petroleum is its extraction from the fluid inclusions without modifying it in any way. The rock sample must first be coarsely crushed to about the size of its component crystals or grains, and then thoroughly oxidized or solvent-extracted. This step must remove all bitumen stains and other extraneous organic

material. Next, the petroleum has to be removed from the inclusions. There are two ways of doing this, decrepitation and crushing, and both have their advantages, disadvantages and adherents. Decrepitation involves heating the sample until the inclusions burst open. This tends to require heating to several hundreds of degrees which may thermally alter petroleum and also produce other volatiles not present in the original inclusions. Crushing is therefore the preferred method for liberating petroleum. Several crushers have been described in the literature (e.g. Andrawes *et al.*, 1984). The crushing needs to be as gentle as possible or there may be substantial adsorption of some components of the petroleum onto newly created mineral surface (this can be further minimized by heating the crusher to around 100°C). Fluid is usually released into an inert carrier gas and trapped by liquid nitrogen prior to analysis.

3.6.3 Gas chromatography

All chromatographic techniques rely on partition of the mixture to be separated − in this case petroleum − between a stationary and a mobile phase. In gas chromatography (GC), the stationary phase is a silicone gum and the mobile phase a helium carrier gas. The petroleum + carrier gas mixture is injected into a column, a capillary lined with the gum. As the mixture travels through the column, molecules partition between gum and gas. The lighter ones spend more time in the gas and take less time to reach the end of the column and detector. To get the heaviest molecules through the column, its temperature is progressively raised. The components of the petroleum are detected as they emerge from the column by a flame ionization detector.

Gas chromatograms are plots of intensity (so that peak height is proportional to concentration) against elution time (the time taken for a component to pass through the column). They record the major components of a petroleum from which *PVT* properties can be calculated and can be used to identify alteration (such as biodegradation and water washing; Horsfield & McLimans, 1984). They also provide limited information about source and maturity. For example, the pristane/phytane ratio can be calculated from a gas chromatogram by simply ratioing the height of the respective peaks. Some features of the shape of the gas chromatogram, relative proportions of light and heavy (waxy) hydrocarbons, even−odd preference in straight-chain alkanes, etc., are also characteristic of both source and maturity. A word of warning here however. Gas chromatograms of 'oils' are generally of petroleum from which low boiling point (usually <200°C) hydrocarbons and also asphaltenes* have been

---

* Asphaltenes are the largest most polar components of an oil and are defined as those which are soluble in a polar solvent (dichloromethane) but insoluble in a non-polar solvent (hexane or heptane). In practice, they too are now usually separated by liquid chromatography.

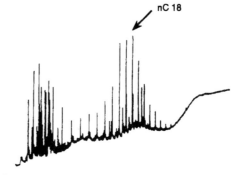

Blue-fluorescing inclusions

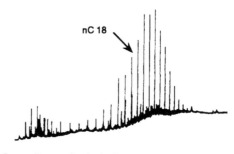

Green-fluorescing inclusions

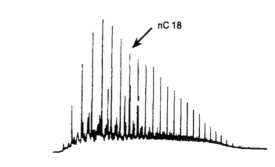

DST 2A free petroleum

Fig. 3.28 GC traces for fluid inclusion petroleums and produced oil, Machar Field, Central North Sea.

removed and which have been separated into alkane and aromatic fractions by liquid chromatography. Alkane gas chromatograms are the most widely used. In contrast, GC analyses of fluid inclusions incorporate everything that was in the inclusion that the crushing process was able to release. The shapes of the gas chromatograms of the two cannot therefore be easily compared.

Figure 3.28 shows gas chromatograms of two groups of petroleum fluid inclusions (defined on the basis of UVF colour; see Fig. 3.26) and that of the alkane fraction of the reservoir oil. The interpretation of these analyses will form part of the Machar case history in Section 6.6.

3.6.4 Gas chromatography–mass spectrometry

The amount of information that GC can provide about the source and maturity of an oil is rather limited. This is because the most informative molecules are present in very small quantities which are hard both to resolve and to identify on a gas chromatogram. These molecules are known as *biomarkers* (or colloquially, 'molecular fossils'). They are those that are clearly derived from specific products of biosynthesis, usually on the basis of the form of their carbon skeletons. For example, steranes, extremely common in oils (though present in small quantities), are derived from sterols in algae (Mackenzie *et al.*, 1982). Some biomarker ratios change with maturity due to changes in the proportion of particular isomers (isomerization), increase in the proportion of aromatic bonds (aromatization) or thermal cracking. When calibrated, the extent of these reactions can be used to measure maturity (Mackenzie & McKenzie, 1983).

Biomarkers are analysed by coupling a GC with a mass spectrometer (MS). As compounds emerge from the GC column, they are fragmented and ionized by collision with electrons and the ions are swept into the mass spectrometer for analysis. Each group of compounds tends to produce fragment ions with characteristic mass/charge (m/z) ratios. A ratio of interest is monitored as a function of elution time in the GC column. For example, steranes, a group of compounds that provide important maturity information, produce a characteristic fragment with $m/z = 217$. On a GCMS plot of intensity against elution time for $m/z = 217$, different steranes can be identified from their elution time and their relative quantities determined by reading off peak heights.

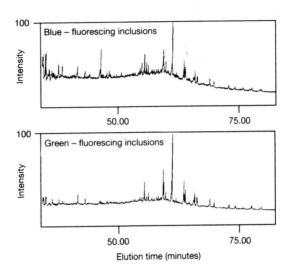

Fig. 3.29 GCMS fragmentograms for $m/z = 191$ (hopanes), fluid inclusion petroleums, Machar Field, Central North Sea.

Application of GCMS to petroleum fluid inclusions is relatively straightforward. Biomarkers are not volatile, so after coarse crushing and thorough cleaning, the rock sample can simply be crushed in a solvent. If possible, liquid chromatography should be used to separate alkanes from aromatics, or signals from the biomarkers being sought may suffer from interference. We have found this to be possible even with minute amounts of fluid. Figure 3.29 shows two GCMS 'fragmentograms' from the same groups of petroleum fluid inclusions whose gas chromatograms were illustrated in Fig. 3.28: $m/z = 191$ represents a characteristic fragment of hopanes, another group of biomarkers useful for maturity determination. Only a few examples of the use of this method have been reported in the literature (e.g. Etminan & Hoffman, 1989).

Chapter 4 **Stable Isotopes**

4.1 Introduction

Isotopes are atoms whose nuclei contain the same number of protons but a different number of neutrons. Most of the 1500 or so naturally occurring isotopes are unstable and undergo radioactive decay to other isotopes. A subset of isotopes, however, is not radioactive, at least within our ability to measure decay times. These are the *stable* isotopes and are the subject of this chapter. The science of stable isotope geochemistry is based on the fact that isotopes of the same element have slightly different thermodynamic and physical properties (Urey, 1947). As a consequence, their behaviour in chemical and physical reactions is slightly different. Isotopes therefore become separated, or *fractionated*, when they, or compounds which contain them, participate in chemical reactions or undergo a change of state. The resulting changes in relative isotopic abundances are often measurable and provide important insights into many geochemical processes.

There are several excellent reasons why most stable isotope geochemistry has been performed on the light elements, hydrogen, carbon, nitrogen, oxygen and sulphur. Firstly, these elements contain at least two stable isotopes in relatively high abundance, facilitating their accurate measurement (Table 4.1). Secondly, the amount of fractionation in a reaction is a function of the relative mass difference between isotopes, so that variations in isotope ratios are greatest for the light elements (and are thus easiest to measure). Thirdly and fortunately, these elements are major components of geological and biological systems, and participate in most geochemical reactions.

A list of the general applications of stable isotopes to problems of petroleum geology is given in Table 4.2. The applications are of two general types: information can be obtained about the source of an element in a mineral or fluid, and also the conditions (temperature, water composition) under which a mineral precipitated. Whilst this information has some intrinsic importance, the power of stable isotopes as a problem-solving tool is greatly enhanced

Table 4.1 Stable isotopes of geological interest.

Element	Isotope	Relative abundance (%)	Terrestrial variation (per mille)
Hydrogen	^1H	99.9844	$\delta D = 700$
	$D\ (^2H)$	0.0156	
Carbon	^{12}C	98.89	$\delta^{13}C = 100$
	^{13}C	1.11	
Oxygen	^{16}O	99.763	$\delta^{18}O = 100$
	^{17}O	0.0375	
	^{18}O	0.1995	
Sulphur	^{32}S	95.02	$\delta^{34}S = 100$
	^{33}S	0.75	
	^{34}S	4.21	
	^{36}S	0.02	

Table 4.2 General application of stable isotopes in studies of diagenetic and petroleum production systems.

Element	Applications
Oxygen	Stratigraphy (carbonates)
	Geothermometry (silicates, carbonates)
	Tracer studies (origin/mixing of waters; water–rock interaction)
Hydrogen	Tracer studies (origin/mixing of waters; origin of hydrous minerals)
Carbon	Stratigraphy (carbonates)
	Tracer studies (origin of carbonate/CO_2)
Sulphur	Stratigraphy (sulphates)
	Tracer studies (origin of sulphates and sulphides)

when used in conjunction with other geochemical techniques. The objectives of this chapter are to outline the principles of stable isotope geochemistry, paying particular attention to the general problems of data interpretation and to describe some applications of different isotopic systems to petroleum geology.

4.2 Principles

4.2.1 Terminology

From an analytical standpoint, it is difficult to determine the rather restricted variations which occur in the absolute concentrations of a given isotope. Stable isotope abundances are therefore measured as *differences* in the isotope *ratios* of two substances. Correspondingly, stable isotope data are reported using the δ, or del notation, in which the isotope ratio of a sample is expressed in terms of its deviation, in parts per thousand (‰, or per mille), from the same ratio in an internationally accepted standard:

$$\delta_A = \{[R_A - R_{STND}]/R_{STND}\} \cdot 10^3 \qquad (1)$$

where R_A is the ratio of the heavy isotope to the light isotope for phase A (D/H, $^{13}C/^{12}C$, $^{18}O/^{16}O$, $^{34}S/^{32}S$) and R_{STND} is the same ratio for a standard material. For example, a $\delta^{18}O$ value of $+15‰$ indicates that the sample is enriched in ^{18}O by 15 parts per thousand relative to the standard, and a δD value of $-50‰$ shows that the sample is depleted in D by 50 parts per thousand relative to the standard.

Although individual laboratories use their own internal standards against which the isotope ratio of a sample is measured, all isotope ratios are eventually reported against an international standard. These are listed in Table 4.3 and by definition, have del values of 0‰. Although V-SMOW is the accepted oxygen standard, $\delta^{18}O$ values of carbonates are often reported relative to the oxygen in PDB*, the carbon standard. The equations for converting between the PDB and SMOW scales are given in Table 4.3.

* PDB is the acronym for the Pee Dee belemnite, a reference standard material which has long been exhausted. All other standards are ultimately calibrated back to this material.

Table 4.3 Internationally accepted standards for stable isotope measurements.

Element	Standard	Comments
Hydrogen	V-SMOW	Vienna Standard Mean Ocean Water; identical to SMOW; $D/H = 155.76 \times 10^{-6}$ (Hagemann et al., 1970); $\delta D = 0.00‰$
Carbon	PDB	Pee Dee belemnite; $^{13}C/^{12}C = 1123.75 \times 10^{-5}$ (Craig, 1957); $\delta^{13}C = 0.00‰$
Oxygen	V-SMOW	As above; $^{18}O/^{16}O = 2005.2 \times 10^{-6}$ (Craig, 1957); $\delta^{18}O = 30.91‰$ relative to V-SMOW. For use in palaeo-temperature studies
Sulphur	CDT	Canyon Diablo troilite; $^{34}S/^{32}S = 449.94 \times 10^{-4}$ (Thode et al., 1961); $\delta^{34}S = 0.00‰$

Note: $\delta^{18}O_{V\text{-}SMOW} = 1.03091\, \delta^{18}O_{PDB} + 30.91$
$\delta^{18}O_{PDB} = 0.97002\delta^{18}O_{V\text{-}SMOW} - 29.98$

The stable isotope fractionation factor between two phases, A and B, is defined as:

$$\alpha_{A-B} = R_A/R_B \tag{2}$$

For example:

$$\alpha^O_{\text{quartz-water}} = (^{18}O/^{16}O)_{\text{quartz}}/(^{18}O/^{16}O)_{\text{water}} \tag{3}$$

The fractionation factor is related to the δ values of the two phases as follows:

$$\alpha_{A-B} = \frac{1 + \delta_A/1000}{1 + \delta_B/1000} = \frac{1000 + \delta_A}{1000 + \delta_B} \tag{4}$$

It is also useful to introduce another term, Δ, as the isotopic difference between two phases:

$$\Delta_{A-B} = \delta_A - \delta_B \tag{5}$$

This in turn can be related to α through a mathematical approximation. Manipulating equations (1) to (3) gives:

$$\Delta_{A-B} = \delta_A - \delta_B \approx 10^3 \ln \alpha_{A-B} \tag{6}$$

This equation is useful because it means that, assuming that phases A and B formed in equilibrium, we can approximate α_{A-B} by simply subtracting the measured isotopic compositions of the phases. The approximation is good when $\Delta < 10‰$, but for values $> 10‰$ it is better to calculate fractionation factors precisely.

4.2.2 Isotope fractionation

Isotopes of a given element fractionate during chemical and physical processes because the strengths of chemical bonds vary slightly with the mass of the isotope. In general, the light isotope forms weaker bonds than the heavier isotope. This gives rise to the three types of isotope fractionation seen in natural systems:

1 kinetic isotope effects, which arise because the lighter isotope reacts more rapidly than the heavier isotope;

2 equilibrium isotope effects, which are controlled by the differing thermodynamic properties of each isotope; and

3 vital or biological isotope effects, observed in certain organisms which produce mineral skeletons which are out of isotopic equilibrium with the water from which they form.

Kinetic isotope effects are important in unidirectional reactions which do not proceed to completion. Examples of reactions which involve a kinetic isotope effect include the bacterial reduction of sulphate and many reactions involving reduced organic carbon. In general, one can describe the isotopic competition as two reactions which proceed at differing rates thus:

$$^{12}R \xrightarrow{k_{12}} {}^{12}P \tag{7}$$

$$^{13}R \xrightarrow{k_{13}} {}^{13}P \tag{8}$$

where R and P are reactants and products containing, for example, ^{12}C and ^{13}C. The ratio of the isotopic rate constants k_{12} and k_{13} is called the *kinetic fractionation factor* and is usually > 1. Take, for example, the bacterially mediated reduction of CO_2 to CH_4. If $k_{12}/k_{13} = 1.05$, then CH_4 will be

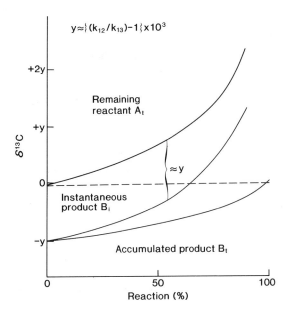

$$y \approx \{(k_{12}/k_{13})-1\} \times 10^3$$

Remaining reactant A_t

$\approx y$

Instantaneous product B_i

Accumulated product B_t

Reaction (%)

Fig. 4.1 Isotope fractionation resulting from a closed-system, unidirectional reaction (Rayleigh distillation). From Krouse and Tabatabai (1986).

about 50‰ lighter (i.e. depleted in ^{13}C) than precursor CO_2. An important corollary of this is that residual CO_2 will be correspondingly *enriched* in ^{13}C. If the reservoir of CO_2 is not replenished (i.e. a closed system), continued reduction will result in CO_2 and CH_4 which both become increasingly enriched in ^{13}C (Fig. 4.1). This is known as the *reservoir* effect (nothing to do with petroleum reservoirs) and can be described mathematically by the Rayleigh distillation equations. In our example, CH_4 produced at a given time will always be 50‰ lighter than the remaining CO_2 reservoir, but the isotopic composition of the accumulated CH_4 will move towards, and eventually reach, that of the initial CO_2. Figure 4.1 shows that, depending on whether or not reaction products are lost from the reservoir, a very wide range of isotope ratios of both reactant and product can result from reservoir processes.

Equilibrium isotope fractionation occurs because the thermodynamic properties of isotopically substituted species differ. When reactions involving isotopically substituted compounds come to equilibrium, isotopes are distributed between the compounds such that the free energy of the system in question is minimized. For example, in the reaction in which CO_2 equilibrates with water,

$$\tfrac{1}{2}C^{16}O_2 + H_2^{18}O = \tfrac{1}{2}C^{18}O_2 + H_2^{16}O \qquad (9)$$

for which the equilibrium constant K is:

$$K = \frac{[C^{18}O_2]^{1/2}[H_2^{16}O]}{[C^{16}O_2]^{1/2}[H_2^{18}O]} \qquad (10)$$

the equilibrium fractionation factor (α) is directly related to the equilibrium constant (K) assuming that the isotopes are distributed randomly within all available sites in phases A and B. In general,

$$\alpha = K^{1/n} \qquad (11)$$

where n is the number of atoms exchanged. As in equation (9), it is usual to write isotope exchange reactions such that only one atom is exchanged. In that case, using equation (9) as an example:

$$K = \frac{(^{18}O/^{16}O)_{CO_2}}{(^{18}O/^{16}O)_{H_2O}} = \alpha \qquad (12)$$

4.2.3 Isotope geothermometry

Like most reactions, the equilibrium constant, or fractionation factor of an isotope exchange reaction, is temperature dependent. This forms the basis of stable isotope geothermometry. Above about 0°C, in most mineral–water or mineral–mineral systems, $\ln \alpha_{A-B}$ varies as an approximately linear function of $1/T^2$, where T is in kelvin. In general, using A and B as constants for a particular mineral–mineral or mineral–water pair:

$$10^3 \ln \alpha = A.10^6/T^2 + B \qquad (13)$$

so that the per mille fractionation ($10^3 \ln \alpha$) decreases with temperature.

Examples of the temperature dependence of mineral–water fractionation factors are shown in Fig. 4.2. In some cases the slopes of the fractionation equations for two cogenetic minerals are very different. Under these circumstances, the difference in isotopic ratios between the two minerals defines the precipitation temperature, *but only if the minerals precipitated coevally, and in equilibrium*. In diagenetic systems, it is extremely difficult to demonstrate that any two minerals fulfil these criteria, so estimation of precipitation temperature invariably requires an assumption about the isotopic composition of the mineralizing fluid (Fig. 4.3). This is a major disadvantage. It means that the accuracy of isotope geothermometry in sedimentary rocks is always dependent on an assumption about the isotopic composition of waters probably long since disappeared.

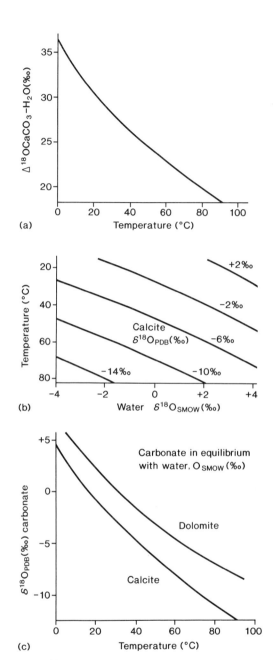

Fig. 4.2 (a) Oxygen isotope fractionation between calcite and water is a function of temperature. (b) Oxygen isotopic composition of calcite is a function of both temperature and the isotopic composition of the water from which the calcite formed. (c) $\delta^{18}O_{H_2O}$ must be specified in order to define the temperature at which calcite formed. If two minerals, such as dolomite and calcite, formed in equilibrium from the same water, then the difference in the isotopic compositions of the minerals ($\Delta = \delta^{18}O_{dol} - \delta^{18}O_{calc}$) defines, uniquely, the precipitation temperature.

4.2.4 Analytical methods

Measurements of the isotopic ratios of the light elements are made in stable isotope ratio mass spectrometers using the following gases: H_2 (hydrogen), CO_2 (carbon and oxygen) and SO_2 or SF_6 (sulphur). The basic analytical procedures are similar in each case and are discussed in detail in standard isotope texts such as Hoefs (1987) and Faure (1986). The principal stages in the analysis are as follows.

1 Separation and purification of the material for analysis. A typical sample size is 5–15 mg, although the sensitivity of modern mass spectrometers is such that, with care, submilligram samples may be analysed routinely.

2 Quantitative conversion of the element to the gas which will be analysed in the mass spectrometer.

3 Purification of the gas by vacuum distillation.

4 Determination of the isotopic composition of the gas by mass spectrometry, relative to that of a laboratory standard.

5 Calculation of the isotopic composition of the sample relative to the relevant international standard (Table 4.3; the isotopic composition of the laboratory standard is known relative to that of the international standard).

Analytical precision is dependent on several factors including sample inhomogeneity, sample preparation and mass spectrometry. Precision may be reported in different ways. In some cases, only errors incurred during mass spectrometry are quoted, generally on the order of $\pm 0.01‰$ for carbon and oxygen, $\pm 0.02‰$ for sulphur and $\pm 0.3‰$ for hydrogen. These are not however a true reflection of total error because most error derives from sample inhomogeneity and preparation. A better (and more common) way of quoting precision is reproducibility, which is the ability of the analyst to obtain the same result from different aliquots of the same sample. For example, an isotope analysis might be reported as: $\delta^{18}O = +18.4 \pm 0.2‰$ (2σ). The second value is the analytical precision quoted as two standard deviations (2σ) from the mean result. This is the analyst's estimate of the likelihood that, should the analysis be repeated, the same result within the quoted limits would be obtained: the analyst is 95% sure that a repeat result between 18.2 and 18.6‰ would be obtained. Typical 2σ reproducibilities are: $\pm 0.1‰$ for CO_2 and SO_2, and $\pm 1‰$ for H_2.

Analytical accuracy is monitored by continuously analysing a standard of known isotopic composi-

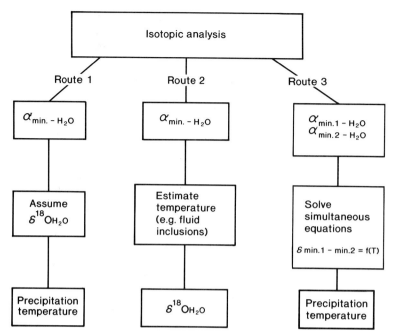

Fig. 4.3 Generating information from the oxygen isotope composition of diagenetic mineral phases. T = temperature; α = fractionation factor.

tion. However, even if we achieve accurate sample analyses, we must remember that the analyses are only average representations of the material that we have separated. If we have failed to separate illite from kaolinite, we cannot expect an accurate analysis of illite. The answer to this most fundamental problem lies in the development of *in situ* methods of analysis. Two analytical methods have shown promise: laser pyrolysis and secondary ion mass spectrometry (SIMS). Lasers have been used to convert sulphides and carbonates in thin-sections to SO_2 and CO_2 for mass spectrometric analysis (Dickson, 1991; Dickson *et al.*, 1991). Reasonable sulphur and hydrogen isotope analyses have also been obtained from SIMS, albeit with lower precision than from conventional mass spectrometry (Eldridge *et al.*, 1988; Deloule *et al.*, 1991).

4.2.5 Data interpretation: general problems

Sensible interpretation of stable isotope data rests on a series of assumptions common to all isotope systems. These will be considered further for each isotope system but their universal importance justifies their collection here.

1 The sample is pure and well characterized. It is notoriously difficult to obtain pure mineral separates from sedimentary rocks. There are three general problems. Firstly, it may be difficult to separate a pure mineral phase. X-ray diffraction can be used to detect contaminants if they comprise more than 5% of the mixture (Section 2.7), but at levels below this, one is obliged to devote a lot of time to the determination of mineral abundances by point-counting (which is itself unreliable for clay minerals; Section 2.2). The second problem arises when the separate consists of both detrital and authigenic forms of a single mineral. Again, it will usually be necessary to resort to point-counting (though see Section 5.3.5 for an attempt to calculate end member K–Ar ages using chemical discriminants). Thirdly, one may obtain a pure separate of the authigenic phase, but find that it is zoned, which may imply growth from chemically distinct fluids. *In situ* analysis may be the only solution to this problem.

2 Isotopic ratios have been preserved since mineral precipitation. In general, isotope exchange at diagenetic temperatures is only effected between minerals and fluids when chemical bonds are broken and reformed; that is, when minerals recrystallize or react to form new minerals (see O'Neil, 1987, for a review). Some minerals, notably clays and carbonates, are more susceptible than others to reaction or recrystallization. Partial or total recrystallization may be revealed by conventional petrography, cathodoluminescence and/or electron microscopy.

Experimental and empirical evidence suggests that in clay minerals, hydrogen and possibly oxygen isotope exchange may occur at temperatures above 100°C (O'Neil & Kharaka, 1976; Bird & Chivas, 1988; Longstaffe & Ayalon, 1990). A checklist for the susceptibility of diagenetic minerals to recrystallization and to isotopic exchange is shown in Table 4.4. As a general rule, the stable isotope composition of a mineral records information about the temperature and fluid with which the mineral last underwent isotopic exchange.

3 Minerals formed in isotopic equilibrium with the fluid from which they precipitated. For minerals precipitating directly from solution, this is generally true unless vital isotope effects are suspected. However, where (i) replacement reactions have occurred, in which part of the isotopic signature of the new phase may be derived from its precursor, and (ii) reactions which may involve kinetic or vital isotope effects are suspected, care must be taken when interpreting the isotopic analyses. Most important amongst the latter are bacterial and thermochemical sulphate reduction.

4 Isotope fractionation factors are accurately known. Because isotope exchange reactions proceed very slowly at the relatively low temperatures characteristic of diagenesis, fractionation factors are often determined at high temperatures and extrapolated to low temperatures. This is justified in some cases (e.g. oxygen in quartz and carbonates), but can introduce significant uncertainty in many others (e.g. oxygen and hydrogen in clay minerals; Table 4.4). A further difficulty is that for some minerals − illite for example − even dubious experi-

mental data are scarce. When using fractionation factors one should bear in mind the conditions under which they were originally determined.

4.3 Oxygen and hydrogen

4.3.1 Water

Reactions of interest to petroleum geologists tend to occur in aqueous media and include precipitation of mineral skeletons from marine waters and diagenetic reactions which may occur in a wide variety of water types from marine and meteoric through to deep basinal fluids. Accordingly, we gain more information about the reactions by studying not only diagenetic minerals, but also waters from which they may have precipitated. Isotopic study of waters in petroleum reservoirs can also provide information about the continuity of different water types in reservoirs and can assist in reservoir correlation and in identifying changes in water composition as the reservoir is produced.

Formation waters (waters within sedimentary basins) are derived from four basic sources, each of which is isotopically (and chemically) distinct (Fig. 4.4).

1 *Meteoric water*. Meteoric waters originate as rain and snow. Their δD and $\delta^{18}O$ compositions vary systematically and define the meteoric water line (MWL), which is expressed as (Craig, 1961):

$$\delta D = 8\delta^{18}O + 10 \tag{14}$$

This regularity results from the systematic behaviour of oxygen and hydrogen isotopes during successive

Table 4.4 Interpretation of the oxygen and hydrogen isotope compositions of minerals in diagenetic systems.

Mineral	Reliability of $^{18}O/^{16}O$ fractionation factor	Reliability of D/H fractionation factor	Susceptibility to O isotopic exchange	Susceptibility to H isotopic exchange	Susceptibility to recrystallization
Quartz	G		G		G
Feldspar	C		G		G
Calcite	G		G		C
Dolomite	C		G		C
Siderite	C		G		C
Kaolinite	C	C	G	X	C
Smectite	C	X	C	X	X
Illite	C	X	C	X	C
Chlorite	X	X	?	?	C

G, Good/low susceptibility
C, Use with care
X, Use with great caution

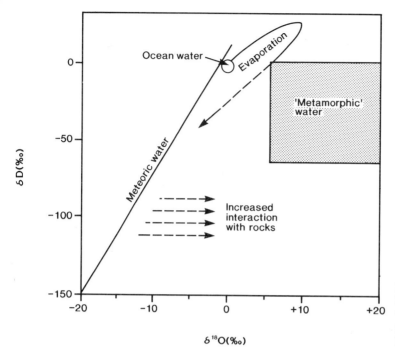

Fig. 4.4 Hydrogen and oxygen isotope compositions of the types of water which may contribute to formation waters in sedimentary basins.

evaporation–condensation cycles. Water evaporated from the oceans is depleted in ^{18}O and D compared to ocean water, becoming increasingly depleted as an increasing proportion of the water vapour condenses as rainfall (Rayleigh fractionation; see Section 4.2.2). Since on a global scale water vapour is transported from low to high latitudes, and from ocean to continental interiors, the isotopic composition of meteoric water changes in a geographically predictable way (Fig. 4.5). With palaeogeographical insight, one can thus estimate the isotopic composition of palaeo-meteoric waters.

2 *Ocean water.* Standard mean ocean water (SMOW) has, by definition, δD and $\delta^{18}O = 0‰$ and most ocean waters fall close to this value. It is less clear whether ocean water has maintained the same isotopic composition throughout the geological past. On an ice-free earth, ocean water probably had a $\delta^{18}O$ value about 1‰ lighter than today's oceans (Shackleton & Opdyke, 1973). There is considerable debate as to whether the $\delta^{18}O$ of the oceans has varied on longer timescales, particularly in the Palaeozoic and earlier. The debate is rooted in the observation that the oxygen isotopic composition of marine carbonates and cherts decreases with increasing age (see reviews by O'Neil, 1986; Veizer *et al.*, 1986). Possible explanations for this are: (i) older

rocks have had more time to equilibrate with isotopically light meteoric waters; (ii) $\delta^{18}O$ of the oceans was lower in the past; and (iii) the oceans were warmer in the past. Of these, (ii) and (iii) have gained most credence, with slight favour towards an ocean which was warmer but isotopically similar to today's.

3 *Evaporation residues* (bitterns). During the evaporation of seawater, the isotopic compositions of the residual brine define a hook on a plot of δD versus $\delta^{18}O$ (Knauth & Beeunas, 1986). Depending on the degree of evaporation, these waters can be enriched or depleted in D and ^{18}O compared to seawater.

4 *Metamorphic water.* Metamorphic waters are defined here as waters which have been expelled into sediments from underlying basement rocks. They have variable isotopic ratios but are characterized by their high $\delta^{18}O$ values.

Most formation waters have isotopic compositions which plot to the right of the MWL but which rarely correspond to the four end-members described above (Fig. 4.6). Waters within individual basins have isotopic compositions which often define restricted regions of the standard $\delta D : \delta^{18}O$ diagram. Nevertheless, a wide range of isotopic compositions occurs in any one basin, and it is dangerous to

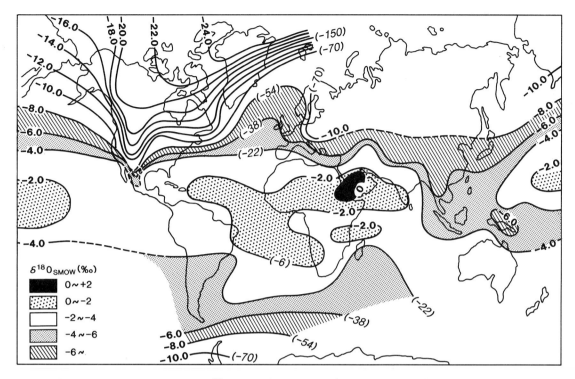

Fig. 4.5 Generalized global distribution of $\delta^{18}O$ and δD (in parentheses) of meteoric water (from Sheppard, 1986).

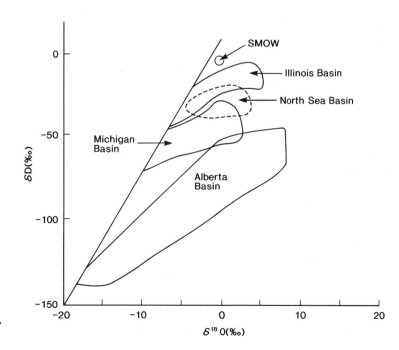

Fig. 4.6 Isotopic compositions of formation waters in a selection of sedimentary basins (from Sheppard, 1986).

assume a single composition for calculating precipitation temperatures from mineral oxygen isotope data.

The chemical and isotopic composition of a formation water is an integrated record of its origin and subsequent diagenesis. The relevant diagenetic processes may be loosely divided into two categories: mixing – between meteoric, metamorphic, evaporative and ocean waters; and reaction – between sedimentary minerals and waters. Comparison of Figs 4.4 and 4.6 shows how simple mixing between the four end-member waters can theoretically account for much of the observed variation in the isotopic composition of formation waters. However, all necessary end-members may well not be available in any one basin. Reactions between water and minerals result in the repartitioning of isotopes and a corresponding change in the isotopic composition of both. If the water–mineral system is closed, then the isotopic composition of the total system must remain constant. However, the isotopic composition of the phases in the system can change. The extent of change partly depends on the fractionation factor between mineral and water, and partly on the relative amounts of the isotopic species in mineral and water, the *water–rock ratio*. Isotopic exchange between a small amount of mineral and a large amount of water will result in a large change in the isotopic composition of the mineral, but very little change in that of the water, and vice versa. These ideas are illustrated in Fig. 4.7a. The closed system initially comprises smectite and water containing equal masses of oxygen. The oxygen isotopic composition of the system, at the fulcrum, is +11‰ and cannot change. If the smectite then recrystallizes to form illite at 100°C, the isotopic separation between illite and water is defined by the fractionation factor at 100°C (12.8‰). The isotopic composition of the system remains constant at 11‰, but the isotopic composition of both water and mineral are substantially changed. A similar situation is shown in Fig. 4.7b, but in this case 90% of the system's oxygen is in the water. Following recrystallization of smectite to illite, the isotopic separation of mineral and water is 12.8‰, but most of the isotopic change has taken place in the mineral, leaving the water largely unchanged.

Taylor (1974) expressed the isotope mass balance approach in terms of some simple equations. In a closed system, in which sediment is simply buried and reacts with its interstitial water, the material balance equation is:

a) – Equal masses of oxygen in smectite and water

(i) Initial conditions

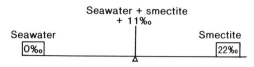

(ii) Final conditions

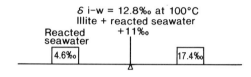

b) – 90% of total oxygen in water

(i) Initial conditions

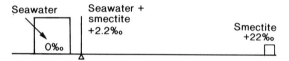

(ii) Final conditions

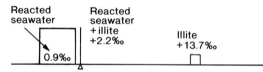

Fig. 4.7 Hypothetical evolution of the oxygen isotope compositions of clay minerals and water during the closed-system conversion of smectite to illite. Conversion of smectite to illite is assumed to take place at 100°C. The isotopic composition of the clay–water system at the fulcrum remains constant, but the composition of each phase changes.

$$W . \delta_{wi} . C_w + R . \delta_{ri} . C_r = W . \delta_{wf} . C_w + R . \delta_{rf} . C_r \quad (15)$$

where W and R are the masses of water and rock minerals in the reaction system, i the initial value, f the final value after isotopic exchange, C_w the atom per cent of oxygen in water, C_r the atom per cent of oxygen in reacting rock minerals and δ is $\delta^{18}O$. If equilibrium occurs between water and rock, then:

$$W/R = \frac{(\delta_{rf} - \delta_{ri})}{\delta_{wi} - (\delta_{rf} - \Delta)} . \frac{C_r}{C_w} \quad (16)$$

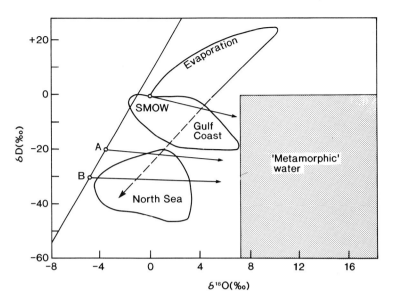

Fig. 4.8 Isotopic compositions of formation waters in the Gulf Coast and North Sea Basins. These formation waters could have formed by: (i) mixing of seawater, meteoric water and metamorphic water, and/ or (ii) water–rock interaction. The arrows indicate the isotopic evolution of mudstone water during the closed-system recrystallization of smectite to illite at 100°C. Initial waters are seawater (SMOW), present-day meteoric water from the Gulf Coast (A) and Mesozoic meteoric water from the North Sea (B).

where $\Delta = \delta_{rf} - \delta_{wf}$ (the fractionation factor between mineral and water at the reaction temperature).

The isotope mass balance approach is relevant to sedimentary basins because, except in areas of hydrodynamically driven recharge, they probably approximate to closed systems. Both Suchecki and Land (1983) and Aplin *et al.* (1993) used this approach to show that the observed ^{18}O enrichment of formation waters in compacting sedimentary basins could be explained by clay mineral recrystallization in mudstones (Fig. 4.8). The hydrogen isotope composition of waters also changes during this process. However, the extent of change is much less than that of oxygen because the hydrogen balance of the sedimentary basin is dominated by water, not rock minerals.

Equation (16) can be used to determine the extent to which $\delta^{18}O_{H_2O}$ changes as a function of water–rock ratio, for a given mineral–water reaction. For reactions relevant to diagenetic systems, such as clay or carbonate recrystallization, water–rock ratios greater than about five will result in little change in $\delta^{18}O_{H_2O}$ (Fig. 4.9). This means that if more than a few pore volumes of water pass through and isotopically exchange with rock minerals, the effect of the reactions will not be recorded in the water. An important corollary of this is that there are finite volumes of ^{18}O-enriched water in sedimentary basins. Whether ^{18}O-enriched waters derive from: (i) reaction with rock minerals; and/or (ii) mixing involving metamorphic or evaporative waters, their total volume is constrained by the volume of reac-

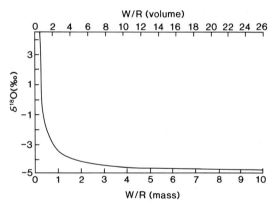

Fig. 4.9 The oxygen isotopic composition of water plotted as a function of effective water–rock ratio. The water–rock system modelled using equation (16) is one in which smectite recrystallizes to illite at 100°C. $\delta^{18}O$ of initial water = −5‰; $\delta^{18}O$ of smectite = 22‰; $\delta^{18}O$ of illite = 19‰. Significant shifts from the initial isotopic composition of the water only occur at low water–rock ratios.

tive rock minerals and/or the volume of evaporative residues plus metamorphic water. In short, the volume of ^{18}O-enriched waters is little more than the pore volume of the basin (Aplin *et al.*, 1993).

4.3.2 Silicates

The most commonly observed authigenic silicate minerals are quartz and the clay minerals illite,

Table 4.5 Oxygen-isotope mineral–water fractionation factors.

Equation		Reference
$1000 \ln \alpha$ (quartz–water) (250–200°C)	$= 3.34 \times 10^6 T^{-2} - 3.31$	Matsuhisa *et al.* (1979)
$1000 \ln \alpha$ (amorphous silica–water) (34–93°C)	$= 3.52 \times 10^6 T^{-2} - 4.35$	Kita *et al.* (1985)
$1000 \ln \alpha$ (Na or K feldspar–water) (350–800°C)	$= 2.91 \times 10^6 T^{-2} - 3.41$	O'Neil and Taylor (1967)
$1000 \ln \alpha$ (calcite–water) (0–500°C)	$= 2.78 \times 10^6 T^{-2} - 2.89$	Friedman and O'Neil (1977)
$1000 \ln \alpha$ (Mg calcite–water) (0 and 25°C)	$= 2.78 \times 10^6 T^{-2} - 2.89 + 0.06M$	Friedman and O'Neil (1977)
$1000 \ln \alpha$ (dolomite–water)	$= 3.14 \times 10^6 T^{-2} - 2.0$	Land (1983)
$1000 \ln \alpha$ (ankerite–water)	$= 2.78 \times 10^6 T^{-2} + 0.11$	Fisher and Land (1986)
$1000 \ln \alpha$ (siderite–water) (33–197°C)	$= 3.13 \times 10^6 T^{-2} - 3.50$	Carothers *et al.* (1988)
$1000 \ln \alpha$ (anhydrite–water) (100–550°C)	$= 3.21 \times 10^6 T^{-2} - 4.72$	Chiba *et al.* (1981)
$1000 \ln \alpha$ (kaolinite–water) (<200°C)	$= 10.6 \times 10^3 T^{-1} + 0.42 \times 10^6 T^{-2} - 15.337$	Savin and Lee (1988)
$1000 \ln \alpha$ (smectite–water)	$= 2.58 \times 10^6 T^{-2} - 4.19$	Savin and Lee (1988)
$1000 \ln \alpha$ (illite–water)	$= 2.39 \times 10^6 T^{-2} - 4.19$	Savin and Lee (1988)
$1000 \ln \alpha$ (illite–water)	$= -2.87 + 1.83 \times 10^6 T^{-2} + 0.0614 \times (10^6 T^{-2})^2 - 0.00115 \times (10^6 T^{-2})^3$	Savin and Lee (1988)
$1000 \ln \alpha$ (illite/smectite–water)	$= (2.58 - 0.19 \times I) \times 10^6 T^{-2} - 4.19$	Savin and Lee (1988)
$1000 \ln \alpha$ (chlorite–water) (Mg, Fe in hydroxide sheet Al, Fe in octahedral sheet) (<150°C)	$= 3.72 \times 10^3 T^{-1} + 2.50 \times 10^6 T^{-2} - 0.312 \times 10^9 T^{-3} + 0.028 \times 10^{12} T^{-4} - 12.62$	Savin and Lee (1988)
$1000 \ln \alpha$ (chlorite–water) (Al, Fe, Mg in hydroxide sheet, Mg in octahedral sheet) (<150°C)	$= 2.56 \times 10^3 T^{-1} + 3.39 \times 10^6 T^{-2} - 0.623 \times 10^9 T^{-3} + 0.056 \times 10^{12} T^{-4} - 11.86$	Savin and Lee (1988)
$1000 \ln \alpha$ (chlorite–water) (Mg, Al in hydroxide sheet Al in octahedral sheet) (<150°C)	$= 6.78 \times 10^3 T^{-1} + 1.19 \times 10^6 T^{-2} - 13.68$	Savin and Lee (1988)

T, temperature in Kelvin
M, mole % $MgCO_3$
I, fraction of illite (0.0 to 1.0)

kaolinite, smectite and chlorite. Feldspar and zeolites are generally less abundant. Preferred oxygen isotope fractionation factors are listed in Table 4.5 (see review by Savin & Lee, 1988).

Quartz

In many ways quartz is the ideal mineral for oxygen isotope analysis. Although the oxygen isotope fractionation between quartz and water has only been determined at high temperatures, extrapolation to lower temperatures yields values which are in broad agreement with those measured on opaline silica in real systems (Matsuhisa et al., 1979; Kita et al., 1985). Furthermore, experiments and empirical observations suggest that post-depositional isotopic exchange is unlikely (with the possible exception of extremely fine-grained [$< 0.5\,\mu m$] quartz at temperatures above 150°C; Yeh & Savin, 1977). The major problem associated with the isotopic analysis of authigenic quartz is separating a pure sample from detrital quartz grains. Two approaches are currently used, neither of which is entirely satisfactory (Brint et al., 1991).

1 Leaching/mass balance approach. Following gentle disaggregation of a sandstone sample, quartz is isolated by chemical dissolution of all other phases (Syers et al., 1968). Two measurements are then made: the isotopic composition of the quartz separate and the proportion of detrital and authigenic quartz in the isolate. The quartz is then leached with HF, preferentially removing the authigenic overgrowths. By measuring the isotopic composition of the detrital quartz residue, the isotopic composition of authigenic quartz can be calculated by mass balance (Milliken et al., 1981). This procedure is subject to several errors which are much greater than the analytical precision. These include: (i) variations in $\delta^{18}O$ of detrital quartz; (ii) the presence of quartz overgrowths from previous sedimentary cycles; (iii) the procedure used to calculate $\delta^{18}O$ of authigenic quartz, which involves significant extrapolation from measured values; and (iv) the occurrence of zonation in the authigenic quartz, implying isotopic heterogeneity. Accuracies of $\pm 2‰$ may be realistically achieved.

2 Separation of quartz overgrowths. In this approach, the quartz isolate is etched in weak HF, and then subjected to ultrasonic agitation (Lee & Savin, 1985). The HF exploits zones of weakness in the quartz separate, which include the boundary between the detrital quartz grain and overgrowth. Authigenic quartz is concentrated in the finer grain sizes which are recovered by sieving. Purity can be checked using cathodoluminescence microscopy. An example is shown in Fig. 4.10. This technique appears to be reasonably successful for samples where there is a marked zone of weakness — for example a dust rim — between the detrital grain and overgrowth, but in the absence of such a zone spalling may occur along fractures within the detrital grain and lead to very significant errors

(Brint et al., 1991). Purity checks are therefore essential.

Clay minerals
The advantage of stable isotope analysis of clay minerals is that both $\delta^{18}O$ and δD can be determined on the same sample, theoretically enabling unique characterization of the isotopic composition of the water from which the mineral formed. Any excitement should however be tempered by uncertainties over clay mineral–water fractionation factors and by the real possibility that some clay minerals may suffer isotopic exchange after they form.

At diagenetic temperatures, mineral–water fractionation factors for oxygen isotopes are only known to ± 1–$2‰$ (Fig. 4.11). This reflects not only the difficulty of determining fractionation factors for clay minerals, but also the fact that they vary according to chemical composition within a particular group of clay minerals. Errors of ± 1–$2‰$ could lead to a similar error in an estimate of $\delta^{18}O$ of the mineralizing fluid, or to an error in the estimated precipitation temperature of about $\pm 20°C$. The best estimates for the hydrogen isotope clay mineral–water fractionation equations are given in Table 4.6; the uncertainties are even greater than those for oxygen. There can be reasonable confidence for kaolinite, since data derived from experimental work and field observations generally coincide (Savin & Lee, 1988). The data for smectite and illite are less certain and are probably not accurate to better than $\pm 10‰$.

In addition to the uncertainties in clay mineral–water isotope fractionation factors, there is also evidence that under certain conditions clay minerals can undergo isotopic exchange with whatever water happens to be around. Isotopic exchange at diagenetic temperatures is usually associated with mineralogical change. The transformation of smectite to illite in Gulf Coast shales is a good example, involving major oxygen and hydrogen isotopic shifts (Yeh & Savin, 1977; Yeh, 1980). However, experimental and field data suggest that hydrogen, and possibly oxygen isotope exchange can occur *independently* of mineralogical change in some clay mineral groups. Smectites are the most susceptible: there is a strong possibility that both oxygen and hydrogen isotope exchange occur at temperatures around 100°C (O'Neil & Kharaka, 1976). The situation is more complex for illite and kaolinite which appear to be robust with respect to

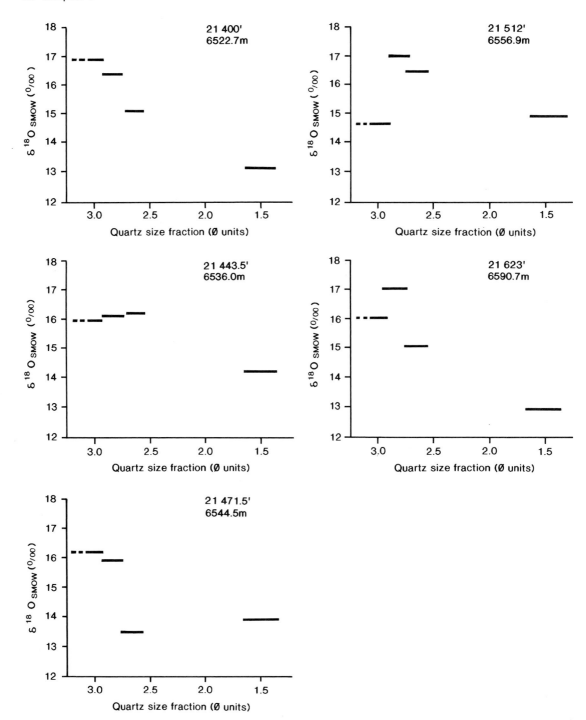

Fig. 4.10 Oxygen isotope analyses of quartz separates, Norphlet Formation (Jurassic), offshore Louisiana. Measured $\delta^{18}O$ is plotted against the size fraction of the separates measured in Φ units ($-\log_2$ [grain size]). Then finer fractions are supposed to tend towards a constant $\delta^{18}O$ value that represents the quartz cement. In this case, authigenic quartz $\delta^{18}O$ would be interpreted to be between about $+16$ and $+17‰$. These samples are from well Mobile 821-1, the subject of the case study in Section 6.3.

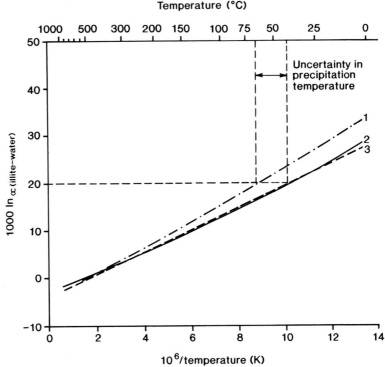

Fig. 4.11 Different estimates of the oxygen isotope fractionation factor between illite and water. Curve 1 is calculated from theoretical considerations (Savin & Lee, 1988). Curve 2 was proposed by Lee (1984) using a combination of theoretical arguments and data for natural samples. Curve 3 is based on data in Eslinger and Savin (1973). Uncertainty in the fractionation factor could lead to a 20°C error in estimated precipitation temperature.

Table 4.6 Hydrogen-isotope mineral–water fractionation factors.

Equation		Reference
$1000 \ln \alpha$ (kaolinite–water)	$= -4.53 \times 10^6 T^{-2} + 19.4$	Lambert and Epstein (1980)
$1000 \ln \alpha$ (illite/smectite–water) $= -19.6 \times 10^3 T^{-1} + 25$ (best estimate for smectite or illite at present)		Yeh (1980)

T, temperature in kelvin

oxygen but susceptible with respect to hydrogen, even at temperatures as low as 40°C (O'Neil & Kharaka, 1976; Bird & Chivas, 1988; Longstaffe & Ayalon, 1990). However, in all the cases where hydrogen isotope exchange could be documented, the clay minerals interacted with large volumes of water which was isotopically very different to that from which they formed. In situations where water–rock ratios are low, isotope exchange may not occur, or may be impossible to identify if the isotopic compositions of new and original mineralizing fluids are similar.

As for quartz, it can be difficult to obtain a pure sample of an authigenic clay mineral. A clay mineral concentrate can usually be isolated from a gently disaggregated sandstone by settling methods because clay minerals are concentrated in the fine ($< 2 \, \mu m$) fraction (Section 2.7.3). It is much more difficult to isolate a specific mineral species from a mixture of clay minerals or to separate an authigenic mineral species from its detrital counterpart. Rules of thumb include the observation that some authigenic clays tend to be concentrated in the finest size fractions ($< 0.1 – 0.5 \, \mu m$), and that illite tends to occur in finer fractions than kaolinite, but there is no substitute for designing each separation process according to the sizes of the minerals to be analysed in the sample in question (these can be determined by examination using a SEM). Attempts have also been made to isolate mineral groups (chlorite, illite, kaolinite) with magnetic separators, with mixed success (Tellier *et al.*, 1988). If, as is often the case, it is impossible to obtain a pure authigenic mineral separate, then it is important to estimate the proportions of each mineral in the mixture. For a multimineralic mixture, this can be done by X-ray diffraction (quick and imprecise) or TEM (slow but more precise).

4.3.3 Example 1: quartz cement in a Pennsylvanian sandstone, West Tuscola Field, north-central Texas

This study was published in 1978 by Land and Dutton and concerns the isotope geochemistry of diagenetic phases in a deltaic sandstone reservoir. It is a benchmark paper in that it was one of the first to use high quality isotope data to investigate the diagenetic history of a sandstone and was perhaps the first to present good quality isotope analyses of authigenic quartz in a reservoir sandstone.

Land and Dutton chose a suite of quartz-rich samples which were variably cemented with quartz cement and used a version of the leaching/mass balance approach to obtain the isotopic composition of authigenic quartz. Following chemical purification (Syers *et al.*, 1968), the isotopic composition of the quartz was determined and plotted against the ratio of authigenic quartz : total quartz, as estimated from point-count data (Fig. 4.12). Linear regression of the data gave a correlation coefficient of 0.91 and estimates of the isotopic composition of detrital and authigenic quartz of +12.6 and +24.5‰ respectively. The value of 12.6‰ for detrital quartz compared with 11.9‰ for a sample of pure detrital quartz. This difference gives some impression of the uncertainty in the estimate of $\delta^{18}O$ of authigenic quartz; this is about ±1‰, much greater than the analytical precision (0.1‰).

Since the isotopic composition of quartz is a function of *both* precipitation temperature and $\delta^{18}O$ of the mineralizing fluid, a single isotopic composition can result from many combinations of $\delta^{18}O_{H_2O}$ and temperature (Fig. 4.13). Constraints on the true conditions of precipitation therefore rely entirely on additional geological and geochemical data or assumptions. The supporting data in this example included a paragenetic sequence determined by thin-section petrography, which showed that quartz was a relatively early phase, and an approximate burial history which suggested that the samples

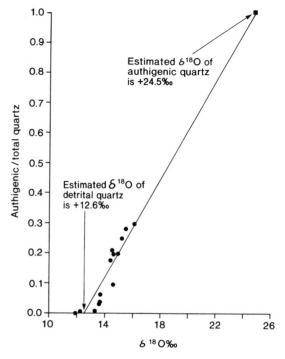

Fig. 4.12 The oxygen isotope composition of quartz in a Pennsylvanian deltaic sandstone from north-central Texas (Land & Dutton, 1978), plotted as a function of the proportion of authigenic quartz in the sample. Extrapolation of the line to 1.0 on the *y*-axis gives an estimate of $\delta^{18}O$ of authigenic quartz of +24.5‰.

reached a maximum burial temperature of less than 80°C before being uplifted in the Tertiary (Fig. 4.13). It was also possible to determine the isotopic compositions of the first and last waters in the formation. The present-day formation water was analysed whilst the isotopic composition of the fresh water in which the sandstones were deposited was estimated from the relationship between latitude and $\delta^{18}O$ of meteoric water (Fig. 4.5). Since meteoric water is the isotopically lightest water that could have passed through the formation, quartz must have formed at temperatures higher than 40°C, and could not have formed from a water enriched in ^{18}O compared to seawater (Fig. 4.13). Beyond that, all is speculation. The cement could have formed at present-day burial temperatures, but from a fluid depleted in ^{18}O compared to the water presently in the reservoir. Such a water could be meteoric water which infiltrated during Tertiary uplift, but the occurrence of quartz relatively early in the

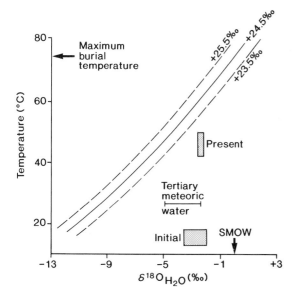

Fig. 4.13 Possible precipitation conditions for authigenic quartz in a Pennsylvanian deltaic sandstone. The curves represent the isotopic composition of quartz cement. Quartz could not have precipitated from a water isotopically heavier than SMOW, nor when the sandstone was first deposited. It could have formed from a Tertiary meteoric water at about 40°C, when the formation was uplifted but petrographic evidence suggests that it was a relatively early cement. Land and Dutton's interpretation was that quartz probably formed at or close to maximum burial temperature from a water with similar characteristics to that presently in the formation.

paragenetic sequence argues against this. Land and Dutton's preferred hypothesis was that the quartz formed from a water similar to that present in the reservoir today, but at the maximum burial temperature of the sequence. The implications would be that: (i) the quartz precipitated from relatively small volumes of basinal water, since the water is isotopically evolved from the original meteoric water; (ii) large-scale fluid flow may not have been the cause of cementation, since the water in the reservoir at the time of cementation is the same as that today; and (iii) cementation was not driven by the ingress of meteoric water due to uplift of the basin margin, but by another mechanism. Reservoir quality would not in this case have been controlled by the access of meteoric water. The conclusions sound plausible and are potentially useful for predicting reservoir quality, but it should not be forgotten that the data are open to alternative explanations.

4.3.4 Example 2: illite cement in fluvial sandstone, Brent Group, Northern North Sea

This example considers the problems of determining the isotopic composition of authigenic illite in a rock that also contains detrital illite. We will return to the same example in Section 5.3.5 where we will look at the difficulties that the presence of detrital illite poses for K–Ar dating. In the well examined in this study, Middle Jurassic fluvial sandstones of the Brent Group have undergone continuous burial and presently reside at their maximum burial depth of 4.3 km, at a temperature of 145°C. An initial SEM study revealed that two morphological types of illite were present: platy illite, which was inferred to be detrital, and fibrous illite, considered to be authigenic.

Clays were separated from core by centrifuging disaggregated samples and then divided into size fractions (<0.2, $0.2–0.5$, $0.5–2$ and $2–5\,\mu m$). Clay separates were examined using TEM in order to estimate visually the relative proportions of detrital and authigenic illite. In general, platy, detrital illite was concentrated preferentially in the coarser fractions. Oxygen isotope compositions of clay separates are listed in Table 5.3 and plotted in Fig. 4.14 as a function of the proportion of authigenic illite in the analysed mixture. If the isotopic compositions of detrital and authigenic illite are homogeneous, then the data should plot on a straight line. To a first approximation, they do: least squares regression and extrapolation give $\delta^{18}O$ of authigenic illite as $+14.5 \pm 0.6\permil$ and $\delta^{18}O$ of detrital illite as $+16.5 \pm 1.2\permil$*. The uncertainties here are, as usual, much greater than the analytical precision.

Without an independent measure of precipitation temperature, even the best isotopic data can only loosely constrain the conditions of illite cement

---

* The calculated $\delta^{18}O$ value is surprisingly low for detrital illite. As a continental weathering product, illite would be expected to have formed in meteoric water at about 10°C. The isotopic composition of water in equilibrium with illite with $\delta^{18}O = +16.5\permil$ at 10°C is about $-9\permil$, compared to -4 or $-5\permil$ for local meteoric water at the time the formation was deposited. This apparent anomaly may be explained in two ways: either the illite inherited some light oxygen from the primary igneous mineral from which it formed or it recrystallized during burial diagenesis. For example, if the illite had recrystallized in meteoric water at 50°C, it would have $\delta^{18}O = +14$ to $+15\permil$. This would leave a niggling doubt as to whether the authigenic illite might also have done the same thing.

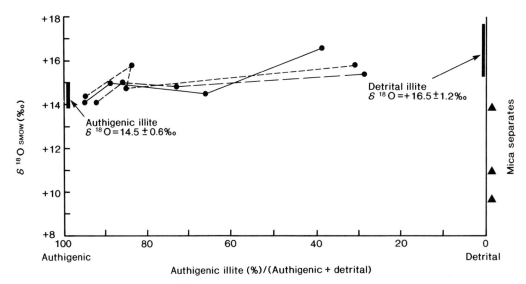

Fig. 4.14 Oxygen isotope compositions of illite separates as a function of the proportion of authigenic illite in the analysed sample, well 3/8b-11, UK North Sea. The diagram shows isotopic analyses of 12 separates, 4 (joined by lines) from each of 3 sandstone samples. Data may be found in Table 5.3. The proportion of authigenic illite in each separate analysed was estimated visually using TEM by a form of point-counting based on the difference in morphology between authigenic (fibrous) and detrital (platy) illite. This procedure is prone to potentially large errors that are hard to quantify. Separates with greater proportions of detrital illite have somewhat higher $\delta^{18}O$ values, but the correlation is not particularly good. The quoted end-member isotopic compositions have an error that does not take into account uncertainty in point-counting. The isotopic compositions of three separates of coarse detrital muscovite are plotted on the right hand axis for reference only; they are quite different from those calculated for detrital illite. See Section 5.3.5. for an attempt to discriminate detrital and authigenic illite on the basis of chemical composition.

formation. Calculated authigenic illite $\delta^{18}O$ is shown on a temperature : $\delta^{18}O_{H_2O}$ plot in Fig. 4.15. In this example, temperature was estimated independently by combining K–Ar ages with a time–temperature graph for the samples based on modelling of their burial and thermal history. Illite grew between about 50 and 72 Ma ago at temperatures between about 90 and 130°C (see Section 5.3.5). Figure 4.15 shows that the isotopic composition of the mineralizing water must therefore have been between about −0.5 and +4.5‰. The present water in the formation has $\delta^{18}O = +1.9$‰, much heavier than local meteoric waters, and has probably acquired its isotopic composition by reaction with sedimentary minerals (Section 4.3.1). *If* illite cementation was driven by fluid flow, then convective or compactional flow is suggested, not hydrodynamic flow from basin margins.

Further insight into the nature of the mineralizing fluid ought to be available from the hydrogen isotope composition of the clay minerals. Figure 4.16 is a plot of δD against proportion of authigenic

illite in the separates analysed. The most striking feature of the data is their variability. The isotopic composition of apparently relatively pure authigenic illite ranges from about −20 to −70‰, far greater than the usual analytical precision (±2‰). This may be interpreted in several ways: the illite may have formed in waters with widely varying isotopic compositions; there may be an unidentified analytical problem, for example the presence of organic contaminants; or the illites may have partially re-equilibrated with the present formation water (δD = −25‰). Based on Yeh's (1980) fractionation equation, illite in equilibrium with this water would have δD = −50‰ at the present reservoir temperature.

4.3.5 Carbonates

Calcite, aragonite, dolomite, siderite and various Fe–Ca–Mg carbonate solid solutions − ankerite (Ca/Fe/Mg), ferroan dolomite (Ca/Mg/Fe) and ferroan calcite (Ca/Fe) − are common as diagenetic

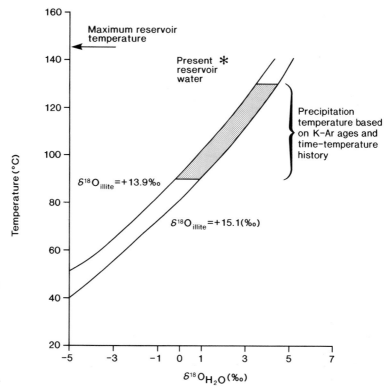

Fig. 4.15 Possible precipitation conditions for authigenic illite in well 3/8b-11, UK North Sea. The curves represent the calculated $\delta^{18}O$ values for authigenic illite (see Fig. 4.13). Precipitation temperatures based on K–Ar ages and burial history modelling suggest that illite formed from a water with $\delta^{18}O$ between −0.5 and +4.5‰: an isotopically evolved, 'basinal' water.

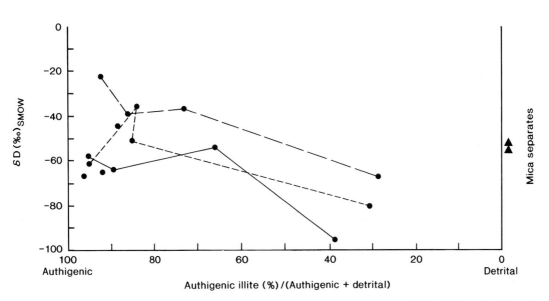

Fig. 4.16 Hydrogen isotope compositions of illite as a function of the proportion of authigenic illite in the analysed sample, well 3/8b-11, UK North Sea. See caption to Fig. 4.14 for an explanation of the horizontal axis.

minerals. Calcite–water and, to a lesser extent, dolomite–water fractionation factors are well known at diagenetic temperatures and can be used with some confidence (Table 4.5). Fractionation factors for siderite are less well constrained, with the most commonly used equation that of Carothers *et al.* (1988). Mineral–water fractionations for intermediates in the solid solution series are not known specifically. In these cases, it is common practice to assume the same fractionation as for the chemically nearest end-member, or to interpolate between the two end-members.

For our purposes, the two main applications of oxygen isotopes in carbonate systems are:
1 the estimation of temperatures of precipitation of diagenetic carbonates; and
2 oxygen isotope stratigraphy from unaltered marine carbonate fossils, which has permitted a very high stratigraphic resolution in the Neogene (Trainor *et al.*, 1988).
Estimation of the temperature of precipitation of carbonate diagenetic phases from oxygen isotopes is identical in principle to temperature estimation using quartz or clay minerals. Carbonate rocks will often comprise a whole range of carbonate mineral components, including original marine calcite, possibly several calcite cement phases precipitated in a variety of diagenetic environments, and other carbonate minerals such as dolomite. There is no point in making a whole-rock isotopic analysis of such a complex mixture of phases and expecting to derive meaningful information about precipitation temperatures. Different phases of interest must be analysed separately to derive useful oxygen isotope ratios (see Hudson, 1977). This can only be done in conjunction with detailed petrography. Once identified, original marine precipitates can be physically separated from cements by the use of stout needles or by modified dentist's drills. Where physical separation is impossible, whole-rock analysis can reasonably be undertaken on samples which appear to comprise simple mixtures whose proportions have been quantified prior to analysis (Emery *et al.*, 1988). When carbonate rocks comprise too many different phases which cannot be physically separated or when they are zoned on a small scale, *in situ* analysis by laser pyrolysis is the best option (Dickson *et al.*, 1991). In sandstones, carbonate cements tend not to present the same problems of separation because it is often possible to identify areas in a thin-section which are cemented by a single carbonate. They may nonetheless show complex zonation, in which case *in situ* analysis

is preferable.

A problem common to carbonate cements in siliciclastic and carbonate rocks is recrystallization. Field and experimental evidence suggests that fine-grained carbonates commonly recrystallize above 100°C, particularly where water–rock ratios are high (Clayton *et al.*, 1968; O'Neil, 1987). The situation is less clear-cut for the relatively coarse crystalline carbonates which commonly cement sandstones. Ayalon and Longstaffe (1988) found that early diagenetic calcite in some deeply buried sandstones in the Western Canada Basin had very variable $\delta^{18}O$ values which they interpreted as indicating partial recrystallization and isotopic resetting of an initially homogeneous cement. On the other hand, comparison of the isotopic composition of carbonates and formation waters in several North Sea reservoir sandstones commonly shows that they are not in equilibrium. These cements could have recrystallized, but not recently. If complex zoning patterns are preserved, one can probably be reasonably happy that recrystallization has not taken place. In general, recrystallization is more likely at higher temperatures and where there are large differences between precipitation and maximum burial temperatures.

Oxygen isotope stratigraphy is based on the record of glaciations and deglaciations preserved by the isotopic signature of marine fossils, which essentially record the changes in the bulk isotopic composition of seawater. Variations in the isotopic composition of seawater are controlled largely by the volume of ice on the Earth's surface. Ice will preferentially contain the lighter isotope, ^{16}O, because most originates as precipitation at high latitudes (Section 4.3.1). During glaciations, the seawater will be correspondingly enriched in ^{18}O, whereas during interglacials the $\delta^{18}O$ value of seawater will approach 0‰ or slightly negative values (Shackleton & Opdyke, 1973). Marine carbonate (chiefly aragonite, calcite or high magnesium calcite) fossils will preserve this isotopic difference, which can be used to provide an isotope stratigraphy of high resolution, particularly over the last 4 Ma or so. The marine carbonate fossils used must come from the same living position in the oceans, otherwise temperature changes through the water column may mask changes in bulk seawater isotopic compositions. They should also ideally be of the same species or genus (otherwise vital effects may mask the isotope stratigraphic record) and should show no diagenetic alteration. Section 8.2 provides more information on the practical problems of oxygen isotope stratigraphy.

4.3.6 Sulphates

Although anhydrite ($CaSO_4$) and barite ($BaSO_4$) are locally important mineral cements, their oxygen isotopic composition is rarely reported. The main reason for this is that at most diagenetic temperatures the oxygen in the dissolved sulphate molecule and in the water itself are not in isotopic equilibrium because the kinetics of the isotope exchange reaction are slow. The rate at which oxygen isotopes are exchanged between dissolved sulphate and water is a function of solution pH and temperature; lower pH and higher temperature facilitate isotopic exchange (Lloyd, 1968; Chiba & Sakai, 1981). Experimental data suggest that at pH 6 isotopic equilibrium between dissolved sulphate and water is achieved within hours or days at 200°C, but takes 10^5–10^6 years at 100°C. Only at temperatures above about 150°C can the oxygen isotope composition of sulphate be used as a geothermometer or to infer the conditions under which the sulphate formed. In seawater, more than 10^9 years are required to achieve isotopic equilibrium between water and dissolved sulphate. However, since the ocean mixing time is around 2000 years, the oxygen isotopic composition of dissolved seawater sulphate is constant throughout the oceans at any time. Because the relative magnitude of the various sources of seawater sulphate has varied through geological time, so has $\delta^{18}O$ of seawater sulphate (Claypool et al., 1980). Despite the significant uncertainty in the precise shape of the secular curve, the oxygen isotope composition of buried evaporites has potential as a stratigraphic tool.

4.4 Carbon

4.4.1 Principles

Equilibrium carbon isotope fractionation factors in the carbonate system are given in Table 4.7. Compared to oxygen, carbon isotope fractionation is relatively insensitive to temperature so that carbon isotopes are of little use as geothermometers. Their main use in diagenetic systems is to unravel the origin of carbon in carbonate cements.

There are two major carbon reservoirs in sedimentary basins: marine carbonate, the product of biological or chemical precipitation from seawater; and reduced organic carbon, derived mainly from marine and land plants buried within fine-grained sediments. Other reservoirs (atmospheric CO_2, soil CO_2, mantle CO_2) may be locally important. The two main carbon reservoirs are readily distinguishable by their isotopic signatures (Fig. 4.17). Most marine carbonates have $\delta^{13}C$ of $0 \pm 4‰$. In contrast, the conversion of CO_2 to organic carbon is accompanied by a series of kinetic isotopic fractionations which produces organic matter with a range of $\delta^{13}C$ values between about -10 and $-35‰$, most typically between -20 and $-30‰$. Petroleum has a similar isotopic composition to the kerogen from which it is generated (see Deines, 1980 for a detailed discussion).

CH_4 is isotopically distinct from both marine carbonate and reduced organic carbon reservoirs. There are two isotopically distinct CH_4 reservoirs: isotopically light (-50 to $-90‰$) biogenic CH_4 ('marsh gas') which is produced at temperatures below 70°C by bacterially mediated processes; and isotopically heavier (-20 to $-55‰$) thermogenic CH_4 which is generated by thermal breakdown of kerogen and oil. Since there is a kinetic isotope fractionation of about $-25‰$ associated with this reaction, early generated (low maturity) CH_4 is isotopically lighter than CH_4 generated at high maturity (Clayton, 1991).

If reduced organic carbon is to contribute to carbonate cements, it must first be converted into carbonate. There are two pathways by which this occurs in sedimentary basins. Firstly, the organic matter itself contains a certain amount of oxygen which is liberated as CO_2 or organic acids, mainly prior to oil generation. Secondly, organic carbon

Table 4.7 Carbon-isotope fractionation factors in the carbonate system (after Deines et al., 1974).

$$1000 \ln \alpha \, (H_2CO_{3(aq)} - CO_2) = -0.91 + 0.0063(10^6 T^{-2})$$

$$1000 \ln \alpha \, (HCO_3^-{}_{(aq)} - CO_2) = -4.54 + 1.099(10^6 T^{-2})$$

$$1000 \ln \alpha \, (CaCO_3 - CO_2) \quad = -3.63 + 1.194(10^6 T^{-2})$$

$$1000 \ln \alpha \, (CO_3^{2-}{}_{(aq)} - CO_2) = -3.4 + 0.87(10^6 T^{-2}) \text{ (derived from limited data)}$$

T, temperature in kelvin

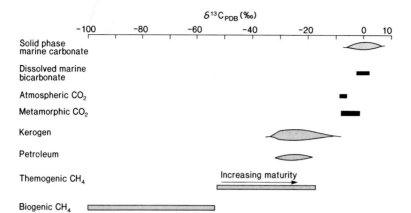

Fig. 4.17 Carbon isotope compositions of major carbon reservoirs in sedimentary basins.

can be oxidized by inorganic oxidants such as molecular oxygen, sulphate and ferric iron. Reactions in both pathways may be purely chemical or, at lower temperatures, catalysed by bacteria. The processes by which organic carbon is converted to HCO_3^- at low temperatures have been well documented from studies of recent muds. During early diagenesis (within metres of the sediment–water interface), bacteria obtain energy by sequentially oxidizing organic matter using oxygen, nitrate, manganese and iron oxides, and sulphate (Fig. 4.18; Froehlich *et al.*, 1979). Bicarbonate is a by-product of all these reactions and has an isotopic composi-

tion similar to the precursor organic matter (-20 to $-25‰$). Given the correct chemical conditions, carbonate precipitates, often as concretions or bands, preserving a record of the oxidation process (Curtis *et al.*, 1972; Coleman & Raiswell, 1981).

If organic matter is buried beyond the sulphate reduction zone, it can be further degraded by methanogenic bacteria at temperatures up to about 70°C. Two main metabolic pathways are involved. *Acetate fermentation* is a disproportionation reaction which yields both CO_2 and CH_4:

$$CH_3COOH \rightarrow CO_2 + CH_4 \qquad (17)$$

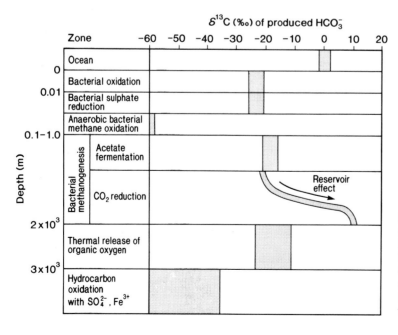

Fig. 4.18 Carbon isotopic composition of bicarbonate introduced to porewaters due to the bacterial and abiotic oxidation of organic carbon. Although expressed as a zonal scheme (see Irwin *et al.*, 1977), zones overlap and are of variable thickness.

Carbon dioxide reduction uses dissolved hydrogen, itself a by-product of bacterial activity, as a reducing agent:

$$CO_2 + 4H_2 \rightarrow CH_4 + 2H_2O \qquad (18)$$

The carbon isotopic composition of CO_2 in the methanogenic zone is a function both of the pathway by which CO_2 is produced and the isotopic fractionation associated with its production. Kinetic isotope effects are important in both processes; for acetate fermentation, early formed CO_2 is isotopically light (-20 to $-30‰$) but approaches the isotopic composition of the carboxyl group as the acetate is fully converted to CO_2 and CH_4 (-5 to $-10‰$; Games *et al.*, 1978). For CO_2 reduction, CH_4 is between 25 and 60‰ lighter than precursor CO_2, which itself may be derived from any of the preceding oxidation/fermentation reactions. As more CO_2 is consumed, residual CO_2 becomes enriched in ^{13}C, reaching values as high as $+15‰$. The end result of all these processes is that the carbon isotopic composition of CO_2 in the methanogenic zone is very variable (Fig. 4.19). Most values are between $+2$ and $-22‰$, not the $+15‰$ which is commonly quoted as representative of methanogenically generated CO_2 (see Irwin *et al.*, 1977). A further complication is that CH_4 may be oxidized, aerobically or anaerobically, and contribute isotopically light CO_2 to the total pool (Whiticar & Faber, 1986; Raiswell, 1988).

Above about 70°C, methanogenic bacteria give up the struggle and CO_2 is produced by chemical processes. Oxygen-bearing compounds are thermally cleaved from kerogen either as CO_2 or low molecular weight organic acids, with an isotopic composition of -10 to $-20‰$. Organic acids may themselves be decarboxylated between 100 and 200°C (Palmer & Drummond, 1986). Above 140°C and perhaps above 100°C, low molecular weight hydrocarbons can be converted to CO_2 by reaction with mineral oxidants such as sulphate or ferric iron. The best known of these reactions is thermochemical sulphate reduction, which can be most simply expressed as the reaction between CH_4 and dissolved sulphate:

$$CH_4 + SO_4^{2-} \rightarrow HCO_3^- + HS^- + H_2O \qquad (19)$$

The isotopic composition of instantaneously produced bicarbonate is similar to that of reactant CH_4. Such light values are, however, rather rarely observed (e.g. Krouse *et al.*, 1988). More commonly, and depending on the particular chemical conditions, the produced carbonate further reacts with pre-existing dissolved or solid-phase carbonate, obscuring its origin. Mixing of dissolved carbonate from various sources is common in diagenetic systems and can severely impede an accurate assessment of the route which led to carbonate cementation. Many carbonate cements in sandstones have carbon isotopic compositions between 0 and $-12‰$ (e.g. Lundegard *et al.*, 1984), which could be the result of several combinations of carbon sources (Fig. 4.17).

There is increasing evidence that inorganic reactions are an important source of CO_2 in the deeper sections of sedimentary basins. Reactions between clay minerals and carbonates are thought to be potentially important sources of CO_2 (Hutcheon *et al.*, 1980), for example the reaction between siderite and kaolinite to form chlorite:

$$5FeCO_3 + SiO_2 + Al_2Si_2O_5(OH)_4 + 2H_2O$$
$$\rightarrow Fe_5Al_2Si_3O_{10}(OH)_8 + 5CO_2 \qquad (20)$$

The isotopic composition of this CO_2 depends on the isotopic composition of the precursor carbonate. However, the average isotopic composition of carbon in metamorphic rocks is around $-6‰$ (Ohmoto & Rye, 1979).

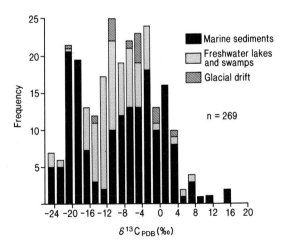

Fig. 4.19 Carbon isotopic composition of CO_2 sampled in the methanogenic zone of recent sediments. Data supplied by C. Clayton.

4.4.2 Example 3: calcite cement in a Miocene carbonate reservoir, Liuhua Field, Pearl River Mouth Basin, offshore China

The Liuhua Oilfield is a Miocene carbonate buildup located in the Pearl River Mouth Basin, some 200 km south-east of Hong Kong in the South China Sea

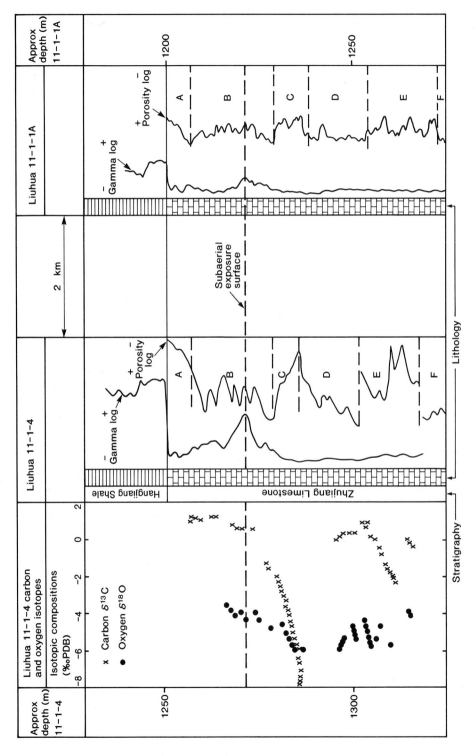

Fig. 4.20 Stratigraphy and isotopic composition of carbonate cements in wells 11-1-4 and 11-1-1A, Liuhua Field, South China Sea (from Turner & Hu, 1991). Zone C corresponds to a lower porosity cemented interval, in which whole-rock $\delta^{13}C$ compositions show a striking fall to ^{13}C depleted values, interpreted to represent the input of isotopically lighter soil-gas carbon. Above zone C is a higher porosity interval (base zone B) in which bioclasts were leached just below a soil horizon, now represented by the deflection to high values on the γ ray log.

(Erlich *et al.*, 1990). The reservoir consists chiefly of shallow water carbonate, comprising coral, algal and other skeletal material and forms an isolated platform a few hundred metres thick, with an oil column height of 75 m. It is this upper, petroleum-bearing portion of the reservoir which shows the most complex relationships between porosity and cementation (Turner & Hu, 1990, 1991).

In order to improve the understanding of the controls on cementation and porosity development within the oil column, several cores were examined by thin-section and cathodoluminescence petrography. There proved to be several petrographically and diagenetically distinct zones (Fig. 4.20). One particular zone (C), at approximately 1280 m in well 11-1-4, showed low porosity caused by abundant pore-filling calcite cement which increased in abundance with depth. This cement appeared to post-date dissolution of some carbonate bioclasts, but pre-dated any significant compaction, suggesting that it was precipitated relatively early. This cement was also present in well 11-1-1A at approximately 1230 m depth. To constrain further the source(s) of carbonate for the cement and its origin and environment of precipitation, isotopic analyses were performed on whole-rock samples taken from the cored interval in well 11-1-4. $\delta^{13}C$ values show relatively little variation in the top portion of the core, ranging between 0 and +2‰. Within zone C, however, $\delta^{13}C$ shows a dramatic shift to values as low as −8‰, tending to decrease as the amount of cement increases (and the amount of porosity falls). No other major carbon isotopic shifts are seen deeper in the core (although a less dramatic shift to −2‰ is seen at approximately 1312 m in 11-1-4) and values range between −2 and +0.5‰. Oxygen isotopic compositions do not show such an obvious shift, with a slight depletion in ^{18}O of slightly less than 2‰ within zone C.

The carbon isotopic composition of the calcite in zone C is interpreted to represent mixing of carbon from two isotopically distinct sources: primary marine carbonate with low positive or low negative values, and an isotopically much lighter carbon, probably originating as soil-gas CO_2, with $\delta^{13}C$ values around −20‰. This interpretation fits well with the evidence of bioclast dissolution prior to (and probably during) cementation in zone C, which supplied carbonate of marine origin and isotopic composition. Dissolved marine carbonate was also supplied from a high porosity interval at the base of zone B which suffered only dissolution and no cementation. The soil profile which supplied the isotopically light carbon is interpreted to have developed above the high porosity interval at the base of zone B, and is represented by a kick on the γ-ray log reflecting the concentration of clay minerals during subaerial exposure. This explanation suggests widespread subaerial exposure of the limestone during a regional or eustatic sea-level fall, and implies that the same cementation and dissolution profile should be widespread. Subsequent wells have shown that this is the case, with distinct carbon isotopic shifts occurring within zone C (N.L. Turner, pers. comm., 1991).

4.5 Sulphur

4.5.1 Principles

In diagenetic systems, the isotopic composition of sulphur yields information about the origin of sulphate and sulphide cements. In produced fluids, $\delta^{34}S$ can be used to infer something about the origin of dissolved sulphate and gaseous H_2S (sour gas). However, the isotope geochemistry of sulphur in low temperature systems is complex and it is rare that sulphur isotope data can be used to define uniquely the origin of a sulphur-bearing compound. Furthermore, many of the important sulphur isotope fractionation factors are variable or poorly known. It is this, rather than the separation of pure phases or analytical precision, that is the limiting factor in the interpretation of sulphur isotope data.

The complexity of sulphur isotope systematics stems from the fact that sulphur exists in several redox states. The mechanisms by which the myriad sulphur species are converted from one to another are not fully understood and neither are the isotope systematics. Nevertheless, a useful generalization is that large (> 10‰) isotope fractionations occur when sulphate is converted to reduced sulphur species (sulphide and elemental sulphur, mainly), whereas small fractionations (± 5‰) occur amongst reduced sulphur species and during sulphide oxidation.

Below 200°C, the rate of isotopic exchange between dissolved sulphate and sulphide is geologically slow and isotopic equilibrium between the species is rare (Ohmoto & Lasaga, 1982). Kinetic effects therefore dominate the isotope systematics of sulphur in sedimentary basins. The main kinetic isotope effect is associated with the reduction of sulphate to sulphide, which can be achieved either chemically, typically above 140°C, or through microbial activity. Experiments at 20–150°C suggest that the kinetic isotope effect associated with chemical

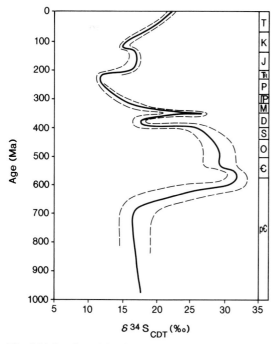

Fig. 4.21 Secular sulphur isotope curve for seawater sulphate, based on analyses of evaporitic gypsum and anhydrite (from Claypool *et al.*, 1980).

sulphate reduction produces sulphide 14–22‰ lighter than the initial sulphate (Harrison & Thode, 1957). However, these experiments were performed using reducing agents which do not occur naturally. The isotope systematics of bacterial sulphate reduction has been studied both in the laboratory and in anoxic marine sediments. Whilst it is clear that bacteria preferentially metabolize $^{32}SO_4^{2-}$, the extent of fractionation is very variable (Goldhaber & Kaplan, 1974). Values between 0 and −50‰ have been measured, although sulphide is commonly 25–50‰ lighter than precursor sulphate.

Sulphur is incorporated into sedimentary basins either as sulphate in evaporitic anhydrite, or as sulphide (pyrite or organically bound) in fine-grained rocks. The isotopic composition of seawater sulphate is constant throughout the world's oceans but has varied between +10 and +30‰ through geological time (Claypool *et al.*, 1980; Fig. 4.21). The sulphur isotopic composition of evaporitic anhydrite therefore has some potential as a stratigraphic tool (cf. oxygen in marine sulphate, Section 4.3.6).

Pyrite and organic sulphur are incorporated into sediments during very early diagenesis by the trans-

port of sulphate across the sediment–water interface; its reduction by bacteria, to sulphide and polysulphide; and the fixation of reduced sulphur by reaction with detrital iron minerals and organic matter. The complexities of these processes and the variable isotope fractionation of the sulphate reduction step lead to the formation of pyrite with widely varying isotope ratios, between −50 and +30‰ with most values between −35 and +10‰ (Kaplan, 1983). A wide range of values (between −17 and +25‰) has also been measured for organically bound sulphur and the petroleum generated from it (A.C. Aplin, unpublished data). Highly negative isotope compositions in pyrite ($< −20‰$) suggest that sulphate reduction occurred in an open system in which sulphate was being transferred across the sediment–water interface more quickly than it could be reduced. In this case the isotope difference between seawater sulphate and sulphide is similar to the bacterial kinetic isotope fractionation. More positive (less negative) values ($> −20‰$) imply that sulphate reduction occurred in a system in which sulphate was reduced more rapidly than it could be supplied, so that reservoir effects increased the isotopic composition of both sulphate and sulphide (see Section 4.2.2). Specific controls on the isotopic composition of organically bound sulphur are poorly understood (Mossman *et al.*, 1991).

Some of the sulphur species incorporated into sediments on deposition participate further in deep burial diagenetic reactions. Pyrite is essentially stable below 200°C, but organic sulphur can be thermally degraded to H_2S, and anhydrite or dissolved sulphate can be chemically reduced to H_2S. Limited experimental work suggests that H_2S generated from organic sulphur may be a few per mille heavier than the organic sulphur (Krouse *et al.*, 1987). Data from deeply buried anhydrite-bearing carbonate reservoirs suggest that the isotopic composition of thermochemically derived H_2S is isotopically similar to, or a few per mille lighter than, the anhydrite (Orr, 1977; Krouse *et al.*, 1988). This is intriguing because the reaction does not appear to be controlled by simple kinetic or equilibrium isotope effects. At 150°C, simple kinetic controls would yield H_2S about 15–20‰ lighter than the sulphate, whilst equilibrium fractionation would give H_2S almost 40‰ lighter than the sulphate. The similarity of the isotopic compositions of sulphate and sulphide suggests that reduction of anhydrite to H_2S is a two-stage batch process in which anhydrite dissolves to give dissolved sulphate (small isotopic

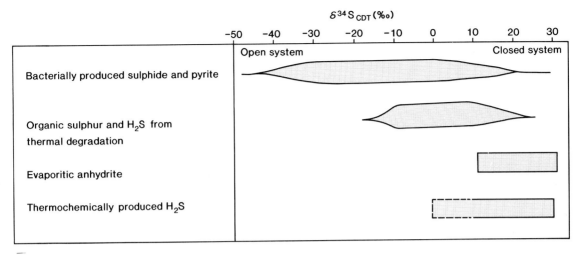

Fig. 4.22 Range of isotopic compositions of sulphur compounds in sedimentary basins.

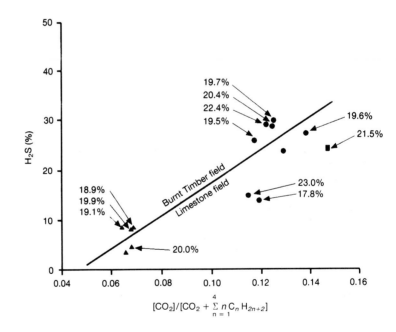

Fig. 4.23 H$_2$S concentration in produced gas from the Limestone and Burnt Timber fields (Alberta, Canada), plotted as a function of the ratio of CO$_2$ to (CO$_2$ + light hydrocarbons). From Krouse *et al.* (1988). The proportion of CO$_2$ in the gas mixture is thought to be a measure of the extent of thermo-chemical sulphate reduction. Numbers are the isotopic composition of H$_2$S.

fractionation), which is then quantitatively converted to H$_2$S. Whilst this may be a reasonable explanation for the anhydrite system, we do not know whether the kinetic isotope fractionation would be more clearly expressed in a sulphate-bearing formation water, where the reducing agent would have easier access to the sulphate.

Let us consider how we might use isotopic data to constrain the origin of H$_2$S in a reservoir fluid. Figure 4.22 shows the range of isotopic compositions which result from particular processes or which occur in particular phases. The ranges are large and overlap so that one cannot generally assign a given isotope ratio to a particular process. Supplementary geological and geochemical data must provide essential constraints. Examples of the required types

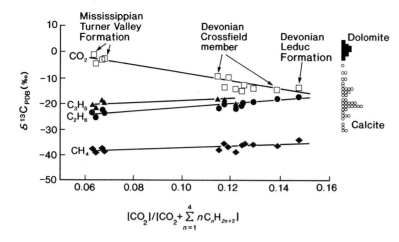

Fig. 4.24 Carbon isotopic compositions of CH_4, C_2H_6, C_3H_8 and CO_2 in produced gas from the Limestone and Burnt Timber fields (Alberta, Canada), plotted as a function of the extent of thermochemical sulphate reduction (see Fig. 4.23). From Krouse *et al.* (1988).

of data would include: samples of formation water to check for evidence of sulphate reduction; samples of organic sulphur to see if the isotopic composition matches that of the H_2S; petrographic studies to look for signs of calcite replacing anhydrite; and maximum burial temperature to determine whether the sediment has reached a temperature high enough for thermochemical sulphate reduction.

4.5.2 Example 4: thermochemical sulphate reduction in a carbonate reservoir, deep Foothills region, Alberta, Canada

H_2S concentrations between 4 and 25% occur in anhydrite-bearing carbonate reservoirs presently at temperatures up to 175°C in southwestern Alberta (Krouse *et al.*, 1988). The isotopic composition of 12 samples of H_2S was 20.1 ± 1.5‰, compared to a range of 22.7–33.0‰ for associated anhydrite. Armed with the knowledge that thermochemical reduction of anhydrite yields H_2S with similar isotopic compositions to the anhydrite, one would be tempted to infer a thermochemical origin for the H_2S. However, the possibilities that the H_2S was generated from the thermal degradation of organic sulphur, or from closed-system bacterial sulphate reduction, could not be ruled out on the basis of the isotopic data alone. In this case, there was abundant geological and geochemical evidence to support a

thermochemical origin. Firstly and most simply, a petrographic study showed that calcite had replaced anhydrite according to the general reaction:

$$CaSO_4 + CH_4 \rightarrow CaCO_3 + H_2S + H_2O \quad (21)$$

Secondly, a strong positive correlation was seen between the H_2S content of the reservoir gas and the ratio of CO_2 to CO_2 plus light hydrocarbons (Fig. 4.23, p. 99). This is consistent with the closed-system destruction of hydrocarbons and the generation of CO_2 and H_2S (similar to reaction (21) above, but with some CO_2 remaining in solution). Thirdly, the isotopic composition of light hydrocarbons increases with increasing extent of reaction (Fig. 4.24). This is consistent with the kinetic isotope fractionation associated with hydrocarbon oxidation in which ^{12}C hydrocarbons react more rapidly than ^{13}C hydrocarbons. The isotopic composition of CO_2 decreases with increasing extent of reaction, indicating an increasing contribution of organic-supplied carbon to the CO_2 reservoir. Finally, the isotopic composition of the calcite replacing anhydrite has isotopic compositions between −5 and −30‰, with most around −20‰ (Fig. 4.23). The more negative values give a clear indication of an organic source for the carbon, whilst the heavier values suggest a mixture of hydrocarbon-derived CO_2 and CO_2 derived from the dolomite reservoir rock.

Chapter 5   Radiogenic Isotopes

5.1 Introduction

The main use of the natural radioactive decay of certain elements has been in dating rocks and minerals. It is possible to determine radiometric dates for a number of diagenetic cements, which for some minerals and under some circumstances can represent the actual ages of mineral growth. Radiometric dating therefore has a clear advantage over both fluid inclusions and oxygen isotopes as a means of establishing a quantitative diagenetic history for a sedimentary rock and integrating diagenesis with geological history: both fluid inclusions and oxygen isotope ratios principally provide information about mineral growth *temperatures*, and *ages* must be derived indirectly, by reference to a modelled thermal history.

Radiogenic isotopes can provide several other kinds of information besides age of mineral growth. Measured dates may not refer to the time of mineral growth at all. Radiometric systems may be reset so that a calculated date can represent the age of a phase of recrystallization or even of a period of fluid movement long after mineral growth. The relative amounts of certain radiogenic isotopes in minerals and waters can also be used as natural tracers for monitoring water–rock interaction and can provide information about the history of formation water chemistry.

We begin this chapter with a refresher in the basics of radiometric dating in which we show how dates are calculated. We then review those isotope systems that can be used for dating components of sedimentary rocks and tracing water–rock interactions in sedimentary basins. Theory is kept to a minimum and enlivened by examples showing the methods in action. The first technique, K–Ar dating, is now almost routinely used to date illite cement, to such an extent that data themselves are sometimes obscured behind comments such as 'K–Ar ages show that illite grew about 110 Ma ago'. We hope that a run through Section 5.3 will convince the reader that such conclusions can be drawn only in certain cases and even then only after much care and deliberation. The details of ^{40}Ar–^{39}Ar dating may be new to many readers. It is based on the same natural decay as the K–Ar method but has some tremendous advantages that can be particularly useful in dealing with diagenetic minerals. Rb–Sr dating will be most familiar to geology students as a technique for dating igneous rocks. Its application to dating diagenetic minerals is in its infancy and has some problems but we will look at how it can be applied to illite. The Rb–Sr system has been far more useful for tracing water–rock interaction and the chemical evolution of natural waters by monitoring mineral and water Sr isotope ratios. Sr isotope stratigraphy is a method for dating sedimentary rocks that relies upon changes in the $^{87}Sr/^{86}Sr$ ratio of seawater through time and the retention of the ratio unmodified in marine carbonate and phosphate. We conclude the chapter with brief descriptions of Sm–Nd and U–Th–Pb dating which are relatively restricted in application but which may be useful for solving specific problems.

5.2 Radiogenic isotope systems

The nuclei of many elements with large atomic numbers are unstable and decay naturally into one or more daughter nuclei. Energy is lost in the form of some combination of helium nuclei, electrons and electromagnetic radiation (α, β and γ radiation). The rate of radioactive decay at any particular time t is proportional to the number of parent atoms present, N:

$$\frac{-dN}{dt} = \lambda N \tag{1}$$

where λ, the constant of proportionality, is called the *decay constant*. Integrating equation (1) gives:

$$N = N_0 \cdot e^{-\lambda t} \tag{2}$$

where N_0 is the number of parent atoms initially present at $t = 0$. The relationship between λ and the more familiar *half-life*, the time it takes for half of the atoms initially present to decay, may be found by setting $N = N_0/2$:

$$T_{1/2} = \frac{\ln 2}{\lambda} \tag{3}$$

The number of radiogenic daughter atoms D^* present at time t is $N_0 - N$; substituting in equation (2) gives:

$$D^* = N_0(1 - e^{-\lambda t}) \tag{4}$$

As N_0 is not usually known in geological systems, it is more convenient to relate the number of daughter

atoms present to the number of parent atoms remaining, N. From equations (2) and (4), we obtain:

$$D^* = N(e^{\lambda t} - 1) \tag{5}$$

The *total* number of daughter nuclei is the sum of the number of non-radiogenic daughter nuclei initially present, D_0, plus the number produced by decay:

$$D = D_0 + N(e^{\lambda t} - 1) \tag{6}$$

This equation is the basis of radiometric dating and the use of radiogenic isotopes as natural tracers and can be found in various forms in several places in this chapter. It can be solved for t which will represent the age of the system provided that:

1 λ is known;
2 D and N can be accurately measured;
3 all changes in the N/D ratio have been due to radioactive decay; in other words, the system has remained *closed* to both parent and daughter atoms since $t = 0$; and
4 D_0 is known.

Though by no means trivial, the first two assumptions are principally analytical problems and need not detain us here. The degree to which the third assumption will hold varies in different geological environments. Argon loss through diffusion can, for example, be a real problem in K–Ar dating of minerals that have been subjected to high temperatures, but the amount of loss will depend on the mineral, partial pressures of Ar in its environment, and the temperature and duration of heating.

The last assumption is that the number of daughter nuclei present in the sample before decay began (D_0) is known. In K–Ar dating, equation (6) is divided by the amount of a non-radiogenic isotope, ^{36}Ar, so D_0 is represented by the ratio $(^{40}Ar/^{36}Ar)_0$. Conventional K–Ar dating involves the assumption that the Ar initially present had the $^{40}Ar/^{36}Ar$ ratio of today's atmosphere, 295.5. In Rb–Sr dating, it is frequently not possible to make a convenient assumption of this kind, in which case a Rb–Sr date cannot be calculated from a single analysis: equation (6) has too many unknowns*. The usual way around this problem is to construct an *isochron diagram*. This involves analysing several cogenetic samples or subsamples — for example, several minerals from a rapidly cooled volcanic rock — and plotting their parent/daughter ratios (Fig. 5.1). The most familiar

---

* Ages calculated assuming an initial ratio are known as *model ages*.

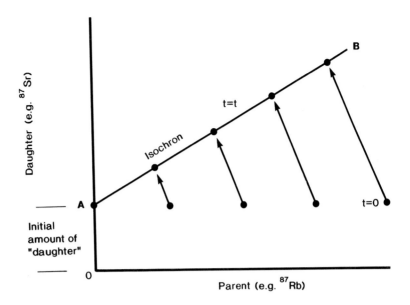

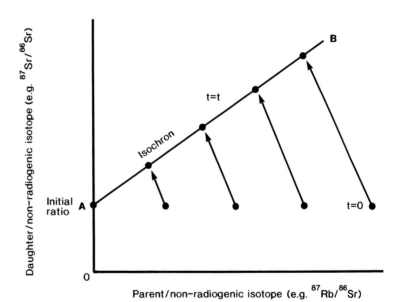

Fig. 5.1 Isochron diagrams. The lower diagram shows the ratio of parent and daughter to a non-radiogenic isotope (such as ^{86}Sr).

of these diagrams is a plot of ^{87}Sr/^{86}Sr against ^{87}Rb/^{86}Sr (though we will consider analogous diagrams involving the K–Ar system in Section 5.3)*. Samples with different Rb/Sr ratios will accumulate radiogenic ^{87}Sr at different rates. At time t, they will

---

* It is analytically convenient to ratio both parent ^{87}Rb and daughter ^{87}Sr to non-radiogenic ^{86}Sr.

plot on an isochron, a line whose slope (from equation 6) is equal to ($e^{\lambda t} - 1$). The intercept of the line is (^{87}Sr/^{86}Sr)$_0$, or D_0 in the notation of equation (6), the ratio of ^{87}Sr to ^{86}Sr in the system before radiaoctive decay began to produce more ^{87}Sr. In this case, the initial ratio does not need to be assumed, but is calculated along with the slope of the isochron, from the array of points. For the method to work, all the samples or subsamples must be of the same age, must have had the same initial ratio $(^{87}Sr/^{86}Sr)_0$ and must of course have a range of parent/daughter (Rb/Sr) ratios.

5.3 K–Ar dating

5.3.1 Principles

Potassium has three naturally occurring isotopes:

^{39}K $93.2581 \pm 0.0029\%$
^{40}K $0.00167 \pm 0.00004\%$
^{41}K $6.7302 \pm 0.0029\%$

Potassium-40 undergoes natural radioactive decay, 88.8% to ^{40}Ca and the remaining 11.2% to ^{40}Ar with a half-life of 1.250×10^9 years (Dalrymple & Lanphere, 1969; Faure, 1986). Although both decay series can be used for dating, the K–Ca method is plagued by interferences from non-radiogenic naturally occurring Ca isotopes, and we will not consider it further. The accumulation of radiogenic ^{40}Ar in a K-bearing mineral as a function of time (t) is given by:

$$^{40}Ar^* = \frac{\lambda_e}{\lambda} \cdot {^{40}K}\,(e^{\lambda t} - 1) \tag{7}$$

$^{40}Ar^*$ is the amount of radiogenic ^{40}Ar produced by decay of ^{40}K; λ is the total decay constant; and λ_e the partial decay constant for ^{40}K to ^{40}Ar decay. Equation 7 can expressed in the form:

$$t = \frac{1}{\lambda} \cdot \left[\frac{^{40}Ar^*}{^{40}K} \cdot \left(\frac{\lambda}{\lambda_e} \right) + 1 \right] \tag{8}$$

This is the basic equation of K–Ar dating. The decay constants and the proportion of ^{40}K in naturally occurring potassium are known (Steiger & Jager, 1977), so equation (2) permits calculation of an 'age' provided that the total amounts of K and radiogenic ^{40}Ar in a sample can be determined and that certain assumptions are met. In the following two sections, we will look first at how these two measurements are made and then at the assumptions implicit in the interpretation of a *geologically meaningful* age.

5.3.2 Analytical methods: precision and accuracy

Calculation of the K–Ar age of a sample requires two separate measurements: the amounts of K and radiogenic ^{40}Ar. The first stage is therefore to pro-

duce a pure separate of the mineral to be dated and to subsample it. A minimum of a few milligrams is presently required and precision will be better if several tens of milligrams or more can be separated. Potassium content can be determined by a number of different methods of which emission spectroscopy and flame photometry are the most popular. The amount of radiogenic ^{40}Ar is determined by *isotope dilution mass spectrometry*. In this technique, the sample is first preheated in vacuum to drive off adsorbed atmospheric gases and then fused to liberate that held in the crystal lattice. To this gas is added a known amount of another isotope, ^{38}Ar. After purification, two isotope ratios are measured by mass spectrometry. The ratio $^{40}Ar/^{38}Ar$ is used together with the known ^{38}Ar volume to calculate the total amount of ^{40}Ar in the sample. This is not however equal to the number that we need to know – the amount of *radiogenic* ^{40}Ar. Around 1% of the atmosphere is ^{40}Ar and no matter how thorough the preheating, a proportion of the measured ^{40}Ar will inevitably be of non-radiogenic origin. The amount of atmospheric ^{40}Ar must therefore be calculated and subtracted from the measured total. This is done using the second of the two measured ratios, $^{36}Ar/^{38}Ar$. This permits calculation of the amount of ^{36}Ar in the gas sample, all of which is of atmospheric origin. In the atmosphere, ^{36}Ar is associated with a known proportion of ^{40}Ar, so the atmospheric ratio of $^{40}Ar/^{36}Ar$ (295.5) can be used to calculate the amount of ^{40}Ar that has come from the atmosphere. The amount of radiogenic ^{40}Ar is then found by subtracting this amount from the total measured ^{40}Ar. The procedure may be summarized as follows

$$\frac{^{40}Ar}{^{38}Ar} \text{[measured]} \quad \rightarrow \quad \text{Total amount of } ^{40}Ar \tag{a}$$

$$\frac{^{36}Ar}{^{38}Ar} \text{[measured]} \quad \rightarrow \quad \begin{array}{l}\text{Total amount of } ^{36}Ar \\ \text{(all atmospheric)}\end{array} \tag{b}$$

$$\frac{^{40}Ar}{^{36}Ar} \text{[known, 295.5]} \rightarrow \quad \begin{array}{l}\text{Amount of non-} \\ \text{radiogenic } ^{40}Ar\end{array} \tag{c}$$

$$(a) - (c) \quad \rightarrow \quad \begin{array}{l}\text{amount of radiogenic} \\ ^{40}Ar\,(^{40}Ar^*)\end{array}$$

K–Ar ages are invariably quoted as plus or minus some number. This is usually the precision or reproducibility of the calculated age, calculated from repeat determinations of K and $^{40}Ar^*$ on single samples or standards, using the method of propagation of errors. The $\pm$ sign denotes one or two standard deviations and means that the analyst estimates that there is a 67% (1σ) or 95% (2σ)

probability that the age calculated is within the quoted range. But how accurate is it? Good precision is a prerequisite for accuracy but it certainly does not guarantee it.

Accuracy may be thought of as having two components. Analytical accuracy is the extent to which the determination is a true reflection of the K and ^{40}Ar contents of the sample. This will depend on any systematic bias involved in the measurements and particularly on sample homogeneity. K and ^{40}Ar analyses have to be performed on separate sub-samples and if these are not homogeneous with respect to both, the precision of the age will be substantially less than the analytical accuracy. Neither good precision nor analytical accuracy are however enough to guarantee that the calculated age will be an accurate reflection of what we would like to think we are dating, usually mineral growth. This may be described as geological accuracy and is ultimately what is of greatest concern. It depends on the validity of a number of assumptions that are implicit in the K–Ar dating method, many of which need to be evaluated separately for different minerals in different geological environments. These are the subject of the following section.

5.3.3 Assumptions

Equation 8 can be used to calculate a true age of mineral growth only if a number of criteria are fulfilled:

1 the sample to be dated must be free from recently introduced K- or Ar-bearing contaminants;
2 there has been no gain or loss of either K or ^{40}Ar* since the mineral grew other than by *in-situ* decay;
3 all non-radiogenic ^{40}Ar − that not produced by *in-situ* decay of ^{40}K − is of atmospheric origin and has the ^{40}Ar/^{36}Ar ratio of today's atmosphere; and
4 potassium in the sample must have 'normal' isotopic composition.

No K- or Ar-bearing contaminants
In order to prepare a sample of any diagenetic mineral large enough for K and Ar analysis, it will be necessary to crush a whole rock sample and isolate the mineral to be dated in some way. It is often difficult to exclude other minerals completely from the separate and if these contain K, as will detrital illite, micas and feldspars, there will be an error in the measured age. The error can be very serious indeed. For example, if a separate consisting largely of 50 Ma old authigenic illite is contaminated

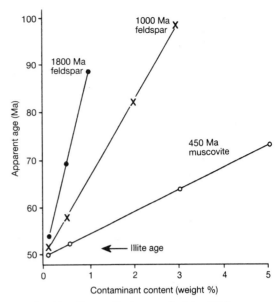

Fig. 5.2 The effect on K–Ar ages of mixing 50 Ma authigenic illite with 450 Ma muscovite and other contaminants (from Hamilton *et al.*, 1989).

with only 2% of a 450 Ma old detrital muscovite, the measured age would be about 59 Ma (Hamilton *et al.*, 1989; Fig. 5.2). The precision of the age might easily be as good as ± 1 or 2 Ma (about 2–4%) but the precise age is wrong by nearly 20%. The older the contaminating phase, the worse the problem will be. Other possible contaminants are soluble K-bearing salts derived from formation water which can precipitate in the pores of subsurface samples as they dry. These salts may be extremely difficult to remove from mineral separates, even with copious washing.

There is no easy solution to the difficulties of preparing a pure, uncontaminated mineral separate. The best way to minimize contamination is to avoid samples with large amounts of detrital K-bearing minerals or that contain more than one generation of the authigenic mineral to be dated. To recognize contamination, there is no real substitute for actually looking at the sample under TEM and analysing selected grains if necessary (Section 2.6). XRD is not capable of detecting contaminants if they are present in amounts of less than a few per cent (Section 2.7). The example given above shows that smaller amounts of old minerals can cause highly inaccurate age determinations. Finally, if, as is common, contamination is likely to be a problem, it is important not to overestimate the accuracy of a set of data.

No gain or loss of Ar or K

Potassium gain or loss is only likely during recrystallization which would in any case involve loss of Ar and resetting of the K–Ar clock. Argon loss or gain is of greater concern. In a subsiding sedimentary basin, authigenic minerals are characteristically subjected to temperatures above those at which they grew, for substantial periods. Diffusion is highly sensitive to temperature so it is important to consider the extent to which ^{40}Ar may diffuse into or out of minerals. Diffusion of ^{40}Ar into a mineral from surrounding pore fluid is unlikely because relatively high partial pressures of Ar would need to be maintained in the fluid. Loss of ^{40}Ar is a more serious concern and would lead to a reduction in the apparent age. If all ^{40}Ar is lost, the K–Ar clock will be reset and the age will measure the time of resetting. If ^{40}Ar loss is partial, the geological significance of the measured age will be difficult to interpret (Damon, 1970).

There are two approaches to evaluating Ar diffusion: studies of natural systems that have been heated and calibration of kinetics by means of experimental work. In a study of the ages of several minerals as a function of distance from an igneous intrusion, K-feldspar was the most susceptible to Ar loss and still showed significant reductions in measured age nearly 7 km from the intrusion (Hart, 1964). Orthoclase and microcline are in fact generally considered unsuitable for K–Ar dating because of their poor Ar retention, even at low temperatures (Dalrymple & Lanphere, 1969). Sanidine does retain Ar and can be used to date volcanic rocks. However, this somewhat bleak picture may be overly pessimistic. Low temperature K-feldspar, usually described as adularia, may retain Ar better than orthoclase and microcline. Girard *et al.* (1988) estimated an age of 98 ± 16 Ma for authigenic feldspar in Lower Cretaceous sandstones, consistent with petrographic evidence for the age of the mineral. Had the feldspar lost a large proportion of its ^{40}Ar, its apparent age ought to have been close to zero.

Studies of Ar retention in illite are more encouraging. The ages of illite separates from a shale intruded by a small Tertiary stock in New Mexico proved to be older than the stock itself, even at distances of only a few feet from the contact (Aronson & Lee, 1986). This suggests that illite is capable of retaining

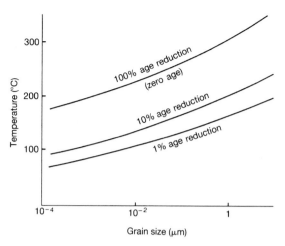

Fig. 5.3 Age reduction in illite due to Ar diffusion (from Hamilton *et al.*, 1989).

Ar when heated for a geologically short period. Authigenic illites may however be heated above their growth temperature for much longer periods, of the order of even 100 Ma. One calibration of diffusion kinetics relies on the use of muscovite as an analogue (Hamilton *et al.*, 1989). Figure 5.3 shows temperatures required to cause age reductions of 1, 10 and 100% as a function of illite crystal size, assuming a heating period of 100 Ma. Heating a typical illite fibre 0.1 µm wide to 180°C causes an age reduction of about 15%. This figure is comparable to an independent estimate obtained by evaluating the reduction in age as a function of depth (temperature) of detrital illites in Plio-Pleistocene sediments (Glasmann, 1987). It appears then that Ar loss may be responsible for limited age reduction in diagenetic illites but that the effect is likely to be serious only under the most extreme diagenetic conditions.

Finally, sylvite. Though it has a high K content, it dissolves and reprecipitates with such alacrity that it does not retain Ar and gives erroneously young ages (Dalrymple & Lanphere, 1969). Neither sylvite nor other K-bearing evaporite minerals are suitable for K–Ar dating.

All non-radiogenic ^{40}Ar is of atmospheric orgin

Equation (7) shows that a K–Ar age is calculated from potassium content and from the amount of radiogenic argon-40, ^{40}Ar*. Unfortunately, we do not measure ^{40}Ar* but a *total* amount of ^{40}Ar that includes not only ^{40}Ar*, but also ^{40}Ar from other sources. Equation (7) may be rewritten to take into account this non-radiogenic component:

$$^{40}Ar_m = {}^{40}Ar_x + {}^{40}Ar^* = {}^{40}Ar_x +$$
$$\frac{\lambda_e}{\lambda} \cdot {}^{40}K \left(e^{\lambda t} - 1 \right) \qquad (9)$$

$^{40}Ar_m$ is the total measured ^{40}Ar and $^{40}Ar_x$ is the non-radiogenic component. Whatever its origin – included into the mineral as it grew, adsorbed onto the mineral as it languished in the core store or during sample preparation, or adsorbed onto the analytical equipment – it is assumed that $^{40}Ar_x$ has the isotopic composition of the atmosphere so that a correction can be applied using the amount of measured ^{36}Ar and the atmospheric $^{40}Ar/^{36}Ar$ ratio (295.5). If the mineral being dated at any time incorporates ^{40}Ar from a non-atmospheric source (*excess argon*) the measured K–Ar age will be older than the true age.

The calculation of a single age provides no information about the presence or absence of excess Ar. In the absence of a reliable independent estimate of the age of the dated mineral, the best way of trying to detect it is by plotting data from a number of cogenetic samples on an isochron diagram (Shafiqullah & Damon, 1974). We will look at two kinds. The first involves plotting $^{40}Ar^*$ against ^{40}K. $^{40}Ar^*$ is calculated assuming that all non-radiogenic ^{40}Ar is atmospheric. If this is true, points representing samples with different ^{40}K contents will lie on a straight line passing through the origin: no ^{40}K, no $^{40}Ar^*$. The slope of the line will be proportional to the true age. If the mineral contains excess argon, the points will plot on a line with a positive intercept on the $^{40}Ar^*$ axis (Fig. 5.4). The true age is still proportional to the slope of the line through the data, but the ages of individual samples calculated assuming that no excess argon is present will be too old. In order to generate a diagram of this kind, it is necessary to have dated at least three cogenetic samples with different K-contents. With igneous rocks, this would be accomplished by dating different minerals. In sedimentary rocks, it is usually difficult even to identify cogenetic minerals, let alone separate three that contain K for dating. We must rely on samples of a single mineral having variable K-content. As we will see in the example below, there does seem to be enough variability in illite K-content to make this approach possible, though the reasons why this should be so are not known.

The second type of isochron diagram is mathematically similar to the more familiar Rb–Sr isochron (see Section 5.5.3) though it is far more

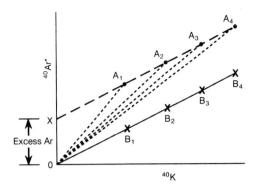

Fig. 5.4 Detection of excess argon on a $^{40}Ar^*-^{40}K$ isochron diagram. Two sets of samples A and B have the same age, given by the slopes of the isochrons OB and XA. However, samples A contain excess Ar. Their individual K–Ar ages will be quite different, proportional to the slopes of the lines OA_1, OA_2 etc., and all will overestimate the true age.

limited in its application. Equation (9) divided by ^{36}Ar gives:

$$\frac{^{40}Ar_m}{^{36}Ar} = \frac{^{40}Ar_x}{^{36}Ar} + \frac{^{40}K}{^{36}Ar} \cdot \frac{\lambda_e}{\lambda} \cdot \left(e^{\lambda t} - 1 \right) \qquad (10)$$

On a plot of $^{40}Ar_m/^{36}Ar$ against $^{40}K/^{36}Ar$, this is the equation of a straight line with intercept $^{40}Ar_x/^{36}Ar$ and slope proportional to age. If the intercept of a line passing through data with different $^{40}K/^{36}Ar$ ratios is not equal to the atmospheric ratio, excess Ar is present. The slope of the line through the data may however not represent a true age because $^{40}K/^{36}Ar$ depends not only on K-content but also on the (variable) amount of atmospheric ^{36}Ar contamination in the sample. The interpretation of the diagram in terms of age also assumes that there has been no diffusive Ar loss.

Potassium has normal isotopic ratio
The amount of ^{40}K in a sample is calculated from the total K-content and an assumed $^{40}K/K$ ratio of 1.167 $\times 10^{-4}$. Nonetheless, a small amount of natural fractionation of K isotopes due to their differences in mass is to be expected. There would be some associated effect on decay constants. Fortunately, even a substantial amount of fractionation will not introduce a serious error: a 5% (50‰) reduction in $^{40}K/K$ will decrease a 100 Ma age by only 0.1% (Hamilton *et al.*, 1989).

Fig. 5.5 SEM photograph of illite cement, Lower Leman Sandstone.

5.3.4 Example 1: illite cement in aeolian sandstone, Rotliegend Group, Southern North Sea

This example involves the assessment of a set of K–Ar ages measured on illite separates from the Lower Leman Sandstone of the Permian Rotliegend Group in the Village Fields area of the Southern North Sea. The Lower Leman Sandstone includes aeolian and fluvial sands and interbedded siltstones and mudstones. Illite occurs as a late, pore-bridging cement (Fig. 5.5). In this section, we are concerned with the validity of the K–Ar ages of these illites. There are two key questions: are the fundamental assumptions of K–Ar dating met and if so, what do the ages represent?

Samples of core were collected from 11 wells in the area, exclusively from aeolian intervals (Robinson *et al.*, 1993). Aeolian sands frequently contain no detrital illite and it was hoped that illite separates would be free from contamination with any older material (Lee, 1984). Those samples with most illite cement were identified using SEM, and clay separates were prepared from them by centrifuging gently disaggregated rock in acetone (Brindley, 1981). Purity was monitored by XRD and by visual examination using TEM. Organic matter was removed by ashing in a low-temperature oxygen plasma but no chemical pre-treatment was used. Finally, the separates were subsampled for K and Ar

analysis. The ages are listed in Table 5.1. Analytical precision is quoted as $\pm 2\sigma$ and does not take into account possible sample inhomogeneity.

The requirements for plausible ages are:
1 the samples must be free of K- or Ar-bearing minerals other than the illite cement to be dated;
2 the illite must have remained a closed system with respect to K and Ar since it grew; and
3 all non-radiogenic Ar incorporated into the sample must have the isotopic composition of the present-day atmosphere.

Detrital illite present in amounts below the detection limit of XRD can easily lead to measured ages substantially older than that of illite cement growth. It is generally coarser than illite cement and concentrates in the coarser separates. In samples where contamination is a problem, these have older ages. The ages of the Village Fields illites show just such a pattern (Fig. 5.6). There are however two reasons why the apparent increase in age in the coarser dated separates is unlikely to be due to contamination in this particular case. Firstly, we attempted to minimize contamination by judicious sample selection and examination of the separates using TEM. Those selected for dating contained no identifiable detrital illite or other K-bearing minerals. Secondly, the ages of the coarsest (> 5 µm) fractions are only around 10% older than those of the finest. Contamination commonly leads to very substantial

Table 5.1 K–Ar ages of illite separates, Lower Leman Sandstone

Well	Field	Depth (mBRT)	Size fraction (μm)	Estimated maximum depth (m)	K–Ar age (Ma)	2σ (Ma)
42/29-4	Cleeton	2852.0	2–5		149	12
		2876.0	<5		150	7
		2897.8	3–8		167	15
42/30-4	Ravenspurn N.	3122.3	<4	4522.3	159	12
		3171.0	1–5	4571.0	156	17
		3171.0	<2	4571.0	166	9
		3172.5	<5	4572.5	172	10
		3172.5	<2	4572.5	170	6
43/26-1	Ravenspurn N.	3137.9	<2	4537.9	147	10
		3148.9	<5	4548.9	165	8
		3153.2	<8	4553.2	150	7
43/26-3	Ravenspurn N.	3111.6	<4	4511.6	177	13
		3113.4	<3	4513.4	168	11
		3153.6	?	4553.6	176	10
47/5A-1		3006.0	?		157	9
47/5A-2	Hyde	3006.5	<2		163	10
		3044.3	<2		151	10
47/5A-4		3057.2		3407.2	124	49
		3057.2		3407.2	85	22
		3071.0	<4	3421.0	167	7
		3105.5	<10	3455.5	170	6
		3127.3	2–8	3477.3	153	7
48/6-25	Hyde	2932.8	<5		157	3
		2961.9	<5		157	3
		2967.7	<5		166	4
		2985.4	<5		165	4
		3000.7	<5		168	4
48/7B-3	Hoton	2977.2	<0.2	4777.2	155	3
		2980.9	<0.2	4780.9	145	3
		2985.8	<0.5	4785.8	154	4
		2996.2	<0.2	4796.2	153	3
		2998.6	<0.5	4798.6	158	5
48/7B-5	Hoton	3145.6	<0.2	4945.6	138	3
		3149.5	0.2–0.5	4949.5	154	3
		3153.3	<0.2	4953.3	167	5
48/7B-4	Hoton	3367.6	<0.2	5167.6	153	4
		3380.9	<0.2	5180.9	153	3
		3436.8	<0.2	5236.8	147	3
		3453.6	<0.2	5253.6	150	4
		3474.6	<0.2	5274.6	146	3
		3498.6	<0.5	5298.6	159	6

increases in the ages of coarser fractions (see for example Fig. 5.11 below).

The second assumption is that there has been no gain or loss of K or Ar to or from the illite since it grew. Burial and thermal history modelling suggest that from the Mid-Jurassic to the Mid-Tertiary, a period of the order of 100 Ma, the samples from the Lower Leman Sandstone are likely to have

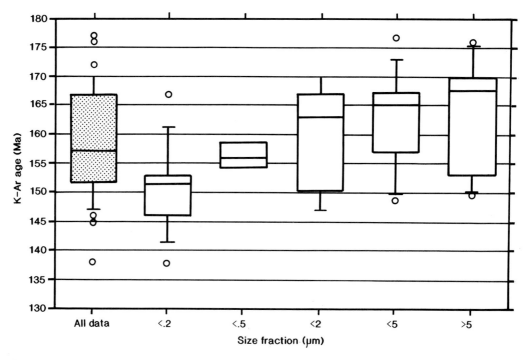

Fig. 5.6 K–Ar age distributions for dated size fractions, Lower Leman Sandstone (Rotliegend Group), Southern North Sea, Village Fields area. The box plots show the 10th, 25th, 50th, 75th and 90th percentiles of each distribution as well as points that lie outside the 10th and 90th percentiles.

been subjected to temperatures as high as 190°C. Estimates of Ar diffusion kinetics suggest that this might lead to age reduction due to Ar loss of up to a few per cent for illite crystals of normal size (Fig. 5.3). This is a possible explanation for the observation that the finest dated size fractions have the youngest ages (Fig. 5.6): the finest illite will have the greatest surface area/volume ratio and will as a consequence be most susceptible to diffusive Ar loss. If however diffusion had caused significant age reduction, measured age might be expected to correlate with maximum burial temperature and thus with estimated maximum burial depth. No such correlation is evident (Fig. 5.7).

The last major assumption is that all non-radiogenic ^{40}Ar has atmospheric isotopic composition. Figure 5.8 shows the illite data from the Village Fields area plotted on a graph of ^{40}Ar* against ^{40}K, one of the isochron diagrams described above (data in Table 5.2). At the 90% confidence level, the unweighted least squares regression line through the data passes through the origin. At the 95% confidence interval, however, the intercept ^{40}Ar$_0$ is

greater than zero, hinting at the presence of excess Ar. The isochron age calculated from the least squares regression line is 138 Ma (Early Cretaceous). The best interpretation is probably that to a first approximation the assumption is true, but that the presence of a small proportion of excess Ar cannot be ruled out. That the line passes close to the origin is also in fact further evidence against diffusive Ar loss since uniform loss of ^{40}Ar* will produce a negative intercept on the graph (Shafiqullah & Damon, 1974).

Figure 5.9 shows the same data plotted on the second type of isochron diagram, ^{40}Ar$_m$/^{36}Ar against ^{40}K/^{36}Ar. The regression line does pass through the atmospheric ^{40}Ar/^{36}Ar ratio at the 90% confidence level but closer scrutiny of the diagram ought to raise doubts about the significance of this observation. The regression line is dominated by one point with a high ^{40}K/^{36}Ar in excess of 120000. This number is high because the sample had a small atmospheric Ar contamination and therefore a low ^{36}Ar content and consequent high ^{40}K/^{36}Ar ratio. In fact, the spread of data on the diagram is largely due

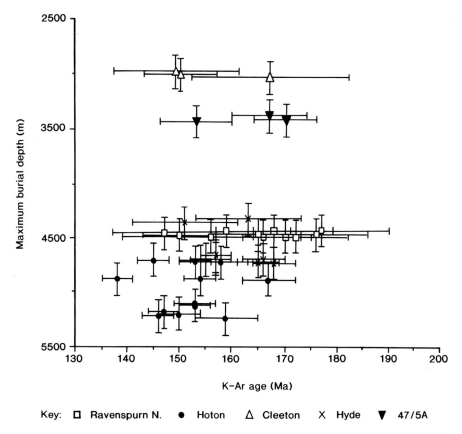

Fig. 5.7 K–Ar age versus maximum burial depth, illite cements, Lower Leman Sandstone (Rotliegend Group), Southern North Sea, Village Fields area. The Village Fields area was inverted during the Mid-Tertiary, so estimates from seismic and from Triassic Bunter shale velocities of the amount of missing section have been used to calculate maximum burial depth. The error bars represent precision (2σ) and an estimate of the likely accuracy of the backstripping. Age and maximum burial depth are not obviously correlated.

to variable atmospheric contamination (Fig. 5.10). This means that the 'isochron' is largely a mixing line with at best cryptic age significance.

It seems then that the Village Fields data do largely meet the assumptions of K–Ar dating and that they are telling us something about when illite cement grew in the Lower Leman Sandstones. But exactly what are they telling us? Even if illite grows over a geologically long period, we will still measure a single K–Ar date which may represent some form of average age. Authigenic illite crystals tend to have relatively large cores and finer, hairy outgrowths (Lee *et al.*, 1985; Fig. 5.5). It might be expected that the cores – which must be older – would lead to older ages in coarser separates. In this example, this does not appear to be the case suggesting that the

duration of illite growth in a sample is of the same order as, or less than, the precision of the method. This may not be true in all cases. As we will see in the next example, a pattern of increasing age in coarser separates may also result from contamination by detrital material and it will usually be difficult to disentangle the two effects. In such cases, a measured age will probably be biased towards the latest time of illite growth (Hamilton *et al.*, 1992).

5.3.5 Example 2: illite cement in fluvial sandstone, Brent Group, Northern North Sea

The interpretation of an age for the growth of illite cement in the Lower Leman Sandstone was relatively straightforward (!) because it was possible to

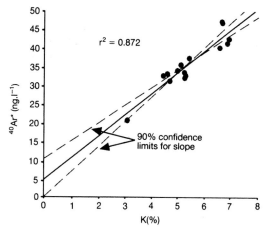

Fig. 5.8 Isochron diagram 1 for Village Fields illites. $^{40}Ar^*$ is plotted against %K (proportional to ^{40}K). In the absence of excess Ar, a line through the data should pass through the origin.

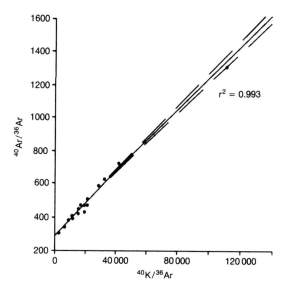

Fig. 5.9 Isochron diagram 2 for Village Fields illites. $^{40}Ar/^{36}Ar$ is plotted against $^{40}K/^{36}Ar$. In the absence of excess Ar, a line through the data should pass through 295.5, the $^{40}Ar/^{36}Ar$ ratio of the atmosphere.

minimize and probably largely avoid contamination of the dated separates by detrital minerals. In this example, we will consider the difficulties in obtaining a reliable age for illite cement in samples which contain detrital illite and mica. Table 5.3 lists the K–Ar ages of different size fractions separated from samples of fluvial sandstone belonging to the Mid-Jurassic Brent Group of the Northern North Sea (these are the same samples whose isotopic ratios were described in Section 4.3.4). There is a clear

trend towards older ages in the coarser separates (Fig. 5.11) and examination by TEM shows that this is due to increasing amounts of relatively coarse detrital illite. Careful sample preparation can reduce the degree of contamination but cannot generally eliminate it (e.g. Liewig *et al.*, 1987). These samples have not been subjected to temperatures

Table 5.2 K–Ar ages of illite separates, Lower Leman Sandstone: analytical data.

Well	Field	Depth (mBRT)	K (%)	Ar (ng)	Rad. ^{40}Ar (ng.l^{-1})	Atm. ^{40}Ar (%)	^{36}Ar (ng.l^{-1})	$^{40}K/^{36}Ar$	$^{40}Ar/^{36}Ar$	Age (Ma)	2σ (Ma)
42/29-4	Cleeton	2852.0	6.948	0.01125	42.10	66	0.27608	16836	449	149	12
		2897.8	3.103	0.01147	21.05	70	0.16588	12514	423	167	16
43/26-1	Ravenspurn N.	3148.9	4.690	0.02904	31.50	56	0.13544	23165	529	165	8
		3153.2	6.605	0.03089	40.04	60	0.20291	21777	493	150	7
		3137.9	6.882	0.03047	41.03	62	0.22612	20360	477	147	10
43/26-3		3111.6	4.453	0.01750	32.84	72	0.28525	10443	411	177	13
		3113.4	5.448	0.02936	37.29	63	0.21447	16994	470	168	11
		3153.6	4.614	0.02666	33.17	69	0.24942	12376	429	176	10
42/30-4		3122.2	6.210	0.04150	40.07	64	0.24064	17264	463	159	12
		3171.0	5.279	0.03516	33.45	79	0.42508	8308	375	156	7
		3172.5	6.674	0.02461	46.94	60	0.23785	18772	493	172	10
		3172.5	6.703	0.02192	46.47	40	0.10465	42849	740	170	6
		3071.0	4.999	0.04963	34.09	46	0.09810	34091	644	167	7
		3105.5	5.114	0.03657	35.46	49	0.11509	29727	604	170	6
		3127.2	5.223	0.03718	32.11	71	0.26556	13157	417	151	8
		3127.2	5.270	0.03188	32.66	63	0.18785	18767	470	153	7

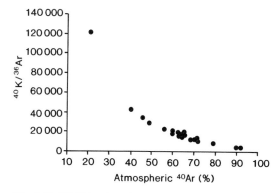

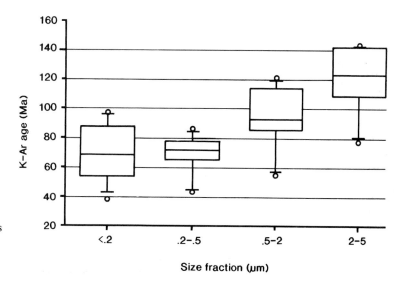

Fig. 5.10 ^{40}K/^{36}Ar plotted against amount of atmospheric Ar contamination. ^{40}K/^{36}Ar is largely determined by the degree of atmospheric contamination. The line through the data in Fig. 5.9 probably has no age significance.

high enough to cause any significant diffusive loss of Ar (which might reduce measured ages), so the most straightforward interpretation would be that the youngest ages are close to, but older than, the true age of illite cement growth. But how much older? In order to calculate isotopic ratio end-members, we tried an approach involving a visual estimate of the proportion of authigenic illite in each analysed separate (Section 4.3.4). An alternative approach is to use some chemical discriminant. Figure 5.12 is a plot of K–Ar age against Al/K ratio. Also plotted on the figure is the average ($\pm 2\sigma$) of Al/K ratios for individual diagenetic illite fibres from the same samples, analysed using TEM. The separates with the youngest ages have Al/K ratios close to the diagenetic illite value of $2.18 \pm 0.08^*$. The older

ages are associated with lower ratios characteristic of phengitic micas. It is possible to calculate a regression line for the ages and calculate where the slope and associated confidence intervals meet the range of TEM authigenic illite analyses. This suggests an age between 50 and 72 Ma and, though not particularly rigorous, does give some idea as to the uncertainty involved.

5.3.6 Example 3: K-feldspar cement, offshore Angola

Just as it is very difficult to isolate a separate of pure authigenic illite from a rock sample that contains detrital illite, it is next to impossible to obtain tens of milligrams of authigenic K-feldspar free from contamination by detrital orthoclase or microcline. The only viable approach is to try to produce separates with a variable proportion of detrital to authigenic material, quantify how much of each is present and then calculate the age of the end-members by extrapolation. Girard *et al.* (1988) have done this for Lower Cretaceous sandstones from offshore Angola. They first produced feldspar separates comprising grains and overgrowths by standard methods like heavy liquids and the ever popular hand-picking. The separates were then etched in acid. The hope was that the grain–overgrowth boundary would be weakened so that when the separates were then shattered ultrasonically, the fine

---

* Values of around 2.5 are reported for illite cements in sandstones by Warren (1987).

Fig. 5.11 K–Ar age distributions for dated size fractions, Brent Group fluvial sandstones, UKCS wells 3/8b-10 and -11. Age increases in the coarser separates due to increasing contents of old detrital illite.

Size fraction (μm)

Table 5.3 K–Ar ages and isotopic analyses of size fractions of fluvial sandstone, Brent Group, Northern North Sea.

Well	Depth (mBRT)	Size fraction (µm)	K–Ar age (Ma)	2σ (Ma)	K/Ar weight ratio	δ¹⁸O (per mille SMOW)	δD (per mille SMOW)	Illite cement (%)*
3/8b-10	4056.35	<0.2	37.7	0.9				63
		0.2–0.5	43.1	1.0				76
		0.5–2	54.6	1.3				52
		2–5	77.5	6.0				47
	4096.20	<0.2	68.9	1.6	1.88			96
		0.2–0.5	73.5	1.6	1.74			88
		0.5–2	91.0	2.0	1.33			42
		2–5	108.0	2.0	0.92			29
	4076.85	<0.2	75.7	1.7	1.94			81
		0.2–0.5	78.1	1.7	1.8			59
		0.5–2	114.0	3.0	1.54			55
		2–5	135.0	3.0	1.02			10
	4076.90	<0.2	89.1	5.5	1.74	16.3	−62	98
	4085.80	<0.2	95.9	12.0		16.2	−70	99
	4100.80	<0.2	96.8	4.3		15.2	−50	97
3/8b-11	4243.60	<0.2	59.7	3.4	2.17	14.2	−24	92
		0.2–0.5	64.5	2.6	2.14	15.0	−39	86
		0.5–2	85.0	5.8	1.39	14.8	−37	73
		2–5	111.0	10.0	1.04	15.4	−67	29
	4257.70	<0.2	6	3.1	2.31	14.4	−61	95
		0.2–0.5	70.4	2.5	2.07	15.8	−36	84
		0.5–2	94.6	4.8	1.54	14.9	−51	85
		2–5	142.0	4.0	1.15	15.8	−80	31
	4257.70	<0.2	46.9	3.6	2.04	16.9	−67	96
	4269.70	<0.2	82.6	2.4		15.4	−45	88
	4326.20	<0.2	51.3	10.0	2.12	14.1	−58	95
		0.2–0.5	85.4	11.0	1.75	15.0	−64	89
		0.5–2	120.0	11.0	1.42	14.5	−54	66
		2–5	143.0	4.0	1.07	16.6	−81	39

* Determined by point-counting using TEM.

overgrowths would fly off and be collected by sieving (the same technique sometimes used to isolate samples of quartz cement for isotopic analysis; see Section 4.3.2). The detrital feldspar luminesced but the overgrowths did not, so the proportion of detrital to authigenic material could be estimated visually using a standard CL microscope. In this study, this was done by point-counting, though image analysis potentially offers greater accuracy (Hearn, 1987).

This is undeniably a laborious procedure requiring samples that contain a large amount of authigenic feldspar (which is rather unusual) as well as substantial care. It does nonetheless appear to work. Girard *et al.* ended up with an age of 98 ± 16 Ma for the feldspar cement, in accord with the inference that cement growth pre-dated compaction. The detrital end-member age of 516 ± 16 Ma was close to the age of the basement in the area. Like the method we used for dating the Brent illite, it involves the assumption that all detrital and all authigenic feldspar can be characterized by single ages.

5.4 ^{40}Ar–^{39}Ar dating

5.4.1 Principles

^{40}Ar–^{39}Ar dating is closely related to K–Ar dating and is based on the same natural decay of ^{40}K to ^{40}Ar (Dalrymple & Lanphere, 1971). It involves the conversion by irradiation of a proportion of one of the naturally occurring isotopes of K, ^{39}K, into an isotope of Ar that does not occur in nature, ^{39}Ar. ^{39}Ar is then a measure of the original K content of the sample and an age can be calculated from the measurement of a single isotopic ratio, ^{40}Ar/^{39}Ar.

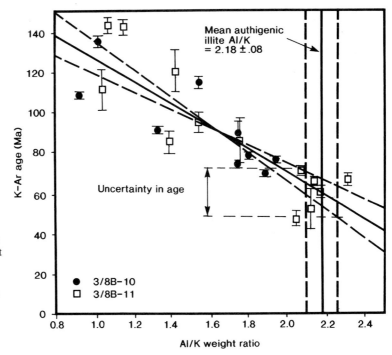

Fig. 5.12 K–Ar age plotted against Al/K ratio of dated separates, Brent Group fluvial sandstones, UKCS wells 3/8b-10 and -11. Error bars represent precision (2σ). The mean Al/K for the authigenic illite end-member is calculated from TEM analyses of individual fibres from both samples.

This is the key difference between ^{40}Ar–^{39}Ar and K–Ar dating. The calculation of a single age in this manner nonetheless involves making exactly the same assumptions as in K–Ar dating: that there has been no Ar loss since mineral growth and that all non-radiogenic ^{40}Ar is of atmospheric origin (no excess Ar). However, ^{40}Ar–^{39}Ar dating can have advantages over the K–Ar method.

1 All of the information needed to calculate an age can be obtained from a single aliquot of gas released from a single sample as small as 1 µg, by heating or total fusion. K–Ar ages require separate determination of K content and Ar isotope concentrations so that sample heterogeneity will introduce an error.

2 Gas to be dated may be released by laser fusion so that individual mineral grains may be dated.

3 Under some circumstances, it is possible to correct for the effects of Ar loss.

4 It is sometimes possible to identify the presence of excess Ar from one sample because an isochron can be produced from a single sample (its identification on a K–Ar isochron diagram requires a suite of data from cogenetic samples).

The first stage is irradiation of the sample in a nuclear reactor. An amount of ^{39}Ar is formed that depends on the amount of ^{39}K (and therefore total K), the length of irradiation, the neutron flux and the capture cross section. The last two are difficult

to calculate from first principles so standards of known age are placed in the reactor and used to monitor the flux. Argon is released from sample and flux monitors by fusion and Ar isotope ratios are measured by mass spectrometry. A factor known as J is then calculated from the ages and ^{40}Ar/^{39}Ar ratios of the standards and is used to determine the age of the samples*:

$$t = \frac{1}{\lambda} . \ln\left[\frac{^{40}\text{Ar}^*}{^{39}\text{Ar}} . J + 1\right] \qquad (11)$$

Note the similarity to equation (8). Just as in K–Ar dating, the radiogenic component, ^{40}Ar*/^{39}Ar, must be calculated from the measured ^{40}Ar/^{39}Ar by subtracting a non-radiogenic component assumed to have atmospheric ^{40}Ar/^{36}Ar:

$$\frac{^{40}\text{Ar}^*}{^{39}\text{Ar}} = \left(\frac{^{40}\text{Ar}}{^{39}\text{Ar}}\right)_m - 295.5 \left(\frac{^{36}\text{Ar}}{^{39}\text{Ar}}\right)_m \qquad (12)$$

A number of Ar isotopes are in fact formed by irradiation of K, Ca and Cl and these must be corrected for (Brereton, 1970), but a serious error is likely only in samples that are very young (< 1 Ma) or that have a higher Ca than K content.

---

*J can be an important source of error as many reactors have severe flux gradients.

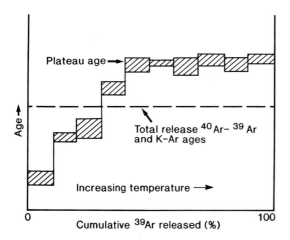

Fig. 5.13 Idealized $^{40}Ar-^{39}Ar$ age spectrum produced by step heating. The width of the boxes parallel to the age axis indicates precision. Gas released at low temperatures comes from less Ar-retentive sites and is younger than the plateau age. This sample has lost Ar by diffusion. Its total fusion $^{40}Ar-^{39}Ar$ and K–Ar ages will be younger than the true age. Note that the plateau age may still underestimate the true age if Ar has been lost uniformly from the more retentive sites.

If all argon is released by completely melting a sample, the measured $^{40}Ar-^{39}Ar$ age (*total release age*) is equivalent to the K–Ar age. However, it is also possible to release aliquots of gas sequentially by heating at progressively higher temperatures and to measure an age for each. This is known as *step heating*. If there has been no Ar loss, the ages of all the gas aliquots should be the same and equal to the K–Ar age. If, however, Ar has been preferentially lost from certain sites in the mineral, there will be a spectrum of ages. The most common pattern, and the easiest to interpret, is illustrated in Fig. 5.13. Gas released at low temperature has young ages because it comes from less retentive sites and from the mineral surface from which Ar is most likely to have been lost. Gas aliquots released at higher temperatures from more retentive sites tend towards a constant *plateau age* which is a minimum estimate of, and frequently close to, the true age of closure. In such cases, the total release and K–Ar ages will be younger than the plateau age. It is also common to observe anomalously old ages in the gas driven off at low temperature. This may indicate the presence of excess Ar or loss of K. We will look at an example of such a pattern below when we discuss $^{40}Ar-^{39}Ar$ dating of chlorite cement.

One further variant of the $^{40}Ar-^{39}Ar$ method involves releasing gas for dating by means of a laser (York *et al.*, 1981). The method has enormous potential because it removes the need to prepare a pure mineral separate: the laser can be directed at individual grains and analyses obtained from spots only a few tens of micrometres across. We will look below at how this method has been used to date feldspar overgrowths. Although there are some problems, laser $^{40}Ar-^{39}Ar$ dating does seem to have a bright future for dating not only authigenic, but also detrital minerals.

There is one problem sometimes associated with $^{40}Ar-^{39}Ar$ dating that we must consider before looking at some examples: the phenomenon of argon *recoil* (Turner & Cadogan, 1974). This happens during irradiation when ^{39}K is lost due to the emission of protons to form ^{39}Ar. Recoil affects a surface layer of the order of 0.8 μm thick and is therefore likely to be a particular hazard with fine-grained minerals. It leads to erroneously old and meaningless ages. We will encounter the problem of recoil in two of the examples below.

5.4.2 Example 4: chlorite cement, Triassic, Central North Sea

In this example, we will look at an attempt to date authigenic chlorite by means of the $^{40}Ar-^{39}Ar$ method. Chlorite usually contains a small amount of K ($<1\%$) even though there is no site in the lattice which it can occupy (Whittle, 1985). Chlorite separates were prepared by disaggregation of sandstone cores followed by repeated high gradient magnetic separation. Purity was checked by XRD analysis and visual inspection using TEM. All step heating profiles show a characteristic saddle shape, with ages considerably older than the Triassic depositional age of the sandstones at both low and high temperature ends (Fig. 5.14). A similar profile has been observed for other minerals (Harrison & McDougall, 1981; Zeitler & Fitzgerald, 1986). There is no clear plateau age although the minimum ages are reasonably consistent, between 65 and 41 Ma (average 52 ± 9 Ma), and compatible with a K–Ar age of 74 ± 3 Ma measured on a mixed layer illite–smectite cement that appears on textural grounds to pre-date chlorite.

In the absence of contamination by another K-bearing phase, there are at least two possible explanations for the saddle-shaped spectra. Excess ^{40}Ar may be present in both the least and most retentive sites so that the gas given off at high

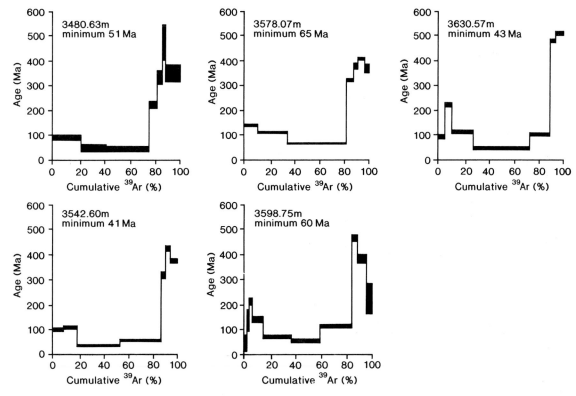

Fig. 5.14 Step heating age spectra for chlorite cement separates from five Triassic sandstone samples, Central North Sea. The profiles show a characteristic saddle shape. Interpretation of a meaningful age is not straightforward.

and low temperatures gives anomalously old ages. Alternatively, there may have been loss of ^{39}Ar in the reactor (recoil) from most and least retentive sites. Though corrections could be applied to compensate for recoil or excess Ar, it is at present difficult to interpret a geologically meaningful age from profiles such as these. More needs to be known about where K is held in the chlorite lattice and how it is released on heating. Note however that total release ^{40}Ar–^{39}Ar and K–Ar ages – weighted averages of the step heating profile ages – would certainly be meaningless.

5.4.3 Example 5: illite cement in aeolian sandstone, Rotliegend Group, Southern North Sea

^{40}Ar–^{39}Ar step heating profiles may sometimes be capable of identifying and correcting for Ar loss or the presence of excess Ar. Might it then be a better method than K–Ar dating for dating illite? Figure 5.15 shows a step heating age spectrum from one of

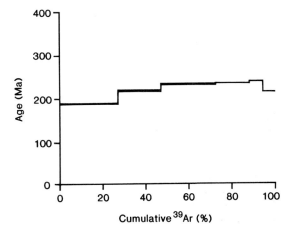

Fig. 5.15 Step heating age spectrum for illite cement from the Lower Leman Sandstone (Rotliegend Group), Southern North Sea, Village Fields area. There is a well defined plateau age at about 240 Ma and a total release age of 217 Ma. Both are older than the K–Ar age of 155 ± 3 Ma.

the samples of illite cement from the Rotliegend Group of the Southern North Sea (Section 5.3.4). There is a fairly well defined plateau at about 240 Ma and the total release age is 217 Ma. However, both are much older than the K–Ar age, 155 ± 3 Ma. This is almost certainly due to approximately uniform loss of ^{39}Ar from all sites through recoil (confirmed by the loss of about 2% K during irradiation). In this case, ^{40}Ar–^{39}Ar step heating cannot provide a meaningful age. Recoil is likely to be a common problem in dating illites because of their fine grain size (Reuter & Dallmeyer, 1987). However, Bray *et al.* (1987) found that the K–Ar and ^{40}Ar–^{39}Ar ages of illite cements from Proterozoic sandstones in Saskatchewan were concordant. They ascribed the lack of recoil to their having left the samples as aggregates. Presumably, ^{39}Ar lost from the illite structure was retained in between grains or adsorbed onto surfaces.

5.4.4 Example 6: K-feldspar overgrowths

Two approaches have been taken to applying ^{40}Ar–^{39}Ar dating to K-feldspar overgrowths. One is to separate bulk samples of feldspar and perform step heating; the other is to use a laser to analyse individual overgrowths. In the first method, step heating profiles are measured for two feldspar fractions (Hearn & Sutter, 1985). One fraction has grains and overgrowths intact and a second comprises only detrital feldspar from which overgrowths have been removed by acid dissolution. The age spectrum of the overgrowths is then calculated from the mass fraction of overgrowths in the mixed sample as estimated using CL methods (see Section 5.3.6). Hearn *et al.* (1987) applied this technique to authigenic feldspars in Cambrian carbonates in the Appalachians. The synthetic age spectrum that they calculated for authigenic feldspar had a well defined plateau age at about 300 Ma with young ages at low release temperatures and old ages at high release temperatures. This suggests that K–Ar ages and total release ^{40}Ar–^{39}Ar ages may not correctly measure the age of authigenic feldspar.

Feldspar overgrowths can also be dated by laser fusion. Though in principle there is no reason why the laser cannot generate step heating profiles, in practice most such ages are total release ages simply because tiny overgrowths do not produce enough gas for several age determinations. Figure 5.16

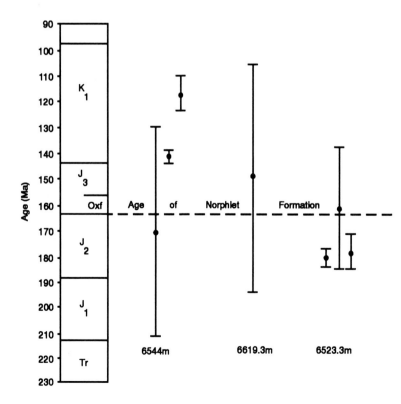

Fig. 5.16 Total release laser-fusion ^{40}Ar–^{39}Ar ages for K-feldspar overgrowths, Norphlet Formation, Gulf of Mexico. Error bars represent 1σ: precision is poor and some ages are older than the age of the Norphlet Formation. There is no suggestion that the feldspar is a late diagenetic mineral but it is not possible to determine the actual age.

shows total release feldspar overgrowth ages from three sandstone samples from the Norphlet Formation, Gulf of Mexico. The ages are variable, their precision is poor (because of the minute amounts of gas analysed) and some are older than the sediment. Other studies have encountered similar problems (e.g. Walgenwitz et al., 1990). The variability and old ages are probably due to release of gas from adjacent detrital feldspars which is hard to avoid and which makes interpretation of the true age of the overgrowths very difficult.

5.5 The Rb–Sr system

5.5.1 Principles

Rubidium has two naturally occurring isotopes:

^{85}Rb   72.1654%
^{87}Rb   27.8346%

Rubidium-87 is radioactive, decaying to stable radiogenic ^{87}Sr with a half-life of 4.88×10^{10} years ($\lambda = 1.42 \times 10^{-11} \mathrm{a}^{-1}$; Steiger & Jager, 1977). Strontium has four natural isotopes:

^{84}Sr   0.557%
^{86}Sr   9.861%
^{87}Sr   6.991%
^{88}Sr   82.59%

The abundance of ^{87}Sr is usually measured and expressed as a ratio, ^{87}Sr/^{86}Sr. In a Rb-bearing rock or mineral, this ^{87}Sr/^{86}Sr ratio will increase with time by an amount proportional to the ^{87}Rb/^{86}Sr ratio of the material and to the time elapsed, such that:

$$\frac{^{87}\mathrm{Sr}}{^{86}\mathrm{Sr}} = \left(\frac{^{87}\mathrm{Sr}}{^{86}\mathrm{Sr}}\right)_{\mathrm{initial}} - \frac{^{87}\mathrm{Rb}}{^{86}\mathrm{Sr}} \cdot (e^{\lambda t} - 1) \qquad (13)$$

This allows the rock or mineral to be dated by solving for t if ^{87}Sr/^{86}Sr and ^{87}Rb/^{86}Sr can be measured, (^{87}Sr/^{86}Sr)$_{\mathrm{initial}}$ (or I(Sr) as it is sometimes written) is known or can be assumed, and certain assumptions (discussed below) are met.

Rb and Sr behave quite differently geochemically because Rb is an alkali element and Sr an alkaline earth. As a consequence, different materials can have vastly differing Rb/Sr ratios, from less than 0.01 for a carbonate to in excess of 100 for a mica. If a cogenetic suite of samples has a range of Rb/Sr ratios, provided that all had the same initial ratio and have remained closed to Rb and Sr since their formation, their ^{87}Sr/^{86}Sr and ^{87}Rb/^{86}Sr will plot on an isochron (Fig. 5.1). The age (t) of the samples is related to the gradient of the isochron:

$$\mathrm{gradient} = e^{\lambda t} - 1 \qquad (14)$$

Isochron dating is especially useful because I(Sr) does not need to be known, but can rather be calculated from the intercept of the isochron on the ^{87}Sr/^{86}Sr axis, fitted to data using least squares regression (York, 1969). As well as calculating confidence limits on the age and I(Sr), most isochron-fitting packages calculate the mean square of weighted deviates (MSWD). This is a 'quality of fit' parameter; values greater than 2.5 indicate scatter in the data above that which would be expected from random analytical error. This would indicate that the assumptions of the isochron method have not been met. In some cases, Rb–Sr isochron dating may provide useful information even if the closed system assumption is not met. If the system to be dated has become 'open' with respect to Rb or Sr over a geologically short time period, related perhaps to a specific geological event (e.g. local heating), then the Rb–Sr clock may be completely reset and the isochron method may be used to date the event rather than the original age of the samples (Morton & Long, 1984).

5.5.2 Analytical methods

Rb–Sr dating requires two analyses: ^{87}Sr/^{86}Sr and Rb/Sr ratios. The first step in measuring ^{87}Sr/^{86}Sr is to pre-concentrate the Sr from the sample and to separate it from Rb (^{87}Rb will interfere with ^{87}Sr during mass spectrometry). This is usually achieved by complete dissolution of the sample followed by an ion-exchange treatment. The Sr separate is then analysed by *thermal ionization mass spectrometry* (TIMS). Precision is monitored in two ways. Internal precision is calculated as the standard deviation (some quote standard error, i.e. $\sigma/\sqrt{n}$) of the results obtained for a number of analytical cycles during the analysis of a single sample. External precision, a better measure of sample reproducibility, is the standard deviation of repeat analyses of a standard over a long period of time. Accuracy is monitored by regular analysis of a standard (this also facilitates interlaboratory calibration). Common Sr standards are: NBS 987 (^{87}Sr/^{86}Sr ≈ 0.71025); Eimer and Amend (0.70800) and Holocene marine carbonate (HMC; 0.70920).

Rb/Sr ratios may be measured by XRF if large enough samples are available (Pankhurst & O'Nions,

1973). More usually, with small diagenetic mineral samples, Rb and Sr concentrations are analysed by isotope dilution mass spectrometry. Usually a 'spike', a solution containing known amounts of Rb and Sr that have been artificially enriched in the minor isotopes ^{87}Rb and ^{84}Sr, is added to a known weight of sample prior to dissolution. The spike isotopes homogenize with the sample during dissolution. Both Rb and Sr are then separated by ion chromatography and analysed by mass spectrometry.

5.5.3 Rb–Sr dating of clay minerals

In principle, several clay minerals may be dated by the Rb–Sr method. These include glauconite (Morton & Long, 1984), illite (Kralik, 1984), kaolinite and smectite (Emery et al., 1987). In particular the K-bearing (and thus Rb-rich) clay minerals glauconite and illite can potentially be dated very precisely. The main problem lies in knowing exactly *what* is being dated. Clay minerals are complicated Rb–Sr systems because they contain two types of Rb and Sr: structural (incorporated within the mineral lattice) and loosely bound (adsorbed onto grain surfaces). There is debate as to whether the loosely bound ions constitute simply troublesome contaminants which must be removed (Morton, 1983) or whether they are better considered as a different part of the clay which may nevertheless remain closed to Rb and Sr and thus contain chronological information. In the first case, an ion-exchange treatment can be used to strip the exchangeable material from the clay, the leachate thrown away and the stripped residue analysed. In the second case, the isotopic analysis of the leachate can provide a point on a leachate–residue, two-point 'isochron' (e.g. Ohr et al., 1991).

Most published Rb–Sr data refer to glauconite. Glauconitic minerals are a family of Fe, K-bearing clay minerals, of which glauconite s.s. is an end member which characteristically occurs as pelletal grains in marine sediments. The pellets form by glauconitization of a pre-existing substrate material at or very close to the sediment–water interface and are thought to evolve gradually towards more potassic compositions over a period of 10^4–10^5 years (Odin & Matter, 1981). Glauconite dating (K–Ar and Rb–Sr) has thus been exploited as a means of calibrating the geological time scale (Odin, 1982).

Glauconitic pellets may be readily separated from the other grains of a disaggregated sandstone using an electromagnetic separator. The development

of highly sensitive mass spectrometers means that individual pellets can now be hand-picked and grouped according to pellet morphology or degree of mineralogical evolution (higher-K pellets are darker in colour) so that isochrons can be developed for individual samples (Fig. 5.17). Ideally, biogenic marine carbonate from the same sample should also be analysed. Carbonate minerals have extremely low Rb/Sr ratios, so their ^{87}Sr/^{86}Sr ratios do not change through time and should be equivalent to glauconite I(Sr). The glauconite isochron age can confidently be assumed to be the age of sedimentation if:

1 measurements from different types of glauconite fall on a single isochron (this indicates initial isotopic homogeneity);

2 biogenic carbonate from the sample plots on the same isochron; and

3 the ^{87}Sr/^{86}Sr of the carbonate is that of contemporaneous seawater (Smalley & Rundberg, 1991).

In several studies, glauconite separates have been found to lie on isochrons defining ages much younger than the known sedimentary age of the host rocks, and with I(Sr) values significantly different from contemporaneous seawater (Laskowski et al., 1980; Morton & Long, 1984; Smalley et al., 1988). In each of these cases, comparison of the isochron age with basin history suggests that ages have been reset by exchange of Sr between the different glauconite types during times of uplift and emergence

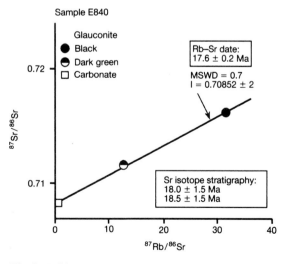

Fig. 5.17 Glauconite dating by the Rb–Sr isochron method (from Rundberg & Smalley, 1989).

above sea level, possibly in response to meteoric water flushing (Morton & Long, 1984). This indicates that Rb–Sr glauconite dates can potentially be used to date diagenesis associated with meteoric water flushing.

5.5.4 Example 7: illite cement in aeolian sandstone, Rotliegend Group, Southern North Sea

In Section 5.3.4 we presented a number of K–Ar ages of illite cements from aeolian sandstones of the Permian Rotliegend Group in the Southern North Sea. These appear largely to fulfil the assumptions of the K–Ar method and show a degree of consistency over a large area and depth range, both of which suggest that the K–Ar ages actually date illite growth. Two of these samples were selected for Rb–Sr analysis in an attempt to obtain concordant Rb–Sr ages. Different size fractions dominated by clay minerals were separated by centrifugation and checked for purity in the now familiar manner. Kaolinite dominated in the coarser fractions separated from sample A; those from B were dominated by illite but contained persistent minor chlorite. This was partially removed by using a high-gradient

magnetic separation technique (Russell *et al.*, 1984), but remained as a minor contaminant. Each of the clay fractions was leached in 1N ammonium acetate to extract loosely bound Rb and Sr. Leachates and residues were then analysed separately for Rb and Sr concentrations using isotope dilution, and for $^{87}Sr/^{86}Sr$. Results are listed in Table 5.4.

Figure 5.18 shows isochron diagrams for the leachate–residue pairs. Two subsamples of the $<2\,\mu m$ fraction of sample A gave similar two-point 'isochron' ages of 140 and 143 Ma but a third was dated at 170 Ma. Illite K–Ar ages for 48/7b-3 range from 145 to 158 Ma (Table 5.1). At least two of the subsamples are therefore extremely close to having concordant K–Ar and Rb–Sr ages. The reason why the third subsample has an older Rb–Sr age is not known. Perhaps the $<2\,\mu m$ fraction is heterogeneous, with the 170 Ma aliquot containing a small amount of old contaminant. The kaolinite-dominated $5–10\,\mu m$ fraction yielded a date of 157 Ma, again within the range of the K–Ar ages. The illite-rich fraction from sample B ($0.5–0.2\,\mu m$) yielded an isochron age of 164 Ma, while the chlorite-rich fraction ($2–5\,\mu m$ magnetic) gave 153 Ma. The intermediate size fraction – a mixture of illite and

Table 5.4 Rb–Sr analyses of illite-rich separates, Lower Leman Sandstone.

Size fraction (μm)	Material	Rb (ppm)	Sr (ppm)	$^{87}Rb/^{86}Sr$	$^{87}Sr/^{86}Sr$*	Residue–leachate date (Ma)
Sample A						
5–10	Leachate	0.521	24.781	0.061	0.710731	
5–10	Kaolinite	82.766	62.855	3.814	0.719085	157
2–5	Leachate	2.083	27.306	0.221	0.711558	
2–5	Kaolinite/illite	166.048	93.845	5.126	0.721226	139
<2	Leachate A	2.784	55.501	0.145	0.711074	
<2	Leachate B	2.872	72.605	0.115	0.711012	
<2	Leachate C	2.805	70.688	0.115	0.711012	
<2	Illite	146.216	48.749	8.699	0.731752	170
<2	Illite	161.088	47.861	9.760	0.730267	140
<2	Illite	171.820	47.918	10.399	0.731907	143
Sample B						
2–5	Leachate	1.527	15.072	0.293	0.711393	
2–5	Mag. frac.	140.469	62.719	6.491	0.724870	153
0.5–2	Leachate	2.377	104.498	0.066	0.710848	
0.5–2	Illite/chlor.	150.713	60.769	7.190	0.728482	174
0.2–0.5	Leachate	1.918	166.483	0.033	0.710752	
0.2–0.5	Illite	166.984	78.867	6.136	0.724948	164

* 2SE for $^{87}Sr/^{86}Sr$ range between 0.000011 and 0.000033.

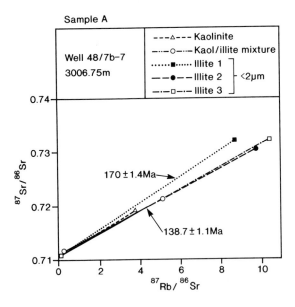

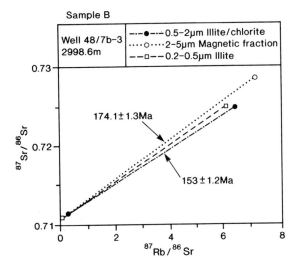

Fig. 5.18 Rb–Sr isochron diagrams for clay mineral residues (high Rb/Sr) and ammonium acetate leachates (low Rb/Sr), illite cement, Lower Leman Sandstone (Rotliegend Group), Southern North Sea, Village Fields area. Sample A is from well 47/7b-7, 3006.75 m and sample B from well 48/7b-3, 2998.6 m. Both contained abundant fibrous diagenetic illite with no detectable detrital illite, and minor dolomite and chlorite. Sample A also contained authigenic kaolinite which texturally appears to be later than the illite. The oldest age is for the illite 1 replicate and the youngest date is for the illite-kaolinite mixture. I(Sr) is about 0.711 for all samples.

chlorite – gave an age of 174 Ma (which is *not* intermediate between the illite and chlorite ages). The K–Ar age of illite from sample B is 158 ± 5 Ma.

These data indicate that leachate–residue pairs give Rb–Sr dates that are in the same range as K–Ar ages. However, in detail there are significant variations in apparent ages, even in one case for different aliquots of the same clay separate. This suggests that the leachate–residue two-point Rb–Sr isochron is a valid means of dating illite but that there are problems with the procedure as it stands at present, possibly due to contamination with small amounts of detrital Sr-bearing minerals. The fact that K–Ar and Rb–Sr dates are so close to being concordant reinforces our belief that the ages are geologically significant.

5.5.5 Sr isotope stratigraphy

Sr isotope stratigraphy is a method for dating marine sediments which is being used increasingly as a supplement to biostratigraphic dating, over which it can have certain advantages. Firstly, it is a numerical method with quantifiable uncertainties and high resolution, and secondly, at least in the marine environment, it is independent of facies.

In today's oceans the $^{87}Sr/^{86}Sr$ ratio of dissolved Sr is constant at 0.70920. This geographical consistency is due to the fact that the residence time of Sr in the oceans is large (≈ 4 Ma) compared with the time that it takes for the oceans to mix ($\approx 10^3$ years). Marine organisms that secrete carbonate or phosphate skeleta incorporate Sr with the marine $^{87}Sr/^{86}Sr$ ratio. Due to the low Rb concentration in carbonates and phosphates, there will not usually be significant *in situ* growth of radiogenic ^{87}Sr so that the fossil material will record the $^{87}Sr/^{86}Sr$ of the seawater at the time the organism was alive provided that it has not been subsequently altered during diagenesis. Studies of the $^{87}Sr/^{86}Sr$ ratio in unaltered marine fossils of known age have revealed that seawater $^{87}Sr/^{86}Sr$ has varied through time in a systematic manner (Fig. 5.19). Figure 5.20 shows a detail of the seawater curve for the Tertiary where the $^{87}Sr/^{86}Sr$ seawater curve is by far the best defined due to the abundance of unaltered and well-dated fossil material from DSDP/ODP cores.

Once a seawater reference curve has been constructed, it may be used as a dating tool. The procedure involves analysing the $^{87}Sr/^{86}Sr$ ratio of unaltered marine carbonate or phosphate in a sample to be dated, locating this value on the

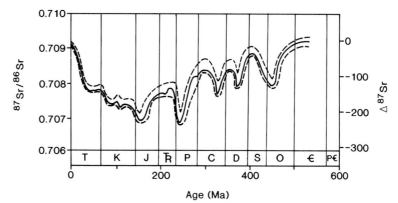

Fig. 5.19 Variations of $^{87}Sr/^{86}Sr$ in seawater during the Phanerozoic (from Burke *et al.*, 1982). The broken lines represent the approximate limits of uncertainty. Data are corrected to a value of 0.70920 for Holocene marine carbonate.

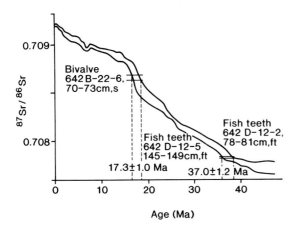

Fig. 5.20 $^{87}Sr/^{86}Sr$ variations in seawater during the Tertiary (from Smalley *et al.*, 1989). Based on the published data of Burke *et al.* (1982), DePaolo and Ingram (1985), Hess *et al.* (1985), DePaolo (1986), and Koepnick *et al.* (1985).

reference curve and dropping a line down to the time axis (Fig. 5.20). The accuracy and precision of dates derived in this way depend on three factors. The first is analytical uncertainty, but modern analytical methods are such that this is relatively insignificant. The second is the gradient of the seawater curve in the age-range of interest (Fig. 5.21). The steeper the curve, the better the resolution attainable. Optimal resolution is theoretically better than 1 Ma for much of the Phanerozoic. The most important influence on accuracy is however the uncertainty in the seawater reference curve itself. The main problems in defining the curve are:

1 interlaboratory analytical bias in published results;

2 contamination of marine carbonate samples with Sr derived from non-marine minerals, e.g. detrital silicates (this is a particular danger where carbonates have been separated from the rest of the rock by acid leaching);

3 post-depositional alteration of the marine carbonate; and

4 uncertainty in the biostratigraphic dating of the samples.

The first problem can be minimized by normalizing all $^{87}Sr/^{86}Sr$ data to a standard value (as were the data in Fig. 5.20)*. The second can be eliminated by excluding from the curve samples with a significant acid-insoluble residue. Ideally, the fossil carbonate should be separated from the host rock physically (i.e. by microdrilling) rather than chemically (acid leaching). It is difficult to assess the importance of the last two problems, let alone correct for them, as the petrography, mineralogy and biostratigraphy of analysed samples usually go undocumented in the published literature.

The uncertainty in a Sr isotope age may be estimated in a variety of ways. Perhaps the simplest is to draw an envelope around the data that make up the seawater curve and to use this together with the accuracy of the $^{87}Sr/^{86}Sr$ analysis to project onto the time axis (Fig. 5.20). While not very rigorous, this does at least give a realistic *maximum* uncertainty in the calculated age. Some authors have used regression techniques to fit a line through the seawater reference data (or through a running average)

*Many workers (e.g. DePaolo, 1986) normalize their dataset to their $^{87}Sr/^{86}Sr$ value for the NBS 987 Sr standard. However, Rundberg and Smalley (1989) found that interlaboratory agreement is improved if the data are normalized using Holocene marine carbonate as the reference.

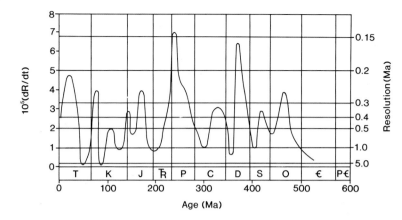

Fig. 5.21 Notional resolution of Sr isotope stratigraphy. dR/dt is the first differential of the seawater curve; the greater dR/dt, the greater the potential resolution of Sr-isotope dating.

and have then calculated confidence limits on the curve. This method is not rigorous either and confidence intervals can grossly underestimate the uncertainty of the age.

5.5.6 Example 8: dating Tertiary sediments, Vøring Plateau, offshore Norway

Sr isotope analyses of samples from ODP Leg 104, Site 642, on the Vøring Plateau, offshore northern Norway are plotted on Fig. 5.20 (Smalley *et al.*, 1989). Three samples were analysed: two phosphatic fish teeth and a calcite bivalve fragment from a glauconitic tuffaceous sand, all from the interval 288–294 m sub-sea floor. The fish teeth (642D-12-2, 78–81 cm and 642D-12-5, 145–149 cm) gave identical $^{87}Sr/^{86}Sr$ values of 0.707828 ± 0.000006 (2 standard errors) and 0.707825 ± 0.000006, corresponding to an age of 37.0 ± 1.2 Ma. The bivalve yielded a statistically indistinguishable $^{87}Sr/^{86}Sr$ value (0.707811 ± 0.000011) and an age of 38.5 ± 1.7 Ma. These consistent Eocene ages were not in agreement with Miocene ages derived from palynomorph biostratigraphy (Manum *et al.*, 1989). While it is conceivable that the dated shell fragments and fish teeth had been reworked from older sediments, the consistency of the Sr ages and a glauconite Rb–Sr age of 40 Ma from the tuffaceous sand, led to a reinterpretation of the palaeontological data involving contamination with younger palynomorphs during drilling and coring (Manum *et al.*, 1989).

5.5.7 Tracing the origin of Sr in subsurface fluids

Most subsurface waters contain appreciable amounts of Sr. The $^{87}Sr/^{86}Sr$ of this dissolved Sr is a reflection of the history of interaction between the water and Sr-bearing minerals and rocks. The ratio therefore contains information about diagenetic reactions and history of fluid movement.

Strontium in formation waters can be derived ultimately from two sources, each with characteristic $^{87}Sr/^{86}Sr$: seawater and detrital minerals. Strontium from seawater can be incorporated into formation waters directly by trapping of seawater in the sediment, or indirectly by dissolution of marine carbonate minerals which will contain Sr with a seawater $^{87}Sr/^{86}Sr$ ratio corresponding to their age, between extreme values of ≈ 0.7069 in the Middle Jurassic and ≈ 0.7092 at present. Most Sr-bearing detrital minerals in clastic sediments have high Rb/Sr ratios and/or are old relative to the depositional age of the host sediment and will have evolved a high ('radiogenic') $^{87}Sr/^{86}Sr$ value. Strontium derived from dissolution of detrital minerals will therefore usually have a much higher $^{87}Sr/^{86}Sr$ than contemporaneous seawater. For example, detrital K-feldspars in the North Sea basin frequently have $^{87}Sr/^{86}Sr$ values > 0.730; high-Rb/Sr micas can have $^{87}Sr/^{86}Sr > 0.800$. In some cases, detrital silicate phases may have lower $^{87}Sr/^{86}Sr$ than contemporaneous seawater. This is particularly true for volcaniclastic sediments which contain young, low-Rb/Sr igneous lithic material with a low I(Sr).

Strontium isotope ratios are particularly useful for characterizing formation waters in marine carbonate rocks where the host rock itself has a distinct and (initially) homogeneous seawater $^{87}Sr/^{86}Sr$ ratio. In some cases, $^{87}Sr/^{86}Sr$ of the present formation water is quite different from that of the host rock. This must mean that the water contains Sr derived from outside the carbonate unit. High $^{87}Sr/^{86}Sr$ is likely to indicate that the Sr comes from clastic sediments, either because the water migrated out of a clastic sequence into the carbonates or possibly due to diffusion of solutes from adjacent formations (e.g. Stueber et al., 1984; Chaudhuri et al., 1987; Emery et al., 1987; Smalley et al., 1992). The presence in formation waters of Sr that is less radiogenic than the host sediment can often be shown to relate to dissolution of volcanic debris (Elderfield & Gieskes, 1982; Egeberg et al., 1990).

Strontium isotopes can also be used to characterize the origin of the strontium contained in diagenetic cements. Unless recrystallized, carbonates and some sulphates such as celestite will retain their initial $^{87}Sr/^{86}Sr$ because of their low Rb/Sr ratios. Carbonate cements from the Smackover Formation of Arkansas, for example, have $^{87}Sr/^{86}Sr$ values far higher than the original marine carbonate of the host rock (Stueber et al., 1984). The implication is that a component of the Sr was derived from an underlying shale through which the formation waters migrated before entering the Smackover. Minerals with high Rb/Sr ratios will have measured $^{87}Sr/^{86}Sr$ greater than the original value due to decay of ^{87}Rb. Their initial ratios must be calculated from present-day $^{87}Sr/^{86}Sr$ and age. Figure 5.18 shows that both of the samples of illite cement from the Rotliegend aeolian sandstones which we attempted to date by the Rb–Sr isochron method have I(Sr) of about 0.711. This is higher than seawater of any age and suggests that the water responsible for illite precipitation had dissolved significant amounts of old detrital minerals.

5.6 The Sm–Nd system

The samarium–neodymium geochronometer is based on the decay of ^{147}Sm to ^{143}Nd with a half-life of 1.06×10^{11} years. ^{143}Nd abundance in geological materials is usually expressed as a ratio, relative to the stable non-radiogenic isotope ^{144}Nd, and the basic equation used for dating is:

$$\frac{^{143}Nd}{^{144}Nd} = \left(\frac{^{143}Nd}{^{144}Nd}\right)_{initial} - \frac{^{147}Sm}{^{144}Nd} \cdot (e^{\lambda t} - 1) \qquad (15)$$

This equation is analogous to that used for Rb–Sr dating and the Sm–Nd system can be used in much the same way, by analysing several cogenetic samples or subsamples (such as mineral separates) and constructing an isochron. Samarium–neodymium is particularly useful for dating basic and ultrabasic rocks which have higher Sm/Nd ratios

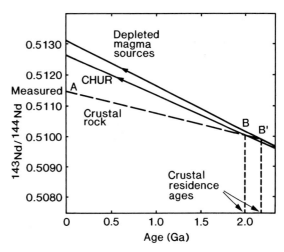

Fig. 5.22 Calculation of a Sm–Nd crustal residence age. AB is the evolution of $^{143}Nd/^{144}Nd$ in the sample calculated from the measured present $^{143}Nd/^{144}Nd$ (intercept) and $^{147}Sm/^{144}Nd$ (gradient). The $^{143}Nd/^{144}Nd$ ratio of CHUR at any time t is given by:

$$\frac{^{143}Nd}{^{144}Nd}CHUR_{(t)} = 0.511836 - 0.1967(e^{\lambda(Sm)t} - 1)$$

The model age calculated from CHUR is given by the intersection of the two curves, B. If the sample had formed from 'depleted' mantle from which magma had already been withdrawn, with higher $^{143}Nd/^{144}Nd$ and $^{147}Sm/^{144}Nd$, its true crustal residence age would be greater (B′).

than granitic rocks and is in this sense complementary to Rb–Sr dating. There are other important differences between the two methods. Both Sm and Nd are light rare earth elements (REE) and display very similar geochemical behaviour, so that the range of Sm/Nd ratios in geological materials is narrow relative to the range of Rb/Sr, from about 0.1 to 0.5. Because Sm and Nd have such similar chemical properties, they are only fractionated significantly by partial melting. Metamorphism, chemical weathering, erosion, transportation and diagenesis are usually considered to have no effect on Sm/Nd and $^{143}Nd/^{144}Nd$ ratios (Chaudhuri & Cullers, 1979; Taylor & McClennan, 1981; Nelson & DePaolo, 1988)[*]. Modern sediments and suspended river sediment retain the same Sm–Nd model ages as the

---

[*] The belief that diagenesis does not cause fractionation of Sm and Nd has been challenged on the basis of a study of the model ages of Tertiary sandstones in the Gulf Coast (Awwiller & Mack, 1991).

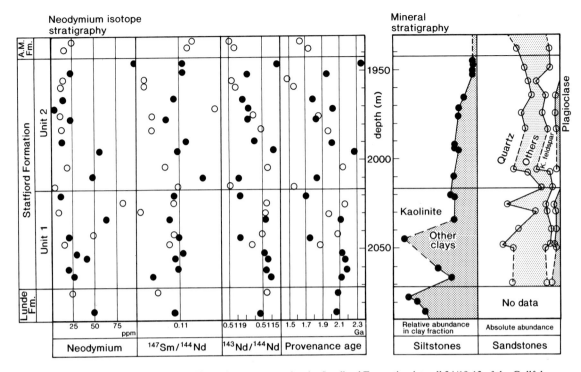

Fig. 5.23 Stratigraphic variations of Nd isotopic parameters for the Statfjord Formation in well 34/10-13 of the Gullfaks Field, Norwegian North Sea (from Mearns, 1989). Open circles are sandstones, closed circles are siltstones and mudstones. Mineral variations in the right hand column are after Malm (1985). There is no clear relationship between kaolinite variations and provenance.

rocks from which they are eroded (Goldstein & Jacobsen, 1988; Mearns, 1988). The main fractionation of Sm from Nd will generally have taken place when a sediment's igneous precursor separated from the mantle by partial melting. Sm–Nd ages even of sediments therefore tend to represent the time since the Nd in the sample separated from the mantle. If $^{143}Nd/^{144}Nd$ and Sm/Nd have been preserved through erosion, transport, sedimentation and diagenesis, then the Sm–Nd age of a sediment sample will represent the weighted mean age of all the sediment source areas that have contributed to the provenance of the sediment.

Samarium–neodymium dating can in principle be used to obtain ages for diagenetic minerals although it will suffer from the same problem as Rb–Sr dating: that of obtaining an isochron. Sm–Nd has in fact been used to date authigenic illite using a procedure closely analogous to that described for Rb–Sr (Section 5.5.4; Ohr *et al.*, 1991). However, the main application of the method to sedimentary rocks has been in characterizing and dating the

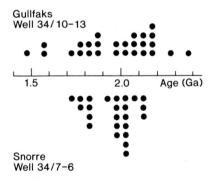

Fig. 5.24 Histogram of provenance ages for the Statfjord Formation from Gullfaks and Snorre fields, Norwegian North Sea (from Mearns, 1989). The majority of the data are comparable (although Gullfaks samples spread to both younger and older provenance ages). This suggests similar source areas for the Statfjord Formation at both localities.

sources of clastic material. This involves the calculation of a model age from a whole rock analysis by assuming an initial $^{143}Nd/^{144}Nd$ ratio.

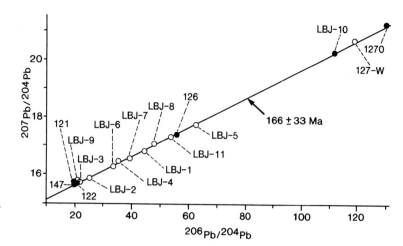

Fig. 5.25 Pb–Pb isochron diagram for marbles from Taiwan (from Jahn, 1988).

The simplest model used for determining initial Nd ratios is known as CHUR (chondritic uniform reservoir), a model for the evolution of the $^{143}Nd/^{144}Nd$ ratio in the mantle which assumes that it has the same $^{143}Nd/^{144}Nd$ and $^{147}Sm/^{144}Nd$ ratios as chondritic meteorites (Fig. 5.22, p. 125). The model age is calculated by using the sample $^{143}Nd/^{144}Nd$ and Sm/Nd ratios to reconstruct the $^{143}Nd/^{144}Nd$ back through time to the point when it was identical to that of CHUR. CHUR is not however an accurate description of the evolution of Nd isotopes in all parts of the mantle (Faure, 1986). Production of magmas by partial melting leaves the mantle locally depleted in light REE and with a higher Sm/Nd ratio than CHUR. More sophisticated models for the $^{143}Nd/^{144}Nd$ evolution path for the mantle underlying a particular crustal segment can be calculated by assuming a time that mantle depletion occurred and a modified mantle Sm/Nd ratio (e.g. Nelson & DePaolo, 1988); or determined empirically by the analysis of mantle-derived rocks of various ages (e.g. Mearns, 1988). $^{143}Nd/^{144}Nd$ ratios are often expressed using an ε (epsilon) notation, which represents the variation in parts per thousand of the $^{143}Nd/^{144}Nd$ ratio in the sample from chondritic reference material such as CHUR (Jacobsen & Wasserburg, 1980).

The provenance information obtained from Sm–Nd crustal residence ages (CRA) may also be used for reservoir correlation. Figures 5.23 and 5.24 show variations in CRA in the Gullfaks field, Northern North Sea (Mearns, 1989). The provenance of the Statfjord Formation is quite distinct from that of the Dunlin and Brent Groups. The very

old CRAs in the Statfjord Formation (2.0–2.4 Ga) indicate a source terrane in the Archean Lewisian Series in Scotland (Mearns, 1989). Some of the younger CRAs (<1.5 Ga) from the Ness Formation, younger even than most of the Precambrian crust on both sides of the North Sea, probably have an input of material from the Tertiary Forties volcanic centre (Mearns, 1989). This study is discussed further in Section 8.5.

5.7 U–Th–Pb dating of carbonates

There are four naturally occurring isotopes of Pb: ^{208}Pb, ^{207}Pb, ^{206}Pb and ^{204}Pb. The first three are radiogenic, the products of radioactive decay of ^{232}Th,

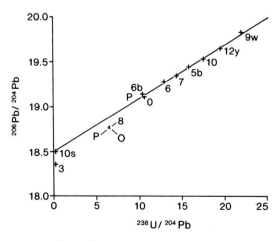

Fig. 5.26 ^{206}Pb–^{238}Pb isochron plot for corals giving an age of 367 ± 5 Ma (from Smith & Farquhar, 1989).

^{235}U and ^{238}U, with half-lives of 14.010×10^9, 0.7032×10^9 and 4.468×10^9 years respectively*. Each of these decay schemes may be expressed as an isochron relationship similar to that described in Section 5.5.1 for Rb–Sr. The two U–Pb decay schemes can also be used in combination in order to date rocks and minerals. This is the Pb–Pb method of dating which makes use of the rate of change of radiogenic ^{207}Pb relative to ^{206}Pb (Fig. 5.25, p. 127). The Pb–Pb method is analytically simple as it only requires measurement of Pb isotope ratios which can be performed using a single Pb chemical separation and a single mass spectrometer run.

The application of U–Pb and Pb–Pb dating to sedimentary rocks is fairly new. The main area of interest is the dating of marine carbonates. Seawater has a high U/Pb ratio, and although U and Pb are only incorporated into marine carbonate in trace concentrations, it appears that the high U/Pb ratio is inherited (Moorbath *et al.*, 1987). This makes U–Pb and Pb–Pb dating of carbonates feasible, and indeed sensible Pb–Pb isochron dates have been obtained for a number of Precambrian limestones (e.g. Moorbath *et al.*, 1987). The uncertainty in the calculation of isochron ages is greater in younger carbonates. Precision may be of the order of $\pm 10\%$ in the Palaeozoic, while in the youngest carbonates for which published Pb–Pb dates exist it is $\pm 20\%$ (Jahn, 1988; Fig. 5.25). One way of obtaining a more precise limestone age is to use U–Pb rather than Pb–Pb dating. While this method is more time consuming (involving isotope dilution measurements of U and Pb concentration as well as Pb isotopic analysis) the precision attainable is far superior, even in rocks with a relatively small range in U/Pb ratios. For example, Smith and Farquhar (1989) obtained a date of 367 ± 5 Ma for Devonian corals (Fig. 5.26, p. 127).

* The various facets of U–Th–Pb dating methods are described exhaustively in Faure (1986), to which the reader is referred for more detail.

Chapter 6 Porosity and Permeability Prediction

6.1 Introduction

This is the first of four chapters in which we will look in detail at some case histories chosen to illustrate the application of geochemical techniques to petroleum geology. In this chapter, we are going to look at its role in contributing to porosity and permeability (*reservoir quality*) prediction. Porosity is the proportion of rock which is available to be filled with fluid – water plus petroleum – and is one of the parameters which must be specified in any prediction of the total amount of petroleum in a structure. Even a high porosity rock will however not be of economic interest if it is impermeable. Permeability and the thickness of net reservoir together largely determine the flow rate at which a well can produce a particular type of petroleum and hence exert a major influence on cash flow and job security. Accurate prediction of porosity and permeability therefore reduces two of the important risks inherent in petroleum exploration, appraisal and development. This chapter is not supposed to be a manual explaining how to predict the porosity and permeability of a petroleum reservoir. Its aim is to use case histories to illustrate how the geochemical methods described in the first half of this book can contribute to the prediction. Nonetheless, so that the reader may better appreciate the nature of this contribution, we will first describe the different approaches to porosity and permeability prediction.

There are two end-member approaches to porosity prediction. One involves the use of a computer program that calculates porosity, usually by coupling

two models* (e.g. Meshri & Walker, 1990). The first
model assumes that chemical equilibrium is maintained between rock and porewater at all times
(Meshri, 1990; Meshri & Ortoleva, 1990). The
second involves calculating the water flux through
the sediment as a function of time (Bethke, 1985;
Bethke *et al.*, 1988; Garven, 1989; Demming *et al.*,
1990; Harrison & Summa, 1991). There are a number of variations on this theme. The simplest assume
that diagenesis and porosity modification take place
in a closed chemical system so that mineralogy and
mineral volume can be calculated without the need
to predict fluid flux. These are clearly inadequate
because the diagenesis of reservoir rocks is not a
closed system phenomenon, at least on the scale of
individual lithological units (Gluyas & Coleman,
1992). Other models incorporate kinetic descriptions of mineral–water reactions. No matter how
sophisticated, all of these models make two key
assumptions: that it is possible to estimate accurately
the fluid flow through a sediment — volumes and
chemistry — over a period of tens to hundreds of
millions of years, throughout the sediment's burial
history; and that as the water passes through the
rock, the nature of the interactions can be predicted
from equilibrium thermodynamic databases. Many
models used to calculate water flow in sedimentary
basins are impressive and physically perfectly sound
but quantitative predictions depend on geological
input which to all intents and purposes has to be
guessed. We would single out the need to specify
horizontal and vertical permeabilities on a regional
scale as a function of time. We do not intend to
belittle the contributions that these models can
make to understanding diagenetic processes. We
simply believe that because they are capable of
precise quantitative predictions, they can be danger-
ously seductive.

The other end-member approach to porosity and
permeability prediction, and that most current in oil
companies, is entirely empirical and relies on the
observation that both are commonly correlated with
other geological parameters. The most commonly
used are maximum burial depth, integrated time–
temperature history, detrital composition, sorting
and grain size (Scherer, 1987; Schmoker & Gautier,
1988; Bloch, 1991). We believe that there is no
realistic alternative to resorting to some form of
empirical prediction and all of the case histories that

we describe make use of empirical observations in
some way.

The most general method of predicting the quality
of an undrilled reservoir involves the construction of
a plot of the mean porosity or permeability of the
target facies (known where it has been drilled)
against depth. In a general sense, rocks tend to be
less porous at depth than they are near the surface,
due to the combined effects of increased compaction and increasing volumes of diagenetic cements.
These plots often therefore show a correlation which
can be used predictively by fitting some kind of
regression equation to the data. This is usually
exponential but over a limited depth range may also
be linear. It is not of course sensible to plot the
porosity of any old rock. The key to making an
accurate prediction is to plot what is to be predicted.
If the aim is to predict the porosity of shallow
marine quartz arenites, it would be unwise to include
volcanogenic sandstones in the database because
these will probably contain unstable minerals which
will affect reservoir quality in some unexpected
way. Just how good a prediction will be is crucially
dependent on the amount and the quality of the data
used to produce the plot. In the most extreme case —
an undrilled basin — there would probably be no
local data at all. The only option would be to use
porosity data from another basin, for a reservoir rock
chosen to be analogous to the hypothetical target.

Regressions of porosity against depth present a
general picture and smooth out irregularities. In Fig.
6.1, some (imaginary) values for mean reservoir
porosity are plotted against depth. In the first case,
the predicted mean porosity for an undrilled structure at a depth of 4 km is $12.0 \pm 5.5\%$. In the
second case, one extra point has been added to
represent a reservoir with 25% porosity at a depth of
4.25 km. After adding this to the dataset, the new
prediction for a structure at 4 km is $14.5 \pm 12\%$.
The main effect of adding the new point has been to
increase the uncertainty of the prediction. Regression of the entire dataset does not therefore permit
the prediction of anomalies such as porous reservoir
at great depth. This is a pity because these are
exactly what we would like to be able to predict.
This is where inorganic geochemistry can make a
contribution.

The geochemical methods decribed in chapters 2 to
5 can be used to describe the conditions under which
rocks became cemented and had their porosities and
permeabilities modified. This information can often
be used to predict where particular mineral cements

---

* See the recent collection edited by Meshri and Ortoleva
(1990).

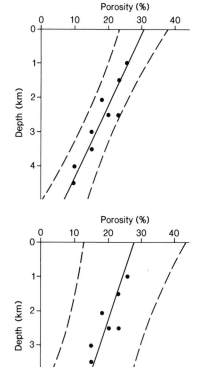

Fig. 6.1 The effect of outliers on porosity predictions. See text for explanation. The broken lines form 95% confidence bands for prediction of porosity from depth.

may be absent and porosity and permeability correspondingly higher. The first case history in this chapter involves exploration for sandstone reservoirs in a virtually undrilled area, the Flemish Pass Basin off the coast of Newfoundland. A porosity–depth plot paints a bleak picture, suggesting that most of the basin is not prospective. However, by determining the age of quartz cementation using fluid inclusions and relating it to that of oil generation, it proved possible to identify parts of the basin where the reservoir might have escaped cementation and retained high porosity and permeability. In the second case history, we describe how a geochemical study of the diagenesis of the Norphlet Formation in the Gulf of Mexico led to an understanding of the factors responsible for preventing quartz cementation and to a quantitative prediction of the thickness of net reservoir (not quartz-cemented). The third

case study involved using geochemical methods for distinguishing between two different types of diagenetic kaolinite in sandstones of the Brent Group. These types must be distinguished if their influence on porosity is to be understood. We then present an example from the Magnus Field in the Northern North Sea which shows how a geochemical study of diagenetic cements in a discovery well might help to produce a porosity map at an early stage and reduce the cost of field appraisal and development. The next case history is a study aimed at understanding the origin of fracture sets in a reservoir in which these exert a major influence on permeability, the Machar Field, Central North Sea. Finally, we describe the use of SEM-based PIA to investigate the origin of apparent high permeability streaks in the Forties Field, Central North Sea.

6.2 Reservoir quality prediction in frontiers exploration: Flemish Pass Basin, offshore Newfoundland

6.2.1 Introduction

The Flemish Pass Basin is located some 400 km east of the Newfoundland coast, close to the margin of the North Atlantic, and is one of a number of rift basins on the Grand Banks (Fig. 6.2). Despite its location just north of the Jeanne D'Arc Basin (which contains at least one giant oilfield, Hibernia; Tankard & Welsink, 1987), Flemish Pass has attracted relatively little exploration interest. This is partly because the water depth over much of the basin is between 1000 and 1100 m and partly because a large number of icebergs pass through the area on their way south from Baffin Bay (witness the fate of the *Titanic* which lies on the continental slope nearby). Exploration drilling in a kilometre of water is expensive but possible. However, the best way to react to a big iceberg is to move out of the way and accept that production will be impossible for a proportion of the year. To compensate for this down-time, high well production rates (governed by permeability–thickness) would be required for any discovery to be commercial, almost regardless of its total reserves. Reservoir quality is therefore a key to the prospectivity of the Flemish Pass Basin.

The problem for the Flemish Pass Basin may be stated as follows: how can we predict where in the basin a particular target reservoir will be capable of sustaining flow rates high enough for a discovery to be worth developing?

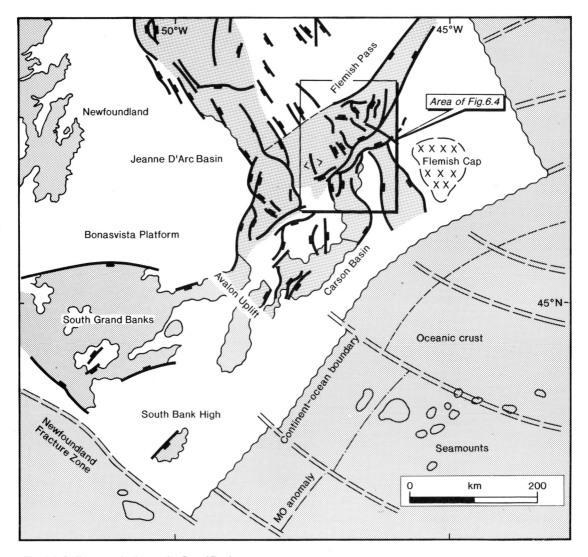

Fig. 6.2 Sedimentary basins on the Grand Banks.

6.2.2 Geological background

The geology of the Flemish Pass Basin is particularly complex. It contains at least three syn-rift megasequences related to three separate cycles of rifting and subsequent sea-floor spreading (Fig. 6.3; Foster & Robinson, 1993). The first rifting event began in the Triassic and was followed by movement of North Africa away from North America south of the Newfoundland Fracture Zone about 175 Ma ago (Aalenian). This transform fault now forms the southern margin of the Grand Banks. The associated megasequence (MS1) is mainly very deeply buried so that it has only proved possible to map and interpret the uppermost sequence, MS1-50 (Kimmeridgian to Early Berriasian) which fills two WNW–ENE trending sub-basins. The second rifting event involved NW–SE extension and was a prelude to spreading of the Grand Banks away from Iberia about 125 Ma ago (Barremian). MS2 fills sub-basins that trend NE–SW and is of Early Berriasian to Mid-Aptian age. The third and final rifting phase preceded movement of Canada away from Greenland 100–105 Ma ago (Albian).

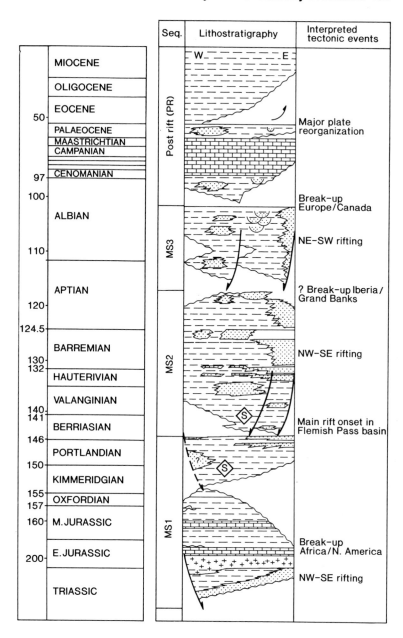

Fig. 6.3 Schematic chronostratigraphy of the Flemish Pass Basin.

MS3 sediments fill sub-basins that trend NW–SE. The last megasequence is a Late Cretaceous to Tertiary post-rift wedge deposited as the continental margin underwent thermal subsidence.

Seven wells have been drilled in the Flemish Pass Basin with one discovery at North Dana (Fig. 6.4). Five of these were however drilled to test what were effectively extensions of the geology of the Jeanne D'Arc Basin; only Baccalieu and Gabriel penetrated thick sequences of syn-rift sediments in Flemish Pass. Several wells showed that a source rock is present in the Kimmeridgian to Early Berriasian MS1-50 sequence and seismic mapping suggested that it would be mature over large parts of the basin. Shallow marine, potential reservoir sandstones in MS1-50 drilled in Baccalieu have very low porosity and permeability because they are cemented by quartz.

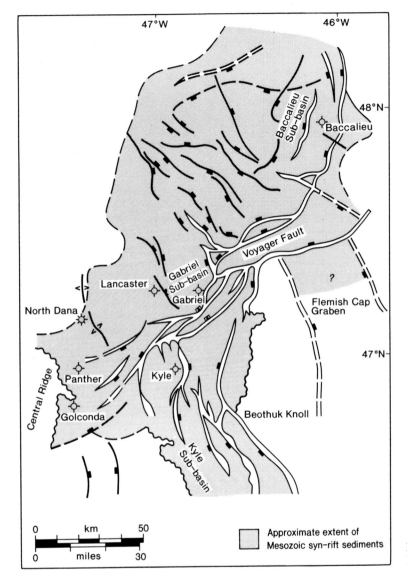

Fig. 6.4 Map of the Flemish Pass Basin.

6.2.3 Approach

Economic constraints dictate that wells from any discovery will need to produce at rates of greater than $5000 \, \text{bbl.day}^{-1}$. Flow rate is determined by petroleum composition (viscosity, gas/oil ratio), the pressure difference between the formation and well during production (*drawdown*) and the product of reservoir permeability × *net thickness**. Making

*Net thickness is the total thickness of reservoir rock that contributes to flow of petroleum into the well. Net : gross is the proportion of net thickness to the total thickness of the reservoir unit.

reasonable assumptions about production conditions, to produce a 30°API oil at $5000 \, \text{bbl.day}^{-1}$ requires a reservoir with permeability–thickness of about $10\,000 \, \text{mD.ft}^{-1}$. This problem can be reduced to that of predicting permeability for different notional values of net reservoir thickness. We tackled this problem in two parts.

1 Calculate a depth below which quartz-cemented shallow marine sands (like those in Baccalieu) are unlikely to produce at $> 5000 \, \text{bbl.day}^{-1}$. The cut-off depth can then be superimposed on a depth map of the target reservoir and areas of acceptable reservoir quality delineated.

2 Predict the distribution of sands protected from quartz cementation (and consequent reduction in porosity and permeability) by early filling with petroleum.

The first approach involves the derivation of a relationship between permeability and depth. The second involves the use of geochemical methods to time quartz cementation, and burial history modelling of the source rock to date petroleum generation. In interpreting the geochemical data, we make a key assumption: that quartz cementation is a regional phenomenon that affects large parts of a basin at about the same time (Robinson & Gluyas, 1992a; Gluyas *et al.*, 1993). It then becomes possible to extrapolate an age determined for quartz cement growth in one well to other parts of the basin and to relate the timing of cementation to that of petroleum generation and migration.

6.2.4 Establishing a relationship between permeability and depth

In an area like Flemish Pass where there are few wells and virtually no core, the first problems in trying to determine how permeability changes with depth are almost always what data to use and where to get it. Permeability measurements are made on core plugs, so are often not available over a range of depths. In such cases, it is necessary to approach the problem in two stages:

1 derive a relationship between permeability and porosity from core analyses; and
2 derive a relationship between porosity and depth using wireline logs (calibrated by porosity measurements made on core if available).

In our example, there is yet another problem: the shallow marine sands have not been cored. The only way around this is to choose an analogue. Fortunately, the Jeanne D'Arc Basin to the south provides a good candidate: the Hibernia Formation is a shallow marine interval of about the same age and has been repeatedly cored because of its importance as a reservoir. Figure 6.5 shows that there is a good correlation between the permeability and porosity of the Hibernia Formation sandstones. A least

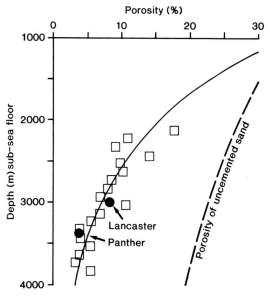

Fig. 6.6 Porosity versus depth; MS1-50 shallow marine sandstones in Baccalieu. Squares are porosities derived from wireline logs and averaged over 100 m intervals. The circles are porosities for the thin correlative intervals in the Lancaster and Panther wells. These plot very close to the exponential curve fitted to the Baccalieu data and give some confidence that this relationship may be applied over the entire basin. The broken line is a curve that we use for predicting the porosity of an uncemented quartz arenite and has the equation $\Phi\% = 50\exp[-z\{km\}/(2.4 + 0.5z)]$ (Robinson & Gluyas, 1992b). The horizontal distance between the two curves is a measure of the volume of mineral cement in the sandstones. Note that depths are plotted below sea-floor, *not* below rotary table. This makes a major difference in Flemish Pass where water depths are very variable.

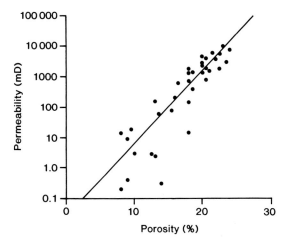

Fig. 6.5 Horizontal permeability versus porosity, Hibernia Formation, Hibernia Field. Permeabilities are for gas, and porosities are calculated using the Boyle's law method (helium porosities; see Archer & Wall, 1986). Both are measured on 1″ plugs of core.

squares fit to the data gives the equation:

$$K \text{ (mD)} = 0.010821 \times 10^{(0.25244\Phi)}$$

(Φ is porosity in %). A porosity–depth relationship for the target shallow marine sands in Baccalieu was obtained by averaging sand porosities calculated from wireline logs over 100 m intervals and regressing these against depth (Fig. 6.6, p. 135):

$$\Phi \text{ (\%)} = 76.76 \times 10^{(-0.34471z)}$$

(z is depth sub-sea-floor in km). A reservoir with 250 ft of net pay needs a permeability of 80 mD to produce at a rate of 5000 bbl.day^{-1}. This corresponds to a porosity of 13.5% and a cut-off depth of 2.2 km. If the amount of net pay were as high as 750 ft, the permeability needed would be only 27 mD, corresponding to a porosity of 15.3% and depth of 2.0 km. The results are displayed in Fig. 6.7. This is a map constructed from seismic data, of the depth to the top of the suggested reservoir interval below the sea-floor. The dark shaded portion shows where the reservoir is deeper than

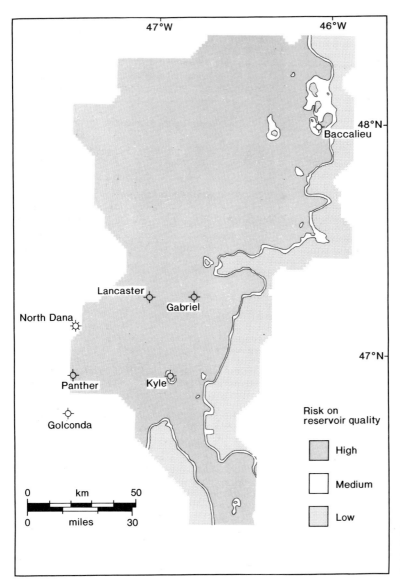

Fig. 6.7 Map of reservoir quality of quartz-cemented MS1-50 shallow marine sands. See text for explanation.

2.2 km. In this area, the risk ascribed to reservoir quality is high because wells are unlikely to produce at rates in excess of 5000 bbl.day^{-1} unless the net reservoir is extremely (and probably implausibly) thick. In the medium shaded parts of the basin, even net reservoir less than 250 ft thick would produce oil at acceptable flow rates. In this area, the risk attached to reservoir quality is low. The area in between, bounded by contours at 2.0 and 2.2 km, represents intermediate risk.

6.2.5 Prediction of uncemented reservoir

The timing of quartz cementation may be estimated by measuring the homogenization temperature of primary fluid inclusions and referring these to a modelled burial and thermal history. Core of MS1 sandstones was available only from the Lancaster and Baccalieu wells. Samples from Baccalieu contained no usable fluid inclusions. In Lancaster, the only core came from shallow marine sands of probable Callovian age. A diligent search through samples of this sandstone revealed only seven primary fluid inclusions in quartz cement, but their

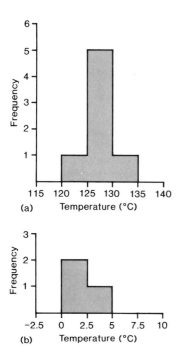

Fig. 6.8 Microthermometric data for primary fluid inclusions in quartz cement, MS1-50 sandstone, Lancaster: (a) homogenization temperatures; (b) solid final melting temperatures.

homogenization temperatures show little scatter and indicate quartz precipitation at temperatures above 124°C (Fig. 6.8). The true temperature of quartz growth will be greater than 124°C by an amount equivalent to the pressure correction (see Section 3.4.3). If CH_4 is present in the inclusion fluids, the pressure correction will be small. The inclusions from Lancaster were too small for analysis by LRS, but final solid melting temperatures above 0°C suggest that a gas is dissolved in the inclusion fluid. This is most likely to be CH_4. When referred to a modelled burial and thermal history for Lancaster, the homogenization temperatures therefore give a maximum age for quartz growth of about 50 Ma ago (Early Eocene) which is probably close to the true age (Fig. 6.9).

The next stage is to identify areas where oil generation commenced before the Eocene. This can be done by producing an isopach map of the interval top-Kimmeridgian (the age of the source rock) to base-Eocene. When decompacted, this map shows the depth of the source rock at the time of quartz cementation which can be calibrated in terms of source rock maturity (Fig. 6.10). The map indicates that oil generation began over large areas of the basin before quartz cement growth reduced reservoir porosity and permeability. Any sealed structures that were charged by these oil kitchens are likely to have been protected from quartz cement growth and to have higher porosities and permeabilities than would be predicted by the empirical depth relationships established for cemented sandstone.

6.2.6 Conclusions

A glance at Fig. 6.7 would be enough to raise doubt in the minds of even the most optimistic of exploration managers about the prospectivity of any play involving MS1-50 shallow marine sands. Over most parts of the basin, it seems at first unlikely that cemented sands will be permeable enough to sustain well flow rates in excess of the required 5000 bbl.day^{-1}. The value of the geochemical data in this case has been that it has led to the identification of substantial parts of the basin that may not be cemented by quartz and which may, as a consequence, buck this trend. The risk on reservoir quality put on any structure close to these areas would be reduced. Unfortunately, we are not able to indulge in a postmortem to check this idea. The Flemish Pass acreage attracted no bids and the prospects in the basin remain undrilled.

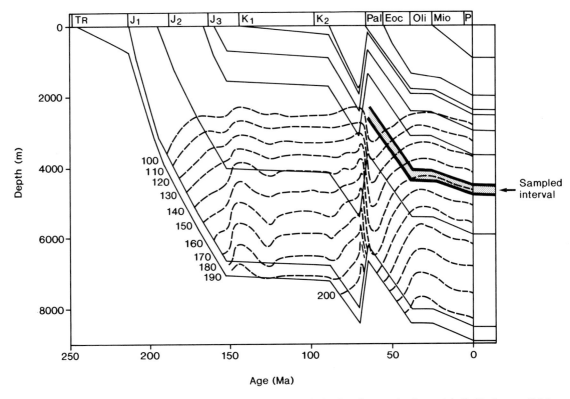

Fig. 6.9 Modelled burial and temperature history for Lancaster. The broken lines are isotherms labelled in degrees Celsius. Two kilometres of Mid–Late Cretaceous burial shown in the figure was put into the model to test the sensitivity of petroleum generation to uncertainties in the amount of pre-Tertiary erosion. In fact, the amount of pre-Tertiary erosion was probably negligible. In this case, the sampled unit would not have reached temperatures of about 125°C until about 50 Ma ago. The methods used to construct the model are described by Allen and Allen (1990).

6.3 Net to gross prediction: Norphlet Formation, Gulf of Mexico

6.3.1 Introduction

The Norphlet Formation is a Middle to Late Jurassic succession dominated by aeolian sediments and is present in the subsurface beneath much of the southern USA, from Texas to Florida. Gas reserves in the Norphlet offshore are of the order of 4.5 tcf (trillion cubic feet). The well that is the subject of this case history, Mobile 821#1, was drilled off the coast of Louisiana in 1987 and tested gas plus minor condensate in Norphlet sandstones (Fig. 6.11). The sandstones showed a distinct vertical zonation in reservoir quality. The upper part had porosities of 1–3% and correspondingly low permeabilities of <0.1 mD and acted as a partial seal. In the lower part, porosities reached 20% or more and perme-

abilities of several hundred millidarcys. The zonation is not due to a change in sedimentary facies since the entire Norphlet in 821#1 is formed of well sorted, medium- to coarse-grained aeolian sandstone. In fact, it is due entirely to variations in the amount of quartz cement: the upper 'Tight Zone' contains around 10% quartz cement while the lower 'pay zones' contain none.

The Tight Zone is a regional phenomenon in the Norphlet and its thickness has a significant influence on the amount of petroleum in a structure. It would therefore be useful to be able to predict thickness before drilling. Unfortunately, it is extremely variable, even within individual prospects. The mineralogical and geochemical work described below was undertaken in order to try to understand the reasons for the differential quartz cementation in the hope that this might provide a means of predicting the thickness of the Tight Zone.

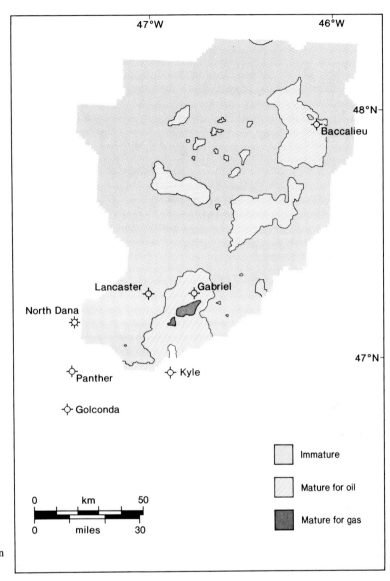

Fig. 6.10 Maturity of Kimmeridgian
source rock in the Early Eocene.

6.3.2 Geological background

The Norphlet Formation was deposited during the
Late Callovian to Oxfordian within broad evaporite
basins on the northern margin of a failed rift zone
and overlies the Mid-Jurassic Louann Salt (Fig.
6.12; Mancini *et al.*, 1985). The centres of the basins
contain thick aeolian sandstone representing dune
fields and these are the target reservoir rocks. To the
south of the dunes, the Norphlet passes into fine-
grained marine clastics and carbonates. The source
rock for the gas in Norphlet reservoirs is the Micro-

laminated Zone (MLZ), the lowest part of the overlying Smackover Formation, which is probably partly
laterally equivalent to the Norphlet. The MLZ is
also the most important top seal and the Louann
Salt forms a bottom seal. Prospective structures are
four-way dip closures associated with salt movement.

6.3.3 Approach

A large proportion of the total thickness of the
Norphlet Formation (about 450 ft) was cored in
821#1, so we had access to a lot of samples and were

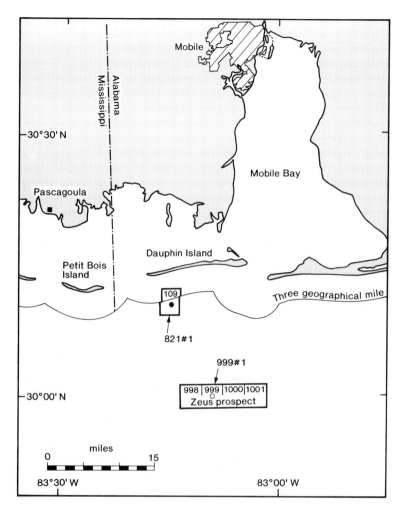

Fig. 6.11 Location map: Mobile Bay and exploration wells 821#1 and 999#1.

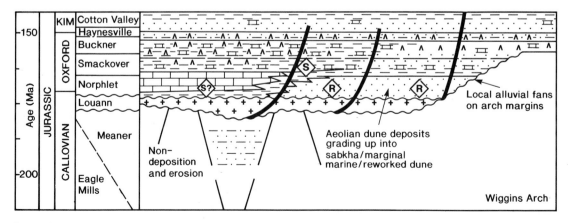

Fig. 6.12 Mid- to Late Jurassic chronostratigraphy of offshore Louisiana.

able to obtain a large amount of quantitative information from XRD and petrography about how mineralogy and certain features of the rock fabric varied between the Tight Zone and the net reservoir below. By relating these data to porosity and permeability measurements made on core plugs, we could see how particular changes in diagenetic mineralogy influenced reservoir quality. Having established a detailed picture of the reasons for the variability of reservoir quality, we began a geochemical project to characterize the diagenetic conditions under which mineral cements grew, principally temperature, time and water isotope chemistry. Our hope was that the mineralogical and geochemical data would

together provide a clue as to why the Tight Zone was so heavily cemented and perhaps even a means of predicting its thickness.

6.3.4 Quantitative mineralogy: controls on reservoir quality

Thin-sections of Norphlet sandstone show that there are several diagenetic minerals. Chlorite is common, forming isopachous coatings on quartz grains, and seems to pre-date quartz cement (Fig. 6.13). Illite cement is present in the Norphlet onshore where it forms characteristic fibres bridging pores (McBride *et al.*, 1987). However, in 821#1, illite takes the

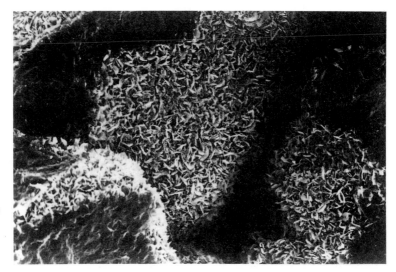

Fig. 6.13 SEM photograph of chlorite cement. Blade-like chlorite crystals completely cover quartz grains. Note the absence of quartz cement. Width of photograph is 200 μm.

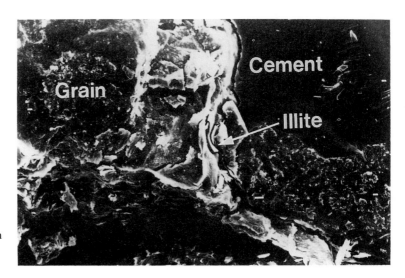

Fig. 6.14 SEM photograph of tangential illite. Plates of illite are sandwiched between a quartz grain (left) and quartz cement (right). Width of photograph is 100 μm.

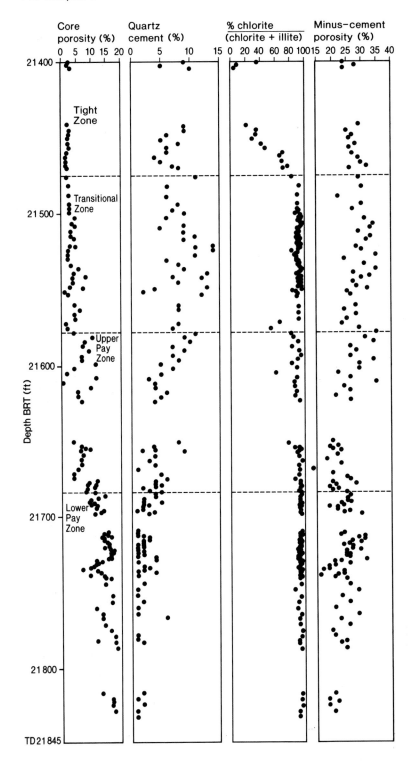

Fig. 6.15 Mineralogical data and core analysis, Mobile 821#1. Porosity is obtained from helium porosimetry on core plugs. Quartz cement volume and minus-cement porosity are measured by point counting. Proportion of chlorite is calculated from XRD analysis of clay separates.

form of tangential grain coatings and is probably not a cement but an early infiltration clay (Fig. 6.14, p. 141). K-feldspar syntaxial overgrowths on feldspar grains are widespread but never abundant. Their position relative to the clays in the paragenetic sequence is unclear, but they pre-date quartz cementation. Before quartz precipitated, petroleum entered the Norphlet, as there is a film of bitumen which stains chlorite and feldspar overgrowths but which is overgrown by quartz. We found no evidence that any of the porosity preserved in the lower part of the Norphlet in 821#1 is secondary.

Figure 6.15 shows how the contents and proportions of some of these minerals vary through the Norphlet formation. The left hand column shows porosity. The Tight Zone is defined operationally as having average porosity consistently below 8%. The amount of quartz cement in the Tight Zone and underlying Transitional Zone varies considerably up to nearly 15%. In the pay zones below, quartz cement content is uniformly low, dropping to less than 5% beneath 21 600 ft. Minus-cement porosity is on average higher in the Tight Zone than in the pay zones (average $28.6 \pm 6.7\%[2\sigma]$ in Tight and Transitional Zones and $23.5 \pm 7.4\%$ in the pay zones). The relative proportions of clay minerals also change vertically through the section. Particularly towards the top of the Tight Zone, illite is dominant over chlorite. This pattern is however reversed near the base of the Tight Zone and from about 21 500 ft down, chlorite is the dominant clay mineral. Relative clay abundances shown in Fig. 6.15 are from XRD analysis of clay separates. XRD cannot however provide reliable measurement of *absolute* clay mineral abundances. To see whether the pattern of absolute abundances also changed, we analysed some samples by thermogravimetry (TG; Fig. 6.16). These show that chlorite content increases from less than 1% at the top of the Norphlet to 5–7% in the pay zones. Over the same interval, illite content decreases from nearly 10% to virtually nothing.

6.3.5 Conditions of mineral cement growth

Laser ^{40}Ar–^{39}Ar ages of K-feldspar overgrowths from the Norphlet Formation are described in Section 5.4.4. The dating was not particularly successful, probably due to a combination of the small size of the overgrowths and Ar release from detrital K-feldspar grains. The most that can be said is that the ages are consistent with K-feldspar growth during early diagenesis.

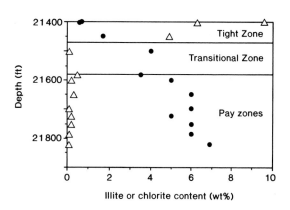

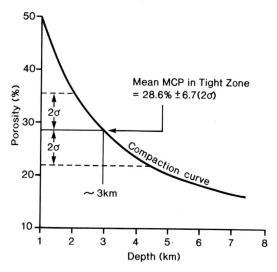

Fig. 6.16 Illite and chlorite abundances versus depth (determined by thermogravimetry. Courtesy E. Warren).

Fig. 6.17 Estimation of cementation depth from minus-cement porosity and a compaction curve. The compaction curve $\Phi\% = 50\exp[-z\{km\}/(2.4 + 0.5z)]$ is from Robinson and Gluyas (1992b).

Though we looked very hard, we failed to turn up any fluid inclusions in quartz cement. We therefore resorted to two other methods to estimate the timing of quartz growth. The first involves calculating minus-cement porosities in the cemented sandstones from point-count data and assuming that this represents the porosity of the sandstone before cementation. The average can then be referred to a compaction curve that describes the porosity of an uncemented sand as a function of depth. The average Tight Zone minus-cement porosity of $28.6 \pm 6.7\%$ corresponds to a cementation depth of about 3 ± 1 km (Fig. 6.17, p. 143).

The other method for estimating the timing of quartz cementation is even more indirect and involves oxygen isotope geothermometry. Quartz cement separates were isolated by the method involving leaching and sonic shattering, with sieving to collect the finest material hopefully richest in authigenic quartz. $\delta^{18}O$ of quartz cement appears to be between $+16$ and $+17‰$ for all of the samples analysed (Fig. 4.10). To calculate a precipitation temperature from mineral $\delta^{18}O$, we have to constrain the isotopic composition of the formation water from which the quartz grew. Since the Mid-Jurassic, Louisiana has not wandered further than $30°$ from the equator. Low altitude meteoric waters in such low latitudes today have $\delta^{18}O$ values between 0 and $-6‰$. Meteoric waters are always the isotopically lightest types of water at a particular latitude and elevation, so $-6‰$ is probably a minimum value for the oxygen isotope ratio of the water

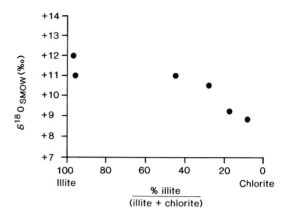

Fig. 6.18 $\delta^{18}O$ of clay separates plotted against proportion of illite.

from which the quartz cement grew. This can be used to calculate a minimum temperature of about 85°C for quartz cement growth.

Finally, chlorite. It may soon be possible to obtain a plausible age for chlorite cementation using the

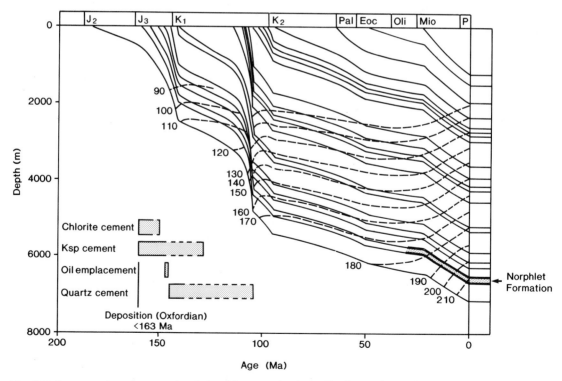

Fig. 6.19 Summary: time of cementation deduced from geochemistry. The diagram is a modelled burial history for well 821#1. Boxes show estimated ranges during which particular cements precipitated.

$^{40}Ar-^{39}Ar$ step heating method (see Section 5.4.2). However when we did this study, we had little option but to try the same indirect method involving oxygen isotopes that we used for quartz, to estimate a minimum temperature for chlorite growth. In Fig. 6.18, $\delta^{18}O$ of analysed clay separates is plotted against the proportion of illite estimated by XRD. This estimate is only semi-quantitative, but it does suggest that chlorite $\delta^{18}O$ is around 8–10‰. We can then use the same minimum water $\delta^{18}O$ of −6‰ to calculate a minimum chlorite growth temperature, remembering as we do so that fractionation factors for chlorite are not accurately known (Table 4.4). It turns out to be about 0°C (we could probably have guessed that chlorite growth from ice is improbable). If chlorite had grown from water as isotopically heavy as seawater (about −1.2‰ in the Jurassic) the isotopic data would be compatible with chlorite growth between about 35 and 65°C.

The results of the geochemical work are summarized in Fig. 6.19. Chlorite and feldspar are both likely to have precipitated before rapid burial during the earliest Cretaceous when temperatures in the Norphlet reached about 100°C. Oxygen isotope data and minus-cement porosities both suggest that quartz precipitated at some time during the Early Cretaceous.

6.3.6 Prediction of Tight Zone thickness

At this point, we have a fairly detailed picture of how diagenetic minerals are distributed vertically through the Norphlet in 821#1 and some fairly

imprecise but probably reasonably reliable estimates of when the various cements grew. How can this be used to predict Tight Zone thickness? The key lies firstly in a recognition of the implications for quartz cementation of the vertical change in the relative and absolute contents of chlorite and illite; and secondly, in an interpretation of the cause of the change.

Early chlorite cement is known to inhibit subsequent precipitation of quartz overgrowths (e.g. Dixon et al., 1989). This appears to have happened in the Norphlet in Mobile 821#1: chlorite content increases from Tight to pay zones while quartz cement volume declines over the same interval. Tangential illite clearly has no such inhibitory effect. The Tight (quartz cemented) Zone is largely developed in Norphlet Formation which contains tangential illite. On the basis of morphology, this illite appears to be a detrital infiltration clay, washed into an unconsolidated sediment through the vadose zone as far as the water table. It seems plausible then that the rapid drop in illite content near the base of the Tight Zone represents a palaeo-water table. In this case, the Tight Zone thickness is equivalent to the depth of the palaeo-water table beneath the ancient land surface.

The top of the Norphlet Formation has substantial relief which was passively infilled by the overlying Smackover Formation. We expected that the distance between the top of the Smackover (originally a horizontal datum) and the base of the Norphlet Tight Zone might change over the region in a regular fashion, reflecting regional changes in the thick-

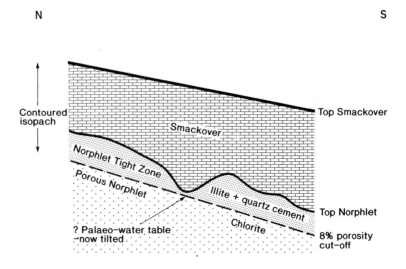

Fig. 6.20 Model for predicting Tight Zone thickness.

ness of the Smackover and any regional dip on the water table (Fig. 6.20, p. 145). We therefore measured this isopach in all the offshore wells and contoured the data (Fig. 6.21). The map shows a systematic regional increase in the isopach towards the south. Tight Zone thickness can be calculated from the contours at any point prior to drilling by subtracting the distance between the top Smackover and the top Norphlet (predicted in the drilling prognosis).

6.3.7 Conclusions

In this study, a combination of quantitative mineralogy and geochemistry of diagenetic minerals provided us with enough information to propose an explanation for the origin of the Norphlet Tight Zone and a way of predicting its thickness. The most important information undoubtedly came from the identification of vertical changes in mineralogy and cement volumes and, arguably, all of the geochemical work on the cements could have been dispensed with. We could have come up with the same explanation for the mineralogical patterns without it (though whether we would have done is another matter). Flushed with success, we went ahead and made a prediction of Tight Zone thickness in the next well, 999#1 (Fig. 6.21). This prediction required extrapolation as no other wells had been drilled so far south. Imagine our disappointment (relief?) when the well penetrated a completely different facies in sediments equivalent to the Norphlet. All bets were off.

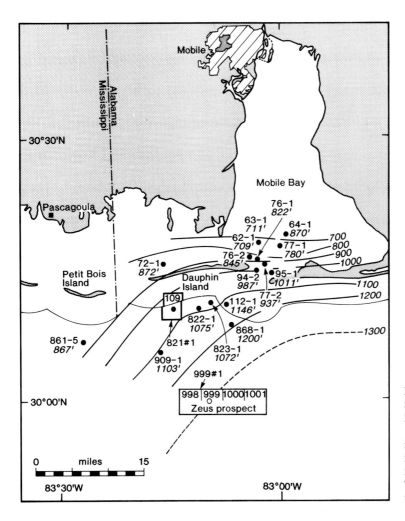

Fig. 6.21 Map of isopach top-Smackover to base Norphlet Tight Zone. The circles are t.Smackover–b.Norphlet Tight Zone thicknesses in feet measured in wells. The isopach increases in a regular fashion towards the south and east. The map can be used to predict the thickness of the Tight Zone.

6.4 Influence of kaolinite on sandstone porosity: Brent Province, Northern North Sea

6.4.1 Introduction

The Brent Province is the collective name for an area located to the north-east of the Shetlands in the Northern North Sea which contains a large number of geologically similar oilfields (Figs 1.1 & 6.22). The main reservoirs are in a succession of Mid-Jurassic terrestrial and shallow marine clastic rocks known as the Brent Group. Kaolinite is a common mineral in Brent Group sandstones and in some areas is an important control on their reservoir quality. It can completely fill and reduce porosity or may be associated with enhanced, secondary porosity immediately adjacent to unconformities (Emery & Myers, 1990). Understanding its origin ought therefore to offer a tool of some value in exploring for good quality reservoirs. Unfortunately, there is kaolinite and there is kaolinite. Its presence is often ascribed to flushing of the sandstone by

meteoric water close to an unconformity and therefore at low temperatures (e.g. Bjørlykke, 1984). Nonetheless, kaolinite can also precipitate as a diagenetic cement without a feldspar precursor, during late burial and at relatively high temperatures (Longstaffe, 1986; Ayalon & Longstaffe, 1988). These two types are difficult to distinguish in thin-section because feldspar precursors are often not obvious. To complicate matters further, kaolinite can form as an alteration product of detrital muscovite and may even be present in sandstones in the form of deformed mudstone clasts (though these types can usually be identified from their textural relationships in thin-sections).

In this section, we describe a geochemical study designed to answer three questions.

1 Does kaolinite in Brent sandstones have a single origin or are different types of kaolinite important in different places?

2 If there are kaolinites of more than one type, how can they be distinguished?

3 What are likely to be the main geological controls on kaolinite distribution?

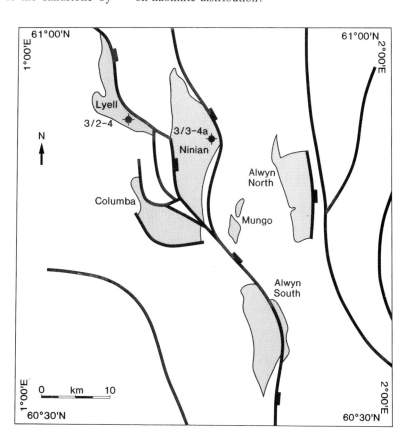

Fig. 6.22 Location of the Ninian and Lyell fields, Brent Province.

6.4.2 Geological background

Coarse clastic sediments of the Middle Jurassic Brent Group overlie Lower Jurassic marine mud-stones of the Dunlin Group and are overlain by Bathonian to Late Jurassic mudstones of the Humber Group. These together form the uppermost pre-rift sediments in the Northern North Sea. A typical

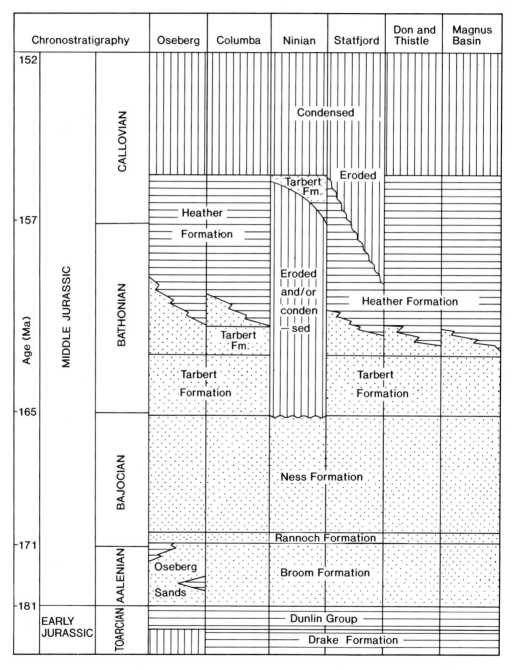

Fig. 6.23 Summary of Brent stratigraphy (after Mitchener *et al.*, 1992).

Brent Province field comprises an eroded, tilted fault block, draped and partially sealed by mudstones of the syn- to post-rift Kimmeridge Clay Formation (which is also the source rock for the oil). The Brent Group has traditionally been subdivided lithostratigraphically into the Broom, Rannoch, Etive, Ness and Tarbert formations (Fig. 6.23; Deegan & Scull, 1977; Graue *et al.*, 1987; see Mitchener *et al.*, 1992 for a sequence-stratigraphic approach to Brent geology). With the exception of the Broom Formation, which appears to represent a separate early phase of basin filling, these formations have been interpreted in terms of the advance and retreat of a wave-dominated deltaic complex. The Rannoch and Etive formations record the advance of a shoreline and form an upward-coarsening unit. The variable facies included in the Ness comprise coastal plain sediments. These are capped by shoreline to shallow marine sandstones of the Tarbert Formation, representing the retreat of the delta.

6.4.3 Approach

Our problem was to see whether it was possible to distinguish one type of kaolinite from another and, if so, to obtain some information about their origins − meteoric alteration of feldspar or precipitation at relatively high temperatures during late diagenesis? Oxygen and hydrogen isotopic compositions are the best parameters for this purpose because they depend on the isotopic composition of the water from which the kaolinite grew and therefore the type of water involved (and also in the case of oxygen, the temperature). We sampled two wells in which Brent Group sandstones contained abundant kaolinite, one each in the Ninian and Lyell fields, chosen to maximize the chances of finding kaolinite of more than one type (Fig. 6.22). Well 3/3-4a is at the crest of the tilted fault block which forms the Ninian Field. The Brent Group there is thin and was probably subjected to one or more phases of subaerial erosion during the Jurassic (Fig. 6.24). We expected that if kaolinite formed by meteoric water flushing was going to occur anywhere, it would be present in this well. Well 3/2-4 in the Lyell Field has a Brent sequence five times thicker than that of 3/3-4a and there are no unconformities above or within the Brent Group. Sandstones in this well are not likely to have been exposed subaerially (although the possibility of meteoric water flow driven along the formation by a

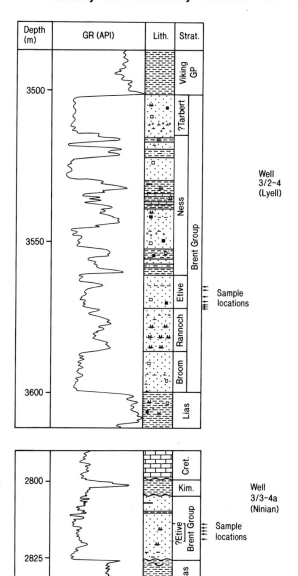

Fig. 6.24 Summary logs of sampled wells.

hydraulic head cannot be excluded). In both cases, samples were collected from the Etive Formation, a massive and homogeneous sandstone with few mud clasts and little detrital mica.

6.4.4 Petrography and isotopic composition of kaolinites

The Ninian kaolinite occurs as clumps of crystal aggregates tens of micrometres long (often described

(a)

(b)

Fig. 6.25 SEM photomicrographs of kaolinite: (a) vermiform kaolinite from Ninian; (b) kaolinite booklets from Lyell.

as *vermiform*) that contain a substantial amount of microporosity (Fig. 6.25). The Lyell kaolinite is not vermiform but consists of densely packed booklets up to about 10 µm across. In both wells, there is a small amount of coarse-grained kaolinite which clearly replaces detrital micas. The oxygen isotope ratios of Ninian kaolinites average

$+15.3 \pm 2.8‰$ (2σ); those from Lyell $+13.9 \pm 1.2‰$ (2σ) (Fig. 6.26). This difference is small but statistically significant at the 95% confidence level. The difference in hydrogen isotope ratios is however very marked. Ninian kaolinites average $-86 \pm 3‰$ (2σ); those from Lyell $-53 \pm 11‰$ (2σ).

We can use these data to determine whether the

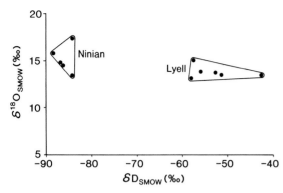

Fig. 6.26 Oxygen and hydrogen isotope analyses of kaolinite separates. For analysis of kaolinite isotopic compositions, sandstone samples were disaggregated and 2–10 µm size fractions separated by centrifuging. These were found by trial and error to contain the greatest proportion of the kaolinites of interest. All but one of the analysed separates contained > 90% kaolinite (determined by XRD).

isotopic compositions of the kaolinites from the two wells are both compatible with growth from meteoric water at low temperatures. This question may be answered in the following way.

1 Estimate the hydrogen isotope ratio of the water from which the kaolinite grew using the hydrogen isotope fractionation equation (though see Table 4.4). Water tends to be about 30‰ more positive than coexisting kaolinite, regardless of temperature. For the mean δD of the Ninian samples of −86‰, the corresponding water isotope ratio will be about −56‰.

2 Calculate the oxygen isotope ratio of a meteoric water with this hydrogen isotope ratio using the relationship between $\delta^{18}O$ and δD for meteoric waters (see Section 4.3.1). This will be a minimum value for the oxygen isotope ratio of the water from which the kaolinite grew, not the actual value.

3 Use this calculated minimum water $\delta^{18}O$ and the measured kaolinite $\delta^{18}O$ to calculate a minimum temperature for kaolinite growth using the oxygen isotope fractionation equation.

In Ninian, the minimum temperature estimated for kaolinite growth is about 20°C; in Lyell, about 60°C. In the case of Ninian, this is perfectly compatible with a near surface origin from feldspar alteration by meteoric water. In Lyell, however, kaolinite growth must have taken place at depths in excess of about 1.5 km. Kaolinites in the two wells seem to have fundamentally different origins.

At this point in the study, we started to acquire that inner glow familiar to all scientists who have set out to test a hypothesis and have just had their prejudices confirmed. This interpretation of the isotopic ratios is however contingent on the assumption that the measured hydrogen isotopic compositions of the kaolinites have not changed since the mineral precipitated. This may not be true. There is a suggestion that hydrogen in clay minerals may undergo geologically rapid isotope exchange with hydrogen in formation waters (see Section 4.3.2). If this is the case, the large difference in δD between Ninian and Lyell kaolinites may reflect differences in the formation waters that are present in the two fields *today*, and *not* differences in the conditions of kaolinite growth. This possibility could and should be tested by analysing the hydrogen isotope compositions of the formation waters in the fields, but we were unable to obtain suitable samples.

6.4.5 Conclusions

Subject to the uncertainty described above, analyses of the hydrogen and oxygen isotope ratios of kaolinite from Ninian and Lyell are compatible with the suggestion from the geological settings of the two fields and from the petrography that the kaolinites in the two wells have fundamentally different origins. That in Ninian probably formed through alteration of feldspar by meteoric water, close to an unconformity and at low temperatures. That in Lyell formed at higher temperatures during burial by a different (and unknown) mechanism. At the very least, this information would stop someone trying to draw a simple contour map of kaolinite content for the Brent Group. At best, it could be used to help understand the distribution of kaolinite. One type should be associated with subaerial unconformities which cut the Brent succession. A second type will not be spatially related to unconformities at all, but might increase in volume with depth.

6.5 Appraisal from a discovery well: Magnus Field, Northern North Sea

6.5.1 Introduction

The Magnus Field is located in Blocks 211/12a and 211/7 of the UK Continental Shelf, about 160 km north-east of the Shetland Islands (Fig. 1.1). It was discovered in 1974 and came into production in 1983 after 9 years of appraisal and development. Oil-in-place is estimated to have been approximately 1650 Mbbl, some 720 Mbbl of which is recoverable. To date, 26 appraisal and development wells have been drilled (17 of them cored) and provide detailed information about the distribution of porosity and permeability in the field. Both are extremely variable. There is a difference in mean porosity of about 7% between crest and oil–water contact, a vertical distance of only about 400 m, with corresponding changes in permeability from less than 30 mD downflank to close to 800 mD near the crest. These

changes in reservoir quality are now well constrained because the field has been so comprehensively drilled. It would however have been extremely useful to be able to map porosity and permeability across the field as soon as possible after its discovery because this information would have influenced the number and locations of appraisal and development wells and possibly even the choice of development scheme.

In this section, we will go back in time to 1974 and pretend that the only information that we have about the discovery comes from seismic data and a

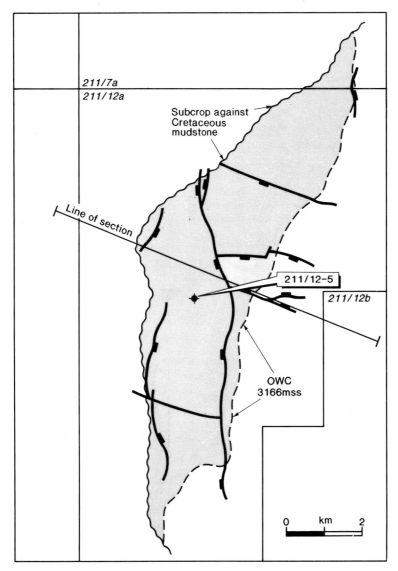

Fig. 6.27 Map of the Magnus Field.

(single) discovery well which is not on the crest of the structure. The well that we are going to use is 211/12-5 (Fig. 6.27). The discovery well was in fact 211/12-1, but the two are in similar positions relative to the crest of the structure and far more data are available for 211/12-5. Using only this small amount of data, how might we have been able to produce maps of mean reservoir porosity and permeability?

6.5.2 Geological background

The Magnus Field consists of a fairly simple east-dipping fault block (Figs 6.27 & 6.28; De'Ath & Schuyleman, 1981; Rainey, 1987; Emery & Myers, 1990). Field limits are defined to the east by the oil–water contact and to the west by erosional truncation of the reservoir. Seal is provided by limestones of the early Cretaceous Cromer Knoll Group and mudstones of the Shetland Group. The main reservoir is the Kimmeridgian Magnus Sandstone Member, made up of four lobes of submarine fan sandstones that interfinger with the Kimmeridge Clay Formation mudstones that further down-dip are the source rock for the oil. The Magnus Member Sandstone comprises 100–200 m of thickly bedded turbidites deposited across the surface of the submarine fan in poorly confined channels or as sheet flows. There are minor slumps and low-density turbidites which do not constitute net reservoir.

6.5.3 Approach

Once the first exploration well on the Magnus prospect had been drilled, we would have at our disposal a depth map of the structure based on seismic data and calibrated by the well and wireline logs, and core of the reservoir rock. How could we predict the distributions of oil-in-place and reservoir permeability? The first step would be to obtain a relationship between porosity and depth so that the depth map of the structure can be converted into a porosity map. The key pieces of information that provide a means of doing this are:
1 identification of quartz cement content as the factor that determines reservoir porosity; and
2 the relationship between the timing of reservoir filling with oil and quartz cement growth.
Both pieces of information can be obtained from the core from a discovery well: the first from the petrography of the reservoir sandstone plus standard core analysis and the second from fluid inclusions in the quartz cement. Together, they permit the porosity at the crest of the structure to be estimated. The porosities at the crest and in the discovery well can then be used to define a porosity–depth relationship. The second step is to obtain a relationship between porosity and permeability so that the porosity map can be transformed into a permeability map. The data required come from

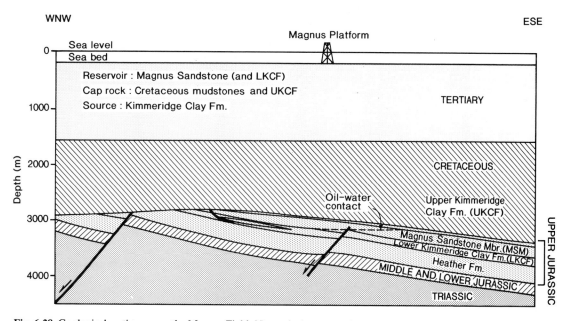

Fig. 6.28 Geological section across the Magnus Field. No vertical exaggeration.

standard core analysis in which both measurements are made on the same core plugs.

6.5.4 Controls on porosity and relationship of cementation to oil filling

Core and sidewall core samples from the 'discovery' well show that the Magnus Member Sandstone is mineralogically simple and that the main control on porosity is the volume of quartz cement. This can be determined by point-counting the amount of quartz cement in thin-sections and relating the amounts to porosities obtained either from the same core plugs (Fig. 6.29) or from wireline logs. The mean porosity in 211/12-5 is $20.8 \pm 0.5\%$ (2σ).

The quartz cement contains both aqueous and petroleum fluid inclusions. Primary inclusions were identified by the usual criterion: isolated occurrence within an overgrowth. Secondary aqueous and petroleum inclusions along healed microfractures are also present and can be observed crossing grains and overgrowths. Measurements of solid final melting and homogenization temperature are shown in Fig. 6.30. The homogenization temperatures of the petroleum inclusions are lower than those of aqueous inclusions because of the greater compressibility of the petroleum (see Section 3.3.3). The difference need not imply that the two sets of inclusions were *trapped* at different temperatures. Indeed, the occurrence of both types as isolated inclusions

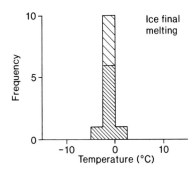

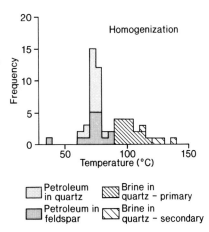

Fig. 6.30 Microthermometric data for primary and secondary fluid inclusions in quartz cement. The data shown come from several wells.

within the quartz overgrowths, sometimes even within the same overgrowth, suggests that quartz grew in the presence of both petroleum and water, as the reservoir filled with oil.

The conclusion that reservoir filling and quartz cementation overlapped in time is an important one. On a gross scale, sealed structures fill with oil from top to bottom. If cement precipitation and filling are partly synchronous, cement is likely to have less time to grow towards the crest of the structure which should therefore have higher porosity. Figure 6.31 shows how this may be used to establish a relationship between porosity and depth. The depth map to the top of the structure can then be transformed into one that allows the prediction of the mean porosity of a well drilled anywhere on the structure (Fig. 6.32). Provided that there is a good correlation between porosity and permeability, it is a fairly

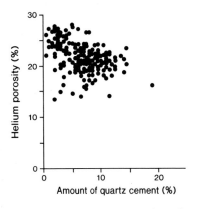

Fig. 6.29 Relationship between porosity and quartz cement content for the main reservoir lithofacies (IV). These data come from all cored wells, not just 211/12-5. There seems to be a weak relationship between the amount of quartz cement determined by point-counting and the helium porosity. It is unusual to obtain better correlations with this kind of data due to the poor accuracy of the point-counting.

Fig. 6.31 Derivation of a relationship between porosity and depth in the discovery well. Two porosities are plotted against depth below sea-floor. One is the calculated mean porosity of net reservoir rock in the discovery well. The other is an estimate of the likely porosity at the crest of the structure which is assumed to be uncemented, calculated using an equation that relates the porosity of a compacted, uncemented sand to depth. A porosity–depth relationship can then be obtained by joining the two points with a straight line. We do not of course know that the relationship must be linear, but there is no justification for trying anything more complex.

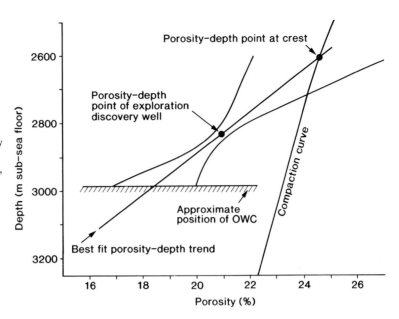

simple matter to transform the map once more so that mean reservoir permeability can be predicted (Fig. 6.33).

6.5.5 Conclusions

There are disadvantages in presenting a case study carried out retrospectively and claiming that it would have been possible to map porosity and permeability after drilling a single discovery well if only we had employed geochemists in 1974. At best, the story tends to lack excitement; at worst, credibility. There is however one advantage: the outcome is known. Having drilled 26 wells, we have an accurate picture of how mean porosity and permeability in Magnus vary with depth. The correspondence is shown in Figs 6.33 and 6.34. The mean porosity for cored wells falls within 1% of the prediction with two exceptions (the anomalous wells had short core sections with unusually low proportions of net:gross reservoir). The prediction of permeability is accurate to a factor of better than 1.5. These predictions are quite accurate enough for planning the locations of development wells and calculating the volume of oil in the field.

The important question that this study raises is not so much could we have produced maps of reservoir quality for Magnus, but rather can we do it for the next discovery and save ourselves some appraisal wells? The applicability of the method will

be limited by a number of constraints. The sedimentology of the reservoir must be fairly simple so that porosity and permeability are determined by diagenetic modification rather than sedimentary facies variations. Equally, the more complex the diagenesis, the less it is likely to be possible to identify one mineral cement as a dominant influence on reservoir quality. The best approach to a new discovery would be to combine the methodology described above with *relative acoustic impedance* (RAI) mapping. AI is the product of rock density × acoustic velocity and therefore depends partly on porosity. With careful processing, it can be obtained from two-dimensional exploration seismic data so there is no reason why RAI cross sections of a prospect should not be available even before an exploration well is drilled. The problem with AI is that it does not depend on porosity alone. It therefore needs to be calibrated against mean well porosity or be supported by other methods to give *absolute* acoustic impedance (AAI). Sections through Magnus show that AAI increases from 6600–7600 at the crest to 8000–9000 AAI units near the oil–water contact. This could be explained by a porosity decrease of about 7% but might also be due to changes in sedimentary facies. If we had already come up with the idea that porosity should decrease from crest to flank due to an increase in the amount of quartz cement, we would probably accept the first explanation.

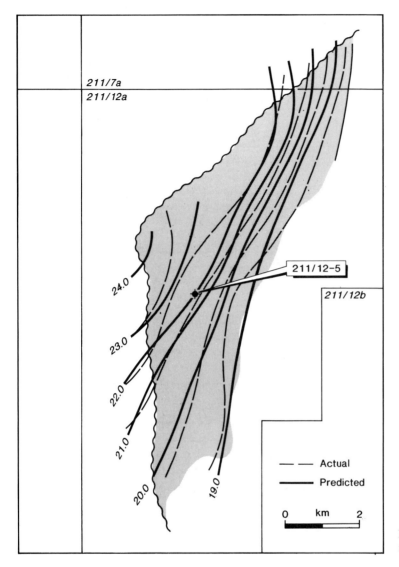

Fig. 6.32 Map of predicted mean reservoir porosity.

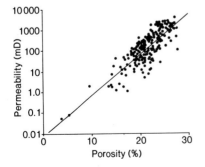

Fig. 6.33 Horizontal permeability versus helium-porosity; core data from well 211/12-5.

6.6 History of fracturing in a Chalk reservoir: Machar Field, Central North Sea

6.6.1 Introduction

The Machar Field is located in the Central Graben of the Central North Sea in UKCS block 23/26 almost due east of Aberdeen (Figs 1.1, 6.35 & 6.36). It has two main reservoirs: a Palaeocene sand and an Upper Cretaceous Chalk, both overlying a salt-cored high at the edge of the graben, which is responsible for the domed structure (Fig. 6.37). An unusual celestite ($SrSO_4$) + siderite rock is found between the Chalk and salt in some parts of the field

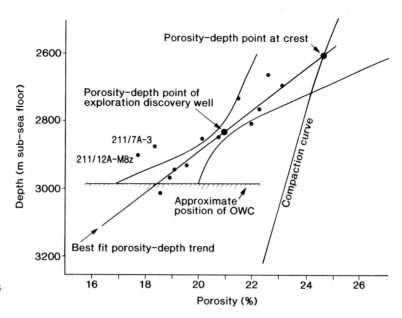

Fig. 6.34 Actual and predicted mean well porosities in the Magnus Field.

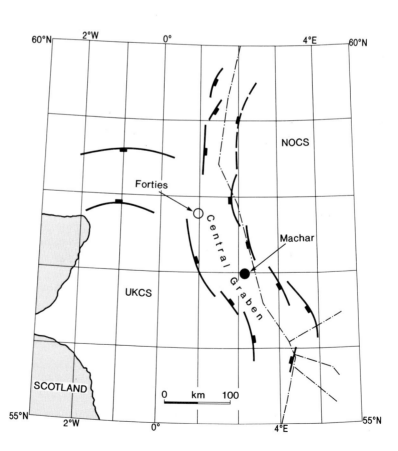

Fig. 6.35 Location of the Machar Field.

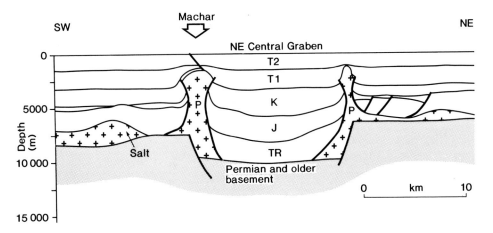

Fig. 6.36 Regional section across the Central Graben.

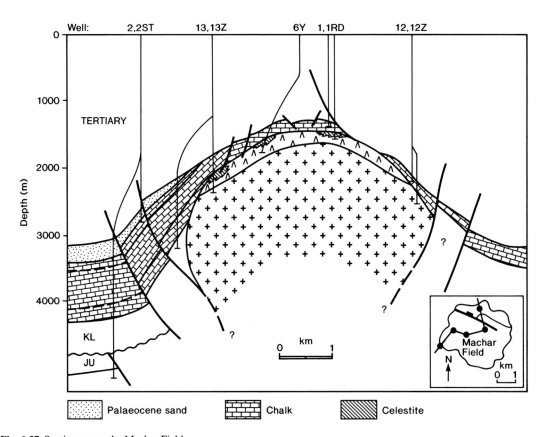

Fig. 6.37 Section across the Machar Field.

and has produced oil at substantial rates (7000 bbl.day^{-1} in 23/26a-6Y). Seal is provided by Early Tertiary mudstones.

In all Chalk reservoirs, matrix permeability is low and oil can be produced at acceptable rates only if the rock is fractured. This is generally the case in Machar where several generations of fractures can be observed in core. However, there is a further complication: many fractures have been filled by mineral cements, mainly calcite but occasionally celestite. Far from contributing to permeability, these filled fractures act as barriers to the flow of oil

into wells. Understanding the development of different types of fractures is therefore important in order to site development wells and also to design efficient schemes for completing wells for production.

In this section, we describe a geochemical study of calcite and celestite fracture fills. The aim of the project was to obtain as much information as possible about the origin of these cements in the hope that we would be able to link them to phases in the evolution of the Machar structure and ultimately predict their distribution. We will return to Machar

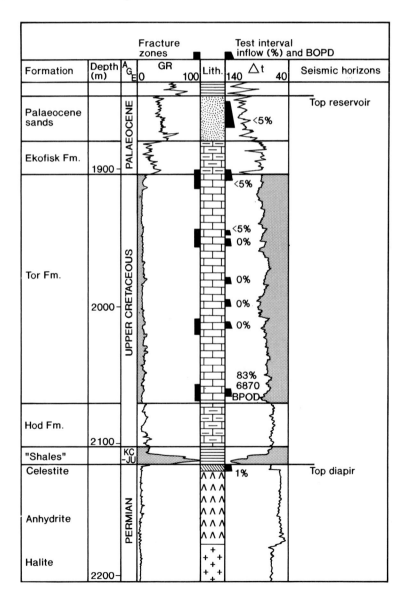

Fig. 6.38 Summary log, well 23/26a-13.

in Chapter 7 when we will look at the origin of the celestite + siderite reservoir rock and the implications for exploration in the Central Graben.

6.6.2 Geological background

The Machar structure is a 4-way dip closure above a dome of Permian Zechstein salt, divided into two main segments by an extensional fault that downthrows to the east (Foster & Rattey, in preparation; Fig. 6.37). There is a condensed Mesozoic and Early Tertiary section on top of the salt indicating that for long periods it remained buoyant relative to adjacent subsiding areas and that it is not principally a piercement structure. Salt pillowing along the graben margins began in the Triassic with erosion of Triassic sediments. The Jurassic and Cretaceous–Palaeocene stratigraphy is condensed: the Chalk is a maximum of 300 m thick over the salt relative to a regional thickness of more than 1000 m. The presence of slumping and resedimented Chalk suggests gravity sliding away from a high on the sea bed at this time. By the Oligocene, the structure was sealed and had begun to receive an oil charge from the Kimmeridge Clay Formation off the flanks of the dome. The only phase of true diapiric activity was associated with Alpine inversion in the mid-Miocene and is reflected in a pronounced unconformity and onlap surface visible on seismic. Since the Miocene, the entire area has subsided and is presently at its maximum burial depth. A 10 m high bump on the sea bed shows that the salt is still buoyant.

The most important oil reservoir is Cretaceous to Early Palaeocene Chalk (which contains > 80% of the oil-in-place) but there are reserves in an overlying Palaeocene sandstone and in the celestite 'cap rock' (Fig. 6.38, p. 159). The Chalk is diagenetically altered and its porosity declines downwards. Matrix permeability is usually less than 1 mD but test permeabilities can reach nearly 1 D due to the presence of fractures. Flow into the wells tends to be restricted to a small number of intensely fractured zones which well tests show to be largely in pressure communication.

6.6.3 Approach

Petrographic study of the calcite fracture fills in well 23/26a-13 revealed the existence of three generations.
1 Type M: (earliest) microspar with no visible zonation occupying very thin hair-line fractures.

2 Type B: blocky calcite, sometimes zoned, occupying thicker fractures that post-date type M calcites.
3 Type T: twinned and zoned calcite occupying the thickest and latest fractures.
We carried out studies of the fluid inclusion contents and stable and radiogenic isotopic compositions of types B- and T-calcite and the rarer celestite fracture fills. Type-M calcite did not contain any fluid inclusions nor did it prove possible to isolate samples for isotopic analysis. We hoped that the results would reveal differences in the temperatures at which these different fractures were cemented and the fluids involved. Our aim was to link these diagenetic events to the geological history of the Machar Field*.

6.6.4 Geochemistry of fracture fills I — fluid inclusions

Types B- and T-calcite and celestite all contain aqueous and petroleum fluid inclusions. Inclusions usually contain either liquid petroleum + vapour or water + vapour but sometimes spectacular three phase liquid petroleum + water + vapour inclusions can be found (Fig. 6.39). Many aqueous and petroleum inclusions were certainly trapped during mineral growth (see Fig. 3.3e) while others are clearly secondary.

Measurements of ice final melting temperatures in primary aqueous inclusions suggest that B- and T-calcites grew from fluids of similar average salinity (about 9–10 eq. wt% NaCl; Fig. 6.40). The range of salinities determined for the T-calcites is however much greater. Celestite seems to have precipitated from a far more saline fluid with an average salinity of about 16%. There are also apparent differences in the UVF characteristics of the petroleum inclusions in the minerals. Those in both calcites fluoresce mainly yellow-green to white; those in celestite fluoresce green-blue to blue and are likely to be more mature (higher API; see Section 3.5.4).

Homogenization temperature distributions for primary fluid inclusions are shown in Fig. 6.41. There is no significant difference between the distributions for aqueous inclusions in the two types of calcite suggesting that the two grew over about the same temperature range. Equally, the distributions for petroleum inclusions are similar to those of aqueous inclusions for both calcites. This means that

---

* For other geochemical studies of the diagenesis of Chalk fields and their fracturing, see Jensenius, 1987; Jensenius *et al.*, 1988; Jensenius and Burruss, 1990.

Fig. 6.39 Three phase fluid inclusions in Machar celestite. Photograph taken in a combination of UV and transmitted visible light. The inclusion cavities are water wet and the oil occurs as a fluorescent meniscus around the vapour bubble. The largest inclusion is about 40 μm across.

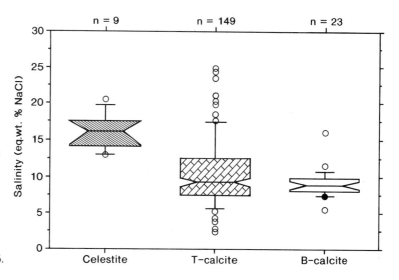

Fig. 6.40 Salinities of fluids from which fracture filling minerals grew. The box plots show distributions of salinity for primary plus pseudo-secondary fluid inclusions. Salinity is expressed as eq. wt% NaCl and is calculated from ice melting point measurements (see Section 3.3.2). Celestite grew from a more saline fluid than did the calcite fracture fills (16.1 ± 5.1% [2σ]). There is virtually no difference in the *mean* salinities of the fluids from which the calcites grew, but T-calcite precipitated from a fluid of more variable salinity (10.4 ± 9.6% relative to 9.1 ± 3.9%). For an explanation of box plots, see Fig. 5.6.

provided that the inclusions have not leaked – a question that we will address in a moment – measured homogenization temperatures are in fact actual trapping (mineral growth) temperatures. The picture for celestite is rather different. Although they overlap with the range for T-calcite, aqueous inclusion homogenization temperatures are significantly higher suggesting growth at higher temperatures.

Before getting too excited about these data, particularly the aqueous inclusion homogenization temperatures, it is important to give some thought to the

question of leakage. Inclusions in both calcite and celestite would normally be susceptible to leakage if heated significantly above their homogenization temperatures. In Machar, this may not have happened since the presence of primary petroleum inclusions shows that both minerals are late diagenetic phenomena. Some clues may also be obtained from the data themselves. Figures 6.42 and 6.43 show that there is no clear relationship between T_h and either depth or inclusion volume (although the absence of such relationships does not *prove* that the inclusions

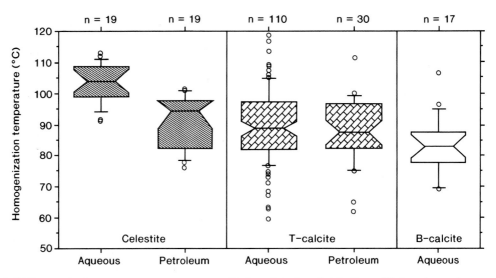

Fig. 6.41 Homogenization temperature measurements. The box plots show distributions of T_h for primary plus pseudosecondary, aqueous and petroleum fluid inclusions. No measurements were made on petroleum inclusions in B-calcite. The only clear difference is that the distribution of T_h for aqueous inclusions in celestite is at higher temperatures.

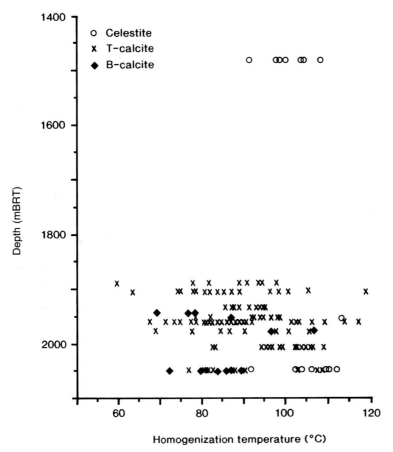

Fig. 6.42 Homogenization temperature versus depth; primary and pseudosecondary inclusions.

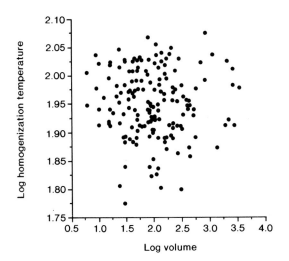

Fig. 6.43 Homogenization temperature versus 'volume'; primary and pseudosecondary inclusions. 'Volume' is calculated by multiplying apparent inclusion long and short dimensions and has the units μm². Note that both scales are logarithmic. There is no correlation between T_h and long dimension either.

have not leaked; see Section 3.4.2). In the case of the calcites, the fact that petroleum and aqueous inclusions have similar T_h distributions argues against leakage*.

Figure 6.44 summarizes the data from the primary aqueous inclusions. Type T-calcite grew over a wide temperature range, from fluids of extremely variable salinity; values range from close to that of seawater to highly saline. Type B-calcite and celestite grew from waters of less variable salinity; celestite grew only at relatively high temperatures.

6.6.5 Geochemistry of fracture fills II — stable and radiogenic isotopes

Figures 6.45 and 6.46 show the oxygen, carbon and strontium isotope ratios of Machar Chalk and fracture filling calcites. $\delta^{18}O$ of Machar Chalk is much

---

*This is a subtle argument. Had the two fluids re-equilibrated to the same *trapping* temperatures, their *homogenization* temperature distributions would be different because of the greater compressibility of oil.

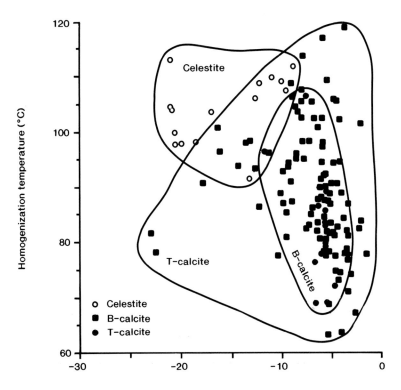

Fig. 6.44 Homogenization temperature versus ice final melting temperature.

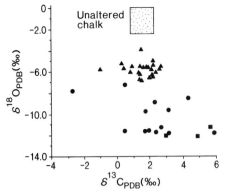

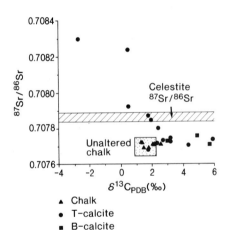

Fig. 6.45 Oxygen and carbon isotopic compositions of fracture filling calcites and matrix Chalk. For data for unaltered North Sea Chalk, see Jørgensen (1987). The precision of both analyses is about ± 0.1‰.

Fig. 6.46 Strontium and carbon isotopic compositions of fracture filling calcites and matrix Chalk. Celestite $^{87}Sr/^{86}Sr$ ranges from 0.70784 to 0.70789; it contains no carbon so cannot strictly be plotted on the graph. The precision of the Sr analyses is better than ± 0.00003.

lower than that of unaltered Chalk indicating a component of diagenetic cement. $\delta^{13}C$ and $^{87}Sr/^{86}Sr$ are however within the range for unaltered chalks indicating that the C and Sr in the cement are derived locally from the Chalk itself. Fracture fills have quite different compositions. Type T-calcite has variable $\delta^{18}O$ and $\delta^{13}C$ suggesting precipitation

from waters of variable composition and/or over a range of temperatures, with carbon coming from several sources. Heavy carbon (> 2‰) might have come from bacterial fermentation of organic matter (though this could not have happened *in situ* as the temperature was too high) or from Permian Zechstein (evaporitic) carbonates; light carbon could have come from decarboxylation of organic matter. T-calcites have variable $^{87}Sr/^{86}Sr$ including some relatively radiogenic values as high as 0.7083. Type B-calcites also contain heavy carbon, but are characterized by rather consistent $\delta^{18}O$ indicating precipitation over a restricted temperature range from a single fluid. $^{87}Sr/^{86}Sr$ lies within the range for unaltered Chalk indicating that the source of the Sr was local.

6.6.6 Conclusions

The earlier of the calcite fracture fills, Type B, grew from a water with a fairly constant salinity of about 9–10% in the presence of a relatively low maturity petroleum and over a substantial range of temperatures (from about 65 to 105°C). Strontium in the calcite was derived from the surrounding Chalk but carbon came from elsewhere. Type T-calcite grew from a water with a wide range of salinity, also in the presence of the same petroleum and over an even wider temperature range (about 60–120°C). Strontium came at least partially from a more radiogenic source and there were at least two carbon sources in addition to the Chalk itself. Celestite grew from a highly saline fluid in the presence of a more mature petroleum, between about 90 and 115°C. At least part of its strontium was externally derived.

Up to a point, the geochemical work was quite successful; certainly, it identified clear differences in the conditions under which the different generations of fracture fills grew. It is however difficult to take the interpretation a step further and link the phases of fracture filling to the evolution of the Machar structure. The wide ranges of homogenization temperatures, the evidence for external sources of strontium and carbon and the presence of petroleum during mineral growth all point to the existence of important convection systems active during fracture filling. With a large amount of water flowing around, it is not safe to use a conductive heat flow model to convert temperature into time as we have in several of the case studies described in this chapter. We cannot easily know when convection cells might have raised temperatures as high as those recorded by the fracture filling minerals. An estimate of absolute mineral age can nonetheless be made, if in a rather roundabout way, by modelling the generation history of the Kimmeridge Clay source rock off the flanks of the Machar salt dome. Oil generation began in the Oligocene and reached a peak in the Miocene, the period of diapiric salt rise. All minerals studied contain primary petroleum inclusions, so they are likely to have grown during this interval. In all probability, most fracturing and fracture filling is linked to this active diapirism.

It might reasonably be argued that this conclusion could probably have been guessed without doing any geochemical work at all. Perhaps the main reason why the value of the work has been rather limited is that the picture is genuinely extremely complex. There are multiple carbon and strontium sources and salinity is highly variable, all indicating that several waters were involved in fracturing and fracture filling, and all probably in a short time during the Miocene. This represents less a failure of applied science than bad geological luck.

6.7 Controls on permeability and the origin of high-permeability streaks: Forties Field, Central North Sea

6.7.1 Introduction

The Forties Field (Figs 1.1 & 6.35) was discovered in the early stages of the exploration of the North Sea in 1970 and originally contained 2480 Mbbl of recoverable oil in Palaeocene sandstones (Lovell, 1990). The main producing reservoir zones are in the main both highly permeable and well connected, so production from Forties has traditionally met forecast levels without any requirement for detailed reservoir studies. Production however reached a peak and began to decline in the early 1980s and in order to meet targets it has proved necessary to improve the description of the reservoir. The most important component of the new description has been the interpretation of a newly acquired three-dimensional seismic survey. Nonetheless, some features of the reservoir that have important implications for production are beyond the limits of seismic resolution and must be tackled in other ways. Amongst these are apparent high-permeability 'streaks' characterized by core plug permeabilities from 3500 to as high as 8000 mD. The previous reservoir model incorporated the streaks but there was no geological explanation for them. Under-

standing their origin and distribution is important because they are likely to act as conduits for injection waters; indeed, they were considered to be important influences on the timing of breakthrough of water into production wells.

This case history describes the use of image analysis of thin-sections of the Forties reservoir to investigate the controls on permeability in 'normal' reservoir and in the high-permeability streaks.

6.7.2 Geological background

The Forties Field is a 4-way dip closure related to drape over a NE–SW trending Mesozoic fault block, the Forties–Montrose Ridge. Seal is provided by Eocene mudstones of the Sele Formation (Figure 6.47; Parsley, 1990). The main reservoir is the Palaeocene Forties Formation which was deposited in the middle to lower parts of a submarine fan system that evolved upwards from a sand-rich type to a mixed sand–mud type. Oil, as ever, is sourced by the Kimmeridge Clay Formation.

Forties Formation sandstones are dominated by monocrystalline quartz with subordinate polycrystalline quartz, feldspars and various lithic fragments, and are mainly subarkoses (Fig. 6.48). Detrital illite, chlorite or haematite coatings are common. Diagenesis has been dominated by compaction and mineral cements generally make up less than 3% of rock volume. Only ferroan calcite may be abundant enough to exert a localized influence on porosity. In general, the permeability of the reservoir is very high and core plug measurements of 1000–2000 mD are not unusual.

6.7.3 Approach

This investigation of the controls on reservoir quality in Forties Formation sandstones was based on PIA. The technique was used to provide a quantitative description of the geometry both of the pore network and of certain features of the detrital rock fabric. In practice, this involved relating PIA parameters to permeabilities measured on core plugs using the standard method of mercury injection (Archer & Wall, 1986) in order to identify the controls on permeability in normal reservoir and in the high-permeability streaks. The expectation was that the controls in the latter would be in some way different and perhaps ultimately predictable.

Thirty samples of sandstone core were collected from three wells (FB55, FA12 and FC22) and

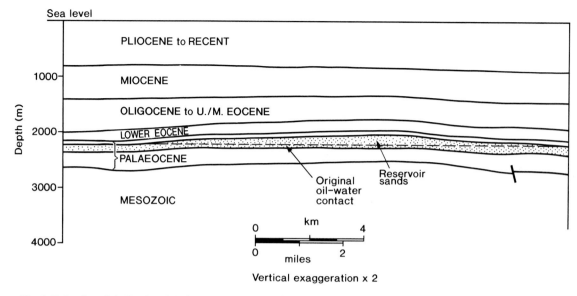

Fig. 6.47 Section of the Forties Field (after Walmsley, 1975).

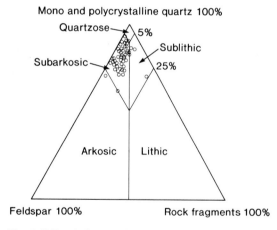

Fig. 6.48 Detrital composition of Forties Formation sandstones.

prepared as polished thin-sections for image analysis using SEM in backscattered electron mode. The samples were selected as representative of the range of normal permeabilities but included some sandstones of abnormally high permeability. Forty fields (*sic*) of view of each thin-section were then digitized at ×100 magnification. The resulting pore images were inverted to give a description of the grain fabric and size distribution (around 10^4 grains per section), a procedure which included the use of an

algorithm to reconstruct grain boundaries at points where two grains touch.

6.7.4 Image analysis of Forties Formation sandstones

Distributions of pore and grain size for a typical sample are shown in Fig. 6.49. Figure 6.50 shows that for the entire sample set, there is a strong correlation between median equivalent circular pore diameter (ECD50) and permeability. ECD50 and the 70th percentile of the grain size are also correlated (Fig. 6.51). The 70th percentile is chosen to represent grain size because it corresponds to the larger grain size mode of the bimodal grain size distributions. The two correlations together suggest that for most sandstones, permeability is strongly influenced by median grain size. Multiple regression of the PIA data and modal analysis shows that 94% of the variance in permeability can be accounted for by a combination of ECD50 and detrital clay content (Fig. 6.52).

A glance at Fig. 6.51 will show that, perhaps rather surprisingly, the high-permeability samples do not have particularly high-variance pore or grain size distributions. One possible explanation is that their pore networks might be better connected in the third dimension than those of the other samples and that this is the feature that renders them anomalous.

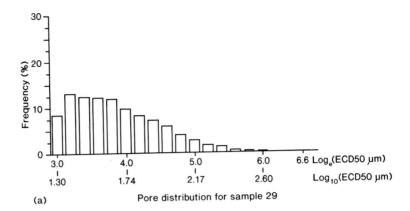

(a)   Pore distribution for sample 29

Fig. 6.49 Pore (a) and grain size (b) distributions for sample 29. Equivalent circular diameter (ECD) is the diameter of a circle having the same area as the analysed pore. ECD50 is the median (or 50th percentile) of the distribution for a sample. The grain size distribution is distinctly bimodal. The mode at larger grain size on the left includes undeformed detrital grains while that at smaller grain sizes includes mainly broken grains and clay-sized detrital particles.

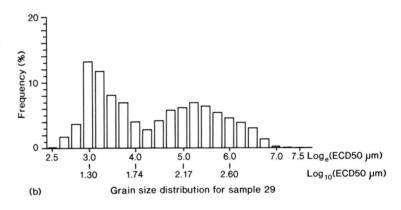

(b)   Grain size distribution for sample 29

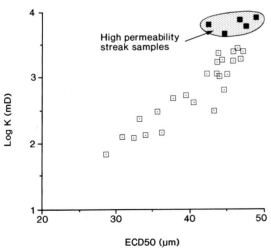

Fig. 6.50 Permeability plotted against ECD50. Equivalent liquid (or Klinkenberg) permeability is calculated from core plug analysis. $r^2 = 0.87$.

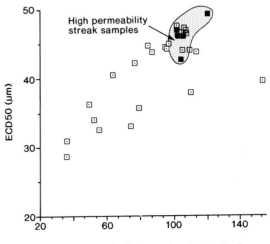

70th percentile imaged grain size (μm)

Fig. 6.51 ECD50 plotted against grain size.

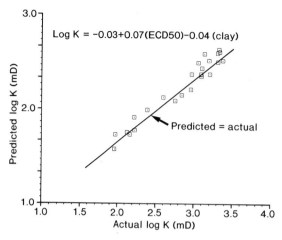

Fig. 6.52 Prediction of permeability from ECD50 and detrital clay content.

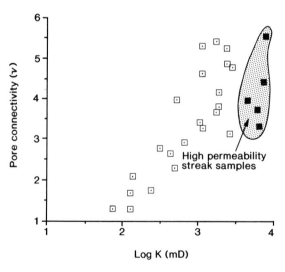

Fig. 6.53 Connectivity plotted against permeability.

Connectivity can be calculated for a sandstone by a method known as series parallel modelling by combining information from image analysis and mercury injection (MacGowan, 1989). The resultant connectivity (v) is a measure of the average number of connected paths between pores. High values of v indicate better connectivity and, in general, higher permeability. For the Forties Formation sandstones, v ranged from about 1 to 6 (Fig. 6.53). The samples from the high-permeability streaks have connectivities from 3 to 6, values which are not in

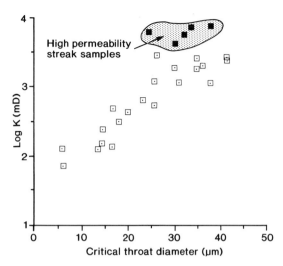

Fig. 6.54 Permeability plotted against critical pore throat diameter.

the least extreme and characteristic more of permeabilities between about 1000 and 3000 mD.

There remains one further possible explanation for the origin of high permeabilities of the streak samples: that for some reason, the critical diameter at which mercury enters the sample upon injection is anomalously large. Figure 6.54 shows that this is not the case; again, the streak samples have critical throat diameters more characteristic of samples with permeabilities of up to 3000 mD.

6.7.5 Conclusions

This study came up with two main conclusions. The first is that permeability in most samples of Forties Formation sandstone is related almost entirely to pore size (related in turn to grain size) with detrital clay content having a small but detectable effect. The second is that so-called high-permeability streak samples do not have pore networks that are compatible with their measured permeabilities. Their grain and pore size distributions are not particularly high, nor do their pore systems appear to be anomalously well connected. One is led to the conclusion that there must be a non-matrix effect on permeability. In this case, it is unlikely that natural fractures are responsible. It is more probable that the core plugs on which permeability was measured were damaged, causing the rock fabric to come apart.

This case study shows how useful PIA can be as a tool for understanding controls on reservoir porosity and permeability. At least part of the first conclusion could have been reached by gathering modal mineralogical analyses by point-counting petrographic thin-sections. However, to make any headway in understanding the more subtle controls on permeability, in particular, requires the large amount of precise, quantitative information that only image analysis can provide (and in any case, who enjoys point-counting?).

Chapter 7 Fluid Migration

7.1 Introduction

In this chapter, we will look at four case histories that show how geochemical studies can be used to characterize episodes of fluid migration through sedimentary basins. Fluid migration is important in petroleum exploration. This is self evident if the fluid in question is oil or gas, but water migration is of concern as well because of the effects that it may have both on oil itself (water washing, biodegradation) and also on the rocks through which petroleum must pass (diagenesis with consequent porosity and permeability modification).

The most obvious manifestations of fluid migration are fluid inclusions. Subject to the limitations described in Chapter 3, these can be analysed to obtain information about the nature and origin of the fluid. Inclusions can also provide an estimate of the temperature and thus, indirectly, the time at which the particular fluid was present. Diagenetic minerals are the product of water–rock interaction and so also have something to say about fluid type and when and at what temperature it migrated through the sample.

The first case history involves a single hand specimen collected in the field (in fact from a ploughed field) in the Aquitaine Basin, France. It shows how water washing can be identified in petroleum fluid inclusions and illustrates how inclusions can be useful even when their homogenization temperatures are probably not recording trapping temperatures. The second case history integrates information from fluid inclusions with mineral isotope chemistry to characterize the conditions under which a rather odd diagenetic celestite reservoir rock formed in the Machar Field in the Central North Sea. So much information came out of this work that it proved possible to specify the actual cause of celestite growth and make a prediction of its regional distribution. The third case history (not our own) describes how the distribution of petroleum fluid inclusions in calcite cement in the Great Oolite in the Weald Basin mirrored the distribution of oil accumulations, the implication being that inclusion distribution can be used to map migration pathways. The final study – on the Waalwijk gas-condensate discovery in the Roer Graben, Netherlands – involves the use of petrography and diagenetic mineral geochemistry to constrain the timing of petroleum migration into the structure.

7.2 History of petroleum migration from outcrop samples: Aquitaine Basin, France

7.2.1 Introduction

The Aquitaine Basin is the most important petroleum producing basin in France (Fig. 7.1; Coustau *et al.*, 1970). Up to 1988, it had produced 75% of the country's total post-war oil production of 500 Mbbl and still produces 40% of her annual oil and 98% of her annual gas output (Weaver & VanDamme, 1988). In parts of the basin, there are few wells and outcropping rocks are not well exposed. A field trip party sent to one of these areas managed to find a single sample of fractured micritic limestone in the middle of a ploughed field. It turned out to be of Albian age and approximately *in situ*. We set out to see if there were any fluid inclusions in the fracture filling carbonate minerals which might provide information about the history of petroleum migration through the Albian sediments.

7.2.2 Geological background

The Mesozoic of the Aquitaine Basin can be divided into four megasequences each of which is related to

a specific phase of relative motion between France and Iberia and which rest on a Hercynian basement (Fig. 7.2; Curnelle *et al.*, 1982; Hiscott *et al.*, 1990). The pre-rift megasequence is of Triassic to Late Jurassic age and comprises continental clastics overlain by thick evaporites and then Jurassic platform carbonates. The syn-rift megasequence of latest Jurassic to Albian age is related to the formation of the Bay of Biscay and includes Lower Cretaceous non- to shallow marine clastics and limestones and thick Aptian to Albian carbonate platform sediments. It is terminated by an intra-Cenomanian regional unconformity associated with inversion along extensional faults. As Iberia moved sinistrally relative to France, a series of transtensional sub-basins formed in which a post-rift megasequence was deposited (Cenomanian to Palaeocene). Extensional tectonics ended in the Eocene when Iberia collided with France. The resulting foreland basin megasequence includes Eocene flysch and Miocene molasse. The present-day configuration of the Aquitaine Basin is dominated by the Lower Cretaceous depocentre bounded to the north by a system of salt ridges and to the south by the Pyrenean thrust front and mid-Cretaceous inversion structures.

Source rocks are present in the Kimmeridgian but the most important are in Lower Cretaceous carbonates which reached maturity over a wide area during development of the foreland basin. Potential carbonate reservoirs are widespread geographically and through the stratigraphy but reservoir quality is usually poor. Lower Aptian mudstones provide a

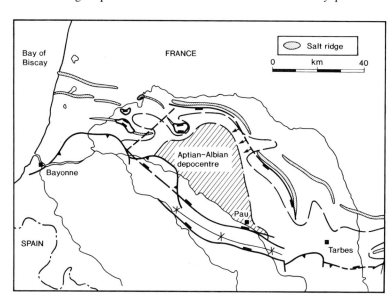

Fig. 7.1 Structural elements, Aquitaine Basin.

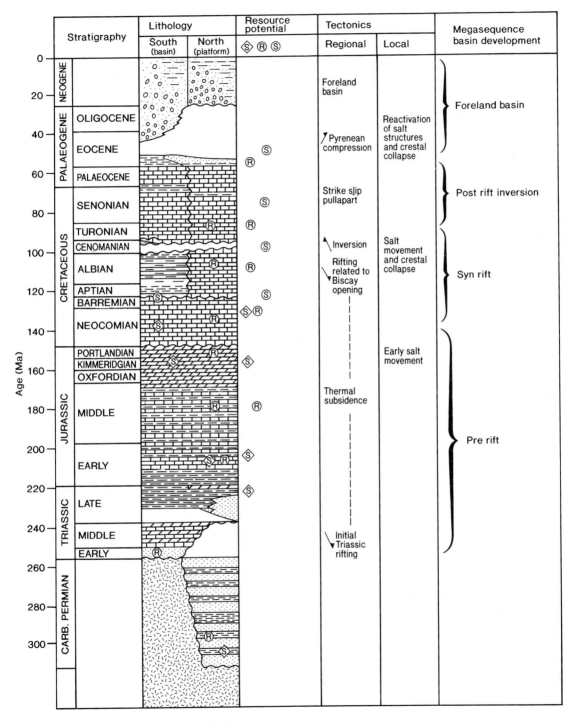

Fig. 7.2 Summary stratigraphy, Aquitaine Basin.

regional seal. The Lacq Field (30 Mbbl oil reserves) is an inversion structure of Cenomanian age. All other known petroleum accumulations are in structures related to salt movement.

7.2.3 Approach

The approach taken in this study was straightforward. The first stage was to characterize the numerous types of fracture-fills in the sample and establish their relative chronology. The second was to search for fluid inclusions and relate groups of inclusions both to one another and also to the episodes of fracture filling. Suitable inclusions were then analysed by microthermometry and, in the case of petroleum inclusions, by GCMS. Bitumen in the pore spaces in open fractures was analysed by GC and GCMS.

7.2.4 Fluid inclusions: petrography, microthermometry and GCMS analysis

A petrographic study of the sample showed that there were five phases of fracturing and filling with ferroan or non-ferroan calcite. The earliest two generations are filled by very fine-grained calcite and contain no visible fluid inclusions. However in the latest three generations of calcite fracture-fills (named C, D and E), there are three main types of fluid inclusions, all containing liquid plus a vapour bubble:

1 aqueous;
2 orange-coloured liquid petroleum + vapour; and
3 colourless liquid petroleum + vapour.

Each of these three groups included inclusions that appeared to be primary and others that were clearly secondary. The colourless petroleum fluoresces strongly under UV light, usually with a blue to blue-green colour, but the orange petroleum shows only very weak fluorescence.

All three generations of fracture-fills contain aqueous inclusions but petroleum inclusions are not uniformly distributed between them (Table 7.1). Type C fracture fills contain colourless petroleum inclusions but none of the orange coloured variety. In contrast, Type D fracture-fills contain orange petroleum inclusions, but only very few colourless petroleum inclusions. Occasionally, a third, liquid water phase could be seen rimming the orange petroleum. The latest (Type E) fracture-filling calcites contain both types of petroleum inclusions. The colourless petroleum was never found associated with liquid water.

A total of 62 fluid inclusions were studied by microthermometry (Fig. 7.3). Orange petroleum inclusions homogenize at temperatures above about 45°C. Primary inclusions show a well defined mode between 45 and 50°C, while the distribution for secondary inclusions is broader with a mode between 75 and 80°C. Colourless petroleum inclusions and aqueous inclusions both show wide ranges of homogenization temperatures with no well defined mode. Ice final melting temperatures however show a more systematic distribution. Types C and E fracture-fills contain relatively saline aqueous inclusions containing about 8 eq. wt% NaCl while inclusions in the intermediate Type D fracture fills contain virtually fresh water (1 eq. wt% NaCl; Table 7.1).

Type D calcite fracture-fills were isolated for analysis of the petroleum in the inclusions. The samples were dominated by the orange petroleum. Unfortunately, it proved impossible to obtain a volume of petroleum large enough to produce a GC trace. However, crushing the calcite under solvent provided sufficient material for limited whole-oil GCMS analysis (Fig. 7.4). Features of the $m/z = 191$ fragmentogram in particular allow the inclusion petroleum to be correlated with other oils from producing fields in the Aquitaine Basin and with the bitumen in the pores of the outcrop sample. All

Table 7.1 Fluid inclusion contents, calcite fracture fills, Albian limestone outcrop sample.

Vein type	Clear petroleum + vapour	Orange petroleum + vapour	Water + vapour	Ice final melting temperature (°C)	Average salinity (eq. wt% NaCl)	Bitumen in pores
C	Present	No	Present	-5.0 ± 2.5 ($2\sigma, n = 5$)	7.9	No
D	Rare	Abundant	Present	-0.6 ± 0.1 ($2\sigma, n = 3$)	1.0	Present
E	Rare	Present	Present	-5.5 ± 1.3 ($2\sigma, n = 9$)	8.5	Present

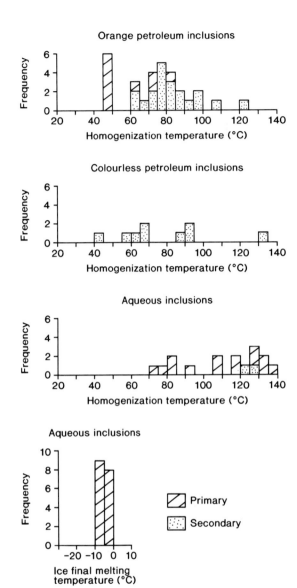

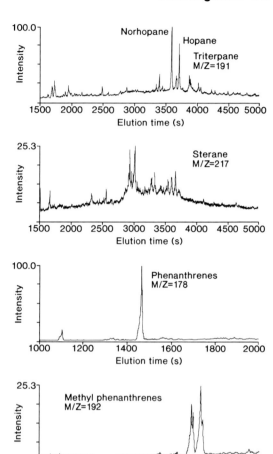

Fig. 7.3 Microthermometric measurements on fluid inclusions in fracture-filling calcites. The upper three histograms show homogenization temperature measurements for the three types of inclusion. The lowest histogram is of ice final melting temperatures in the aqueous inclusions.

Fig. 7.4 GCMS fragmentograms of petroleum in fluid inclusions.

are likely to have come from the same Mesozoic carbonate source rock. Molecular maturity parameters calculated from the GCMS fragmentograms do not constrain the temperature of oil expulsion from the source rock with any great accuracy, but suggest a figure between 110 and 140°C.

What can we make of these data? It would be a brave person who would ascribe much significance to the homogenization temperatures. The wide spread of the data and the fact that the mineral host is calcite suggest that the aqueous inclusions may well have leaked. This may also be true of the petroleum inclusions. Two features of the data in Fig. 7.3 however suggest that the homogenization temperatures of at least the orange petroleum inclusions *may* still be reflecting trapping temperatures: the fairly tight distribution of T_h for primary inclusions and the different T_h distributions for primary and secondary inclusions. Even if this were true, trapping temperatures would be difficult to estimate because the (probably substantial) pressure correction cannot be accurately determined. If the aqueous inclusions have leaked, ice final melting temperatures may still faithfully record salinity. Two features of the data suggest an association between low salinity (indeed virtually fresh) water and the orange petroleum:

1 the restriction of low salinity aqueous inclusions to Type D fracture-fills which contain abundant orange petroleum inclusions; and

2 the existence of water + orange petroleum + vapour inclusions (water was never observed in the colourless petroleum inclusions).

The orange colour of the petroleum in these inclusions may well be caused by removal of light alkanes by fresh water (a process known as *water washing*). This could also dissolve the aromatic hydrocarbons that are responsible for fluorescence, explaining the low intensity of fluorescence shown by this petroleum when illuminated by UV light.

7.2.5 Conclusions

Several conclusions can be drawn from the fluid inclusion study. The most obvious is that the carbonate source rock that was responsible for filling other producing oilfields in the basin was also producing oil in the area in which the sample was picked up, where there was virtually no rock exposed at surface and no drill core. In this particular case, the same conclusion could have been drawn simply from the presence of bitumen in the pores. However, pore-filling bitumen can sometimes be so heavily biodegraded in outcrop samples that correlation with other oils is impossible; it may even be completely removed by weathering. Fluid inclusions would then have been the only way of demonstrating a working oil source system. The second significant conclusion is the recognition of water washing from the association of orange (probably altered) petroleum inclusions with fresh water. This suggests that at least one migration pathway was connected to surface and raises concerns about the quality of any petroleum that might be found in the area and also about seal effectiveness. The presence of bitumen in the pores does not demonstrate water washing during some past phase of oil migration because it might equally well be, and probably largely is, a product of recent near-surface weathering.

As is sometimes the case, the postscript is somewhat disappointing. BP is no longer exploring in the Aquitaine Basin.

7.3 Prediction of the occurrence of diagenetic celestite cap rock: Central North Sea

7.3.1 Introduction and approach

Those who have read Chapter 6 will remember the Machar Field in the Central North Sea. Those who have not should read about the geology of the field before embarking on this case history. It is based on a geochemical study of an unusual diagenetic reservoir rock composed almost entirely of celestite (strontium sulphate). Although the bulk of the reserves in Machar are in fractured Chalk, some oil is contained in a highly porous and permeable celestite 'cap rock' which is found between the Chalk and a unit of Zechstein anhydrite that rests above the salt dome (see Figs 6.38 & 6.39). DST no.1 in well 23/26a-6Y tested oil at $7000 \, \text{bbl.day}^{-1}$ from the celestite.

The high permeability of the celestite cap rock together with the presumed large area of its interface with the Chalk led to the suggestion that by completing production wells within the cap rock, it might be possible to drain the Chalk reservoir more effectively than would be possible by completing within the zones of fractured Chalk. The celestite therefore assumed an importance disproportionate to the small proportion of the field's reserves that it contains.

We set out to predict which of the similar untested prospects within the Central North Sea might have a celestite cap rock. Our approach was to analyse samples of cap rock for strontium, oxygen and sulphur isotope compositions and to study its fluid inclusion contents in the hope that we might be able to identify the cause of celestite cap rock formation.

7.3.2 Conditions and cause of celestite precipitation

Table 7.2 lists strontium, sulphur and oxygen isotope analyses of celestite from well 23/26a-1. The isotopic ratio of the strontium provides information about the source of the strontium and the isotopic ratios of sulphur and oxygen do the same for the sulphate. $\delta^{18}O$ lies between 13.3 and 13.9‰ and $\delta^{34}S$ between 14.8 and 15.6‰. $^{87}Sr/^{86}Sr$ is between 0.70784 and 0.70789. Also listed in the table are analyses of samples of anhydrite dissolved out of core samples of halite. These are Permian Zechstein evaporites and have quite distinct isotope ratios. $\delta^{18}O$ lies between 9.6 and 10.7‰ and $\delta^{34}S$ is about

Table 7.2 Isotopic composition of celestite cap rock and anhydrite in salt.

Depth (mBRT)	$\delta^{34}S$ (‰ CDT)	$\delta^{18}O$ (‰ SMOW)	$^{87}Sr/^{86}Sr$
Celestite			
1481.1	15.6	13.7	0.70788
1481.7	15.2	13.4	0.70788
1481.85	14.9	13.4	0.70787
1482.3	15.4	13.9	0.70784
1482.5	14.8	13.3	0.70789
Anhydrite			
1520.0	10.9	10.7	0.70708
1525.0	10.8	9.6	0.70706
1534.0	10.9	10.4	0.70712

Celestite samples from well 23/26a-1; anhydrite from well 23/26a-5.

10.9‰; $^{87}Sr/^{86}Sr$ is between 0.70706 and 0.70712.

The strontium in the celestite is much more radiogenic than Zechstein strontium and must have largely come from another source with $^{87}Sr/^{86}Sr > 0.70789$. This value also precludes unaltered Chalk which has $^{87}Sr/^{86}Sr < 0.70774$ (see Fig. 6.47). The most likely candidate is a formation water that had passed through clastic rocks and obtained its strontium by dissolving old and/or Rb-rich detrital minerals such as micas or feldspars. The sulphate in the celestite cannot have come entirely from Zechstein evaporites either, as both $\delta^{18}O$ and $\delta^{34}S$ are 4–5‰ heavier (more positive) than Late Permian seawater sulphate (Claypool et al., 1980).

The fluid inclusions found in celestite fracture fills and cap rock were described in Section 6.7.4. There are primary aqueous and petroleum fluid inclusions that suggest growth of the mineral from a highly saline water (average 16 eq. wt%NaCl) in the presence of a mature petroleum, at temperatures between about 90 and 115°C.

The geochemical evidence shows then that celestite grew at high temperatures from a saline water that contained radiogenic strontium and isotopically heavy sulphate, neither of which could have had a local origin. The presence of primary fluid inclusions containing oil sourced by the Kimmeridge Clay Formation indicates a connection between the top of the salt dome and the deeply buried Jurassic section and suggests that the water that caused celestite cap rock to form could have originated in the Mesozoic, off the flanks of the dome. But what caused the

mineral to precipitate? Celestite solubility may be written:

$$SrSO_4 = Sr^{2+} + SO_4^{2-}$$

$$K_{sp} = [Sr^{2+}] \cdot [SO_4^{2-}] \text{ at equilibrium}$$

K_{sp} is the solubility product; $[Sr^{2+}]$ and $[SO_4^{2-}]$ are the ionic activities of strontium and sulphate. Celestite will precipitate from solution if the actual ionic activity product (IAP) $> K_{sp}$. This can come about in only a limited number of ways.

1 Change in K_{sp}. Pressure or temperature variation can cause K_{sp} to fall below IAP and precipitation will ensue.

2 Change IAP. (a) Changing activity coefficients (factors which relate activity to concentration) by altering ionic strength (equivalent to salinity).

3 Change IAP. (b) Increasing the concentrations of dissolved sulphate and/or dissolved strontium so that IAP increases.

Simply moving a formation water to a structurally higher position above the salt dome at lower P and T (**1**) cannot cause celestite precipitation because, for celestite, K_{sp} increases with decreasing temperature. Dissolution would be more likely. Increasing the ionic strength of the formation water through salt dissolution (**2**) is not a plausible precipitation mechanism either as this too would increase celestite solubility. There is however a geologically plausible and simple way in which a formation water arriving at the top of the Machar dome could have acquired additional sulphate and some strontium (**3**). If such a water were undersaturated with anhydrite, it would begin to dissolve the anhydrite that caps the halite: Ca^{2+} and SO_4^{2-} as well as some Sr^{2+} would go into solution and cause celestite − orders of magnitude less soluble than anhydrite − to precipitate. This mechanism is compatible with the isotopic constraint that at least some of the strontium and the sulphate of which the celestite is composed was not locally derived.

7.3.3 Simulation of celestite precipitation

To evaluate the suggested cause of celestite precipitation, we simulated reaction of 20 Central North Sea Mesozoic formation waters with anhydrite, using a computer program for the calculation of chemical equilibria*. The calculations were carried out ·at

* We used PHREEQE (Parkhurst et al., 1980). The 20 waters were selected from a total database of 113 using criteria specified in advance for identifying credible analyses.

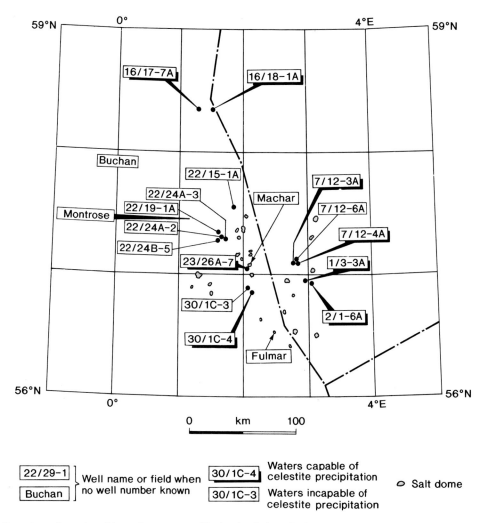

Fig. 7.5 Locations of samples of formation water used in the simulation of celestite precipitation.

100°C to honour the fluid inclusion data and at 1 bar total pressure (for convenience only; pressure has little effect on the results). The first stage in the simulation was equilibration of the water with anhydrite and barite. This is necessary to remove barium from solution: barite is so insoluble that if any Ba^{2+} is present in the water, as soon as anhydrite dissolves, barite will precipitate instead of celestite. The second stage is reaction of the water with anhydrite and celestite.

Of the 20 waters used in the simulation, 12 precipitated celestite upon reaction with anhydrite (Fig. 7.5; Table 7.3). The formation water that precipitated most celestite per unit volume was from

well 23/26a-7, the closest well to the Machar dome, which dissolved 2077 mg.l^{-1} of anhydrite and precipitated 2810 mg.l^{-1} of celestite. Celestite is considerably denser than anhydrite so this reaction in fact creates porosity (which may explain the high porosity of the celestite cap rock).

7.3.4 Conclusions

The geochemical study of the Machar celestite cap rock has identified the probable mechanism by which this bizarre diagenetic reservoir rock formed. By simulating the reaction using chemical analyses of real formation waters from the area of explora-

Table 7.3 Results of celestite precipitation simulations.

Well	Area	Field	Ionic strength	Celestite precipitated (mg.l^{-1})	Anhydrite dissolved (mg.l^{-1})
1/3-3a	NOCS		6.3	197	146
2/1-6	NOCS		6.0	660	489
7/12-3a	NOCS		7.1	625	463
7/12-4a	NOCS		5.3	992	732
7/12-6a	NOCS		1.9	−10	−7
16/17-7a	UKCS		2.3	1117	800
16/18-1a	UKCS		1.3	68	39
16/3-?	UKCS	East Brae	2.0	0	0
21/1-?	UKCS	Buchan	4.0	1387	1018
22/15-1a	UKCS		4.9	−299	−221
22/18-?	UKCS	Montrose	2.0	1162	861
22/18-?	UKCS	Montrose	2.0	1415	1017
22/19-1a	UKCS		8.9	−715	−529
22/24a-2	UKCS		3.6	−345	−641
22/24a-3	UKCS		0.1	−18	−10
22/24b-5	UKCS		10.2	−2300	−1703
23/26a-7	UKCS		4.7	2810	2077
30/16-?	UKCS	Fulmar	3.0	1061	769
30/1c-3	UKCS		3.7	125	92
30/1c-4	UKCS		0.2	−15	−934

tion interest, it has proved possible to begin to delineate parts of the basin where a cap rock of this type could, and where it could probably not, have formed. Although we do not have retrospective knowledge about the actual distribution of celestite cap rocks in the Central North Sea, in our view this represents an almost uniquely successful outcome to a geochemical project. Unfortunately, we cannot claim all the credit. The success is due in large measure to the fact that celestite is composed entirely of geochemical tracers, a geochemist's dream. In the case of celestite, strontium and sulphate are not just present in trace amounts but are the major components of the mineral. Furthermore, celestite chemistry is simple so that there is only a small number of processes that can cause it to precipitate. Pulling the same rabbit out of the hat for other diagenetic minerals is far from easy.

7.4 Regional mapping of migration pathways: Weald Basin, onshore UK

7.4.1 Introduction and approach

The Weald Basin is located in south-east England, south of the London–Brabant Massif (Fig. 7.6). It contains a number of small oil accumulations

sourced by mudstones of the Lower Jurassic Lias and mainly reservoired in the Bathonian Great Oolite. During the late 1980s, there was a considerable amount of exploration interest in the Weald even to the extent of drilling close to the author's home in Guildford and in Winnie the Pooh's wood in Sussex (Milne, 1928). Oil accumulations do not however appear to be distributed throughout the basin. The case study that we will describe here involves an attempt to explain the distribution of oil pools by defining migration pathways based on the presence or absence of petroleum fluid inclusions in late diagenetic ferroan calcite cement (McLimans & Videtich, 1987, 1989).

7.4.2 Geological background

Most of the oil in the Weald Basin is contained in carbonate grainstones of the Great Oolite Formation. This unit comprises a number of facies of which the best reservoirs are oolitic grainstones which are widely distributed across the basin (Fig. 7.6; Sellwood *et al.*, 1989). Early diagenesis in the meteoric environment is thought to have influenced porosity by stabilizing the ooids through precipitation of a fringe cement and by causing alteration of aragonite and high-Mg calcite grains to low-Mg

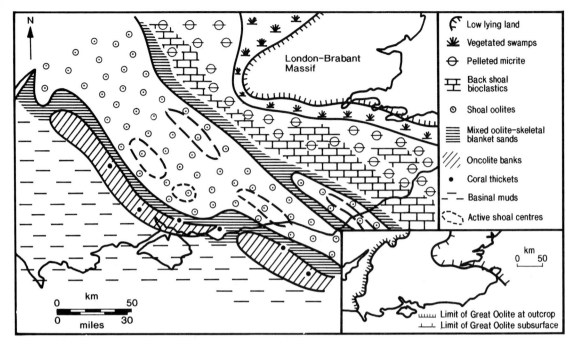

Fig. 7.6 Great Oolite facies distribution in the Weald Basin (from Sellwood *et al.*, 1989).

calcite. Porosity is however largely controlled by a sparry calcite cement which is slightly ferron and which is the latest cement to have formed.

7.4.3 Fluid inclusions in ferroan calcite cement

McLimans and Videtich (1987) collected 75 grainstone samples from 11 wells. The latest, ferroan calcite cement contained aqueous and petroleum fluid inclusions. Most inclusions were either aqueous or petroleum but a few contained both liquids (as well as the customary vapour bubble). Sometimes, the inclusions were clearly secondary; sometimes they appeared to be primary. The preferred interpretation of the textural relationships was that precipitation of the calcite cement and oil migration were broadly contemporaneous.

Homogenization temperatures for fluid inclusions ranged from below 50 to over 130°C (Fig. 7.7). The authors presented a smoothed frequency distribution representing nearly 600 T_h determinations, but without making it clear which inclusions were represented (aqueous *and* petroleum in one histogram?) and at the same time admitting that the measurements differed from well to well. This makes it impossible to evaluate the subsequent

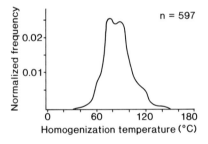

Fig. 7.7 Homogenization temperature measurements made on fluid inclusions in late calcite cement (from McLimans & Videtich, 1987).

claim that the inclusions suggest calcite precipitation at temperatures between 80 and 105°C (those who have read Chapter 3 may be unwilling to take this on trust). Nonetheless, an important conclusion did come out of this comprehensive survey and was drawn simply from the presence or absence of petroleum inclusions. It proved possible to map areas with and without these inclusions in the late calcite cement. Only in the south-western part of the basin did the calcite consistently contain petroleum inclusions (Fig. 7.8). This part of the basin was mature

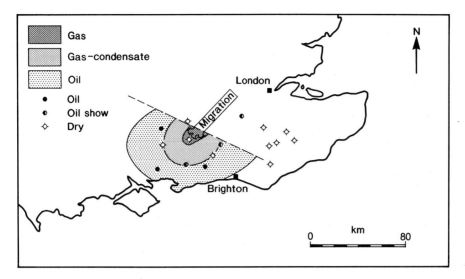

Fig. 7.8 Map of Weald Basin showing location of sampled wells and migration path based on fluid inclusion distribution (from McLimans & Videtich, 1989).

for oil at the end of the Cretaceous and deeper than the eastern part and is where virtually all the oil accumulations have been found.

7.4.4 Conclusions

The coincidence of oil-producing wells with late calcite cements that contain petroleum inclusions suggests that it is possible to identify petroleum migration pathways on a regional scale by mapping the occurrence of the right type of fluid inclusions. The exploration significance of this study in the Weald Basin was presumably limited by the need to drill a large number of wells to collect the samples to make the map. The technique might however be far more useful in surface mapping of poorly explored basins where inclusions may provide the only samples of migrated petroleum not oxidized by weathering (see also the Aquitaine Basin example in Section 7.2).

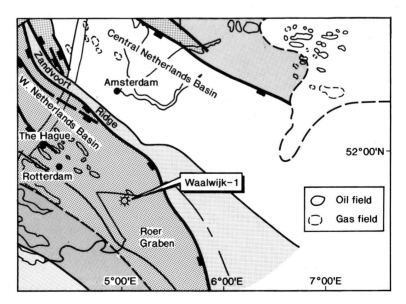

Fig. 7.9 Location map and structural elements, Roer Graben.

7.5 Filling history of a reservoir: Waalwijk, onshore Netherlands

7.5.1 Introduction

The Roer Graben is in the southern part of The Netherlands (Fig. 7.9, p. 181). One of the keys to prospectivity in the graben is the timing of structure formation relative to oil generation. The age of oil generation is usually determined by combining a burial history for a well or depth-converted seismic shot point with a kinetic description of kerogen breakdown (Quigley & Mackenzie, 1988). In the Roer Graben, this is not straightforward for two reasons. Firstly, increased heat flow associated with rifting in the mid-Jurassic about 160 Ma ago may have raised temperatures enough to cause petroleum generation from the Lias source rock. Secondly, Early Tertiary inversion led to the erosion of a substantial amount of Cretaceous sediment so that it is difficult to estimate the highest temperatures attained prior to erosion.

Waalwijk-1, an exploration well drilled in 1987

discovered gas-condensate reservoired in Triassic Bunter sandstone. The uncertainties associated with modelling the timing of oil generation meant that the reservoir could have received its oil charge in the Middle Jurassic during rifting, in the Late Cretaceous before inversion or in the Late Tertiary to Recent. We set out to determine whether there was any feature of the diagenesis of the Bunter sandstones in Waalwijk that could constrain the timing of petroleum filling.

7.5.2 Geological background

The Roer Graben is the south-easterly extension of the West Netherlands Basin. During deposition of the Triassic Bunter sandstones, the graben was bounded to the south by the Brabant Massif and to the east by the Rhenish Massif. The Bunter Formation, which includes the main reservoirs, comprises conglomerates, sandstones and mudstones deposited by laterally extensive sheet-floods. The structure drilled at Bunter level is an elongate tilted fault block (Fig. 7.10). Upper Triassic mudstones, dolomites

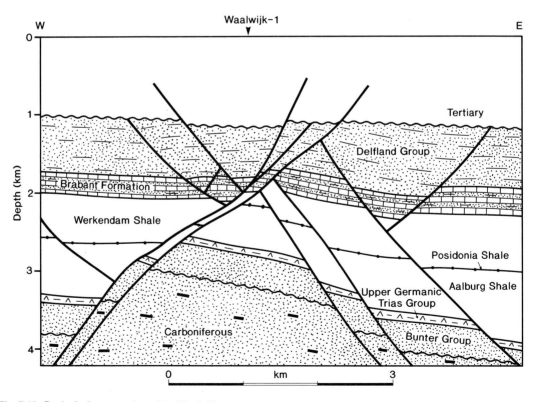

Fig. 7.10 Geological cross section of the Waalwijk structure.

and anhydrites of the Muschelkalk Formation provide a top seal with cross-fault seal provided by the Jurassic Aalburg Shale. Gas is sourced by underlying Carboniferous Coal Measures.

7.5.3 Approach

We first carried out a detailed petrographic study to identify the diagenetic minerals and events represented in the Bunter sandstone reservoir and to establish their timing relative to one another. We then decided which geochemical methods should be applied to which mineral cements in an attempt to affix some absolute ages to the relative ages of the paragenetic sequence in the hope that this would in turn constrain the age of phases of oil migration. Illite cement turned out to be quite abundant in the sandstone and could be dated using the K–Ar method. No other cement could be dated directly so we attempted to constrain precipitation temperatures by analysing the stable isotope ratios of illite, non-ferrous and ferroan dolomite cements and present-day formation water.

7.5.4 Petrography, K–Ar illite ages and dolomite stable isotope ratios

Figure 7.11 shows the paragenetic sequence for the Bunter sandstone established by thin-section petro-

graphy. Filling of the pores with gas-condensate post-dates all of the features on the diagram. The most significant observation from the point of view of petroleum filling is the identification of bitumen at a consistent point in the paragenesis. The bitumen post-dates illite cement growth but is overgrown by ferroan carbonate cements. Bitumen emplacement cannot itself be dated but its age can be bracketed if earlier illite and later carbonate cements can be dated.

Clay mineral separates were prepared from three samples for K–Ar dating and oxygen and hydrogen

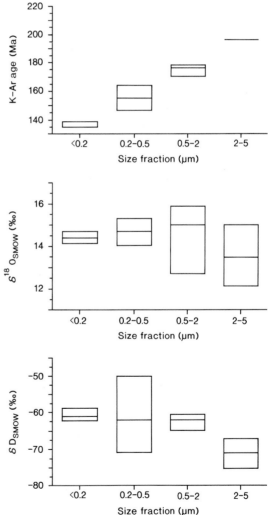

Fig. 7.12 K–Ar age, oxygen and hydrogen isotope ratios of clay separates from Bunter sandstones.

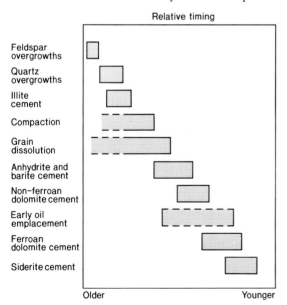

Fig. 7.11 Paragenetic sequence for the Bunter sandstone reservoir.

Table 7.4 Analytical data for clay separates, Waalwijk-1.

Depth (mBRT)	Size fraction (μm)	K–Ar age (Ma)	2σ (Ma)	δ¹⁸O (‰ SMOW)	δD (‰ SMOW)
3482.6	<0.2	134	3	14.8	−58
	0.2–0.5	155	4	13.8	−74
	0.5–2	168	4	11.9	−66
	2–5			15.5	−71
3491.4	<0.2	135	3	14.4	−63
	0.2–0.5	167	15	14.7	−46
	0.5–2	179	4	16.2	−62
	2–5	179	17	11.6	−77
3511.5	<0.2	140	3	14.0	−61
	0.2–0.5	143	3	15.5	−62
	0.5–2	176	3	15.0	−60
	2–5	214	21	13.5	−66

isotope analysis and the results are shown in Fig. 7.12 (p. 183) and Table 7.4. Sample purity was monitored by TEM. The <0.2 μm size fractions contained only elongate illite fibres that appeared to be diagenetic while coarser fractions were invariably mixtures of the same fine fibres with platy and ragged illite crystals, probably detrital. Coarser separates have distinctly older K–Ar ages which also indicates the presence of old detrital minerals. The highly precise ages of the <0.2 μm size fractions from the three samples are close together and may reasonably be interpreted to indicate an age of about 130–140 Ma for illite cement growth (early Cretaceous). This represents a maximum age for early oil emplacement.

To estimate a minimum age for early oil emplacement, we had to adopt a rather different approach involving interpreting mineral oxygen isotope ratios in terms of temperature. To do this, we need to know or be able to constrain the oxygen isotope ratio of the water or waters from which the minerals precipitated. This cannot be done for any one mineral in isolation. Instead, we need to build up a picture of how the porewater isotope ratio has evolved through time using information from both hydrogen and oxygen isotope ratios of diagenetic minerals and of present-day formation water.

Dolomite oxygen and carbon isotope ratio measurements were made on small samples of whole rock selected for their high dolomite content. All samples contained both dolomite and ferroan dolomite so that measured ratios are of mixtures, intermediate between the true values of the minerals (Fig. 7.13). A plot of δ¹⁸O against Fe content

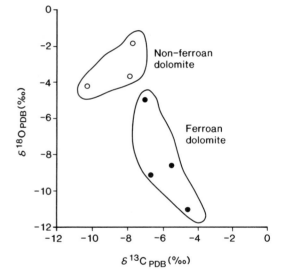

Fig. 7.13 Carbon and oxygen isotopic compositions of dolomite cements.

suggests that a ferroan dolomite end-member has δ¹⁸O < −11.1‰ and a non-ferroan end-member δ¹⁸O > −1.9‰ (Fig. 7.14). Figure 7.15 is a plot of water oxygen isotope ratio against temperature. Mineral oxygen isotope ratio is a function of these two variables so a measured mineral δ¹⁸O value plots as a curve. The curves represent the range of illite and ferroan and non-ferroan dolomite δ¹⁸O measurements. There are two other points on the graph. The present-day formation water (δ¹⁸O = −3.5‰) is plotted at the present reservoir temperat-

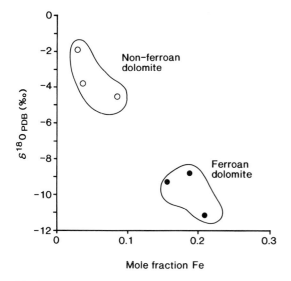

Fig. 7.14 Oxygen isotopic composition versus mole fraction Fe for dolomite cements.

ure of 118°C. This is the end point of any formation water evolution path. The original $\delta^{18}O$ of the water from which illite precipitated is calculated from the illite hydrogen isotope ratio of about −60‰. Waters in equilibrium with illite are about 30‰ heavier; a meteoric water with δD = −30‰ would have $\delta^{18}O$ = −5‰ (see Section 4.3.1). The area on the graph representing the conditions of illite cement growth is further constrained by estimating the temperature in the Bunter during the Lower Cretaceous (70–80°C). Waters in sedimentary basins tend to evolve towards more positive $\delta^{18}O$ because they undergo isotope exchange with minerals which are relatively isotopically heavy. The simplest path for water isotopic evolution is therefore some kind of curve which moves towards the right from waters present during early diagenesis and passing through areas representing mineral growth conditions, ending at present-day formation waters. This path is shown as a broad arrow on Fig. 7.15.

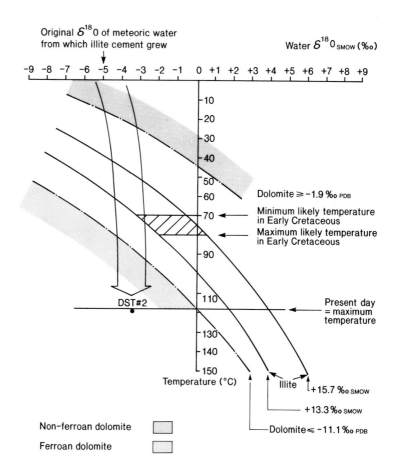

Fig. 7.15 Isotopic evolution of porewaters. See text for explanation.

The water evolution path can be used to estimate precipitation temperatures for minerals whose oxygen isotope ratio has been measured. For ferron dolomite, a figure above about 80°C is indicated; for non-ferroan dolomite, a figure less than about 25°C. Note that the latter is at odds with the position of non-ferroan dolomite in the paragenesis (Fig. 7.11). In fact, the path for water isotope evolution cannot be as simple as that shown on Fig. 7.15. For a start, there was inversion in the Early Tertiary which would have caused cooling so that the evolution path would have to backtrack at some point and move vertically back up the diagram. Furthermore, in an area with a complex geological history like the Roer Graben, we cannot be at all certain that the same formation water has been in the Bunter Sandstone, gently evolving for 200 Ma. These uncertainties seriously limit the value of oxygen isotope ratios as palaeo-temperature indicators in this and many similar cases.

7.5.5 Conclusions

K–Ar dating of illite cement in the Bunter sandstone of the Waalwijk gas-condensate reservoir plausibly indicates a maximum age of Early Cretaceous for early filling of the structure by petroleum. This information partly answers the question set at the start of the study. An Early Cretaceous age for the illite – which itself pre-dates early oil emplacement and carbonate cementation – shows that the gas-condensate now in the reservoir was not generated during the mid-Jurassic. It cannot distinguish between possible Late Cretaceous and Late Tertiary episodes of generation. The status of the bitumen remains somewhat unclear. It may represent oil generated in the Late Cretaceous that was subsequently flushed out only for the structure to be filled by gas-condensate in the Late Tertiary, after dolomite cementation (this is however somewhat speculative).

Attempts to constrain the diagenetic and filling history of the Bunter sandstone more accurately by stable isotope analysis of mineral cements were not particularly successful. As is so often the case, this was due to the uncertainty associated with estimating water isotope composition. If we were undertaking the analytical programme today, we would use a laser for isotopic analysis of the carbonates. This would provide a much more detailed and accurate picture of the isotopic composition of the several minerals present in the sandstone. Nonetheless, it would probably not provide better estimates of cement growth temperature because however accurate $\delta^{18}O$ measurements may be, there will remain the unquantifiable uncertainty about water $\delta^{18}O$. The accuracy of the temperature estimates depends entirely on assumptions that are virtually impossible to assess.

Chapter 8 Correlation

8.1 Introduction

Correlation can be thought of as the linking of rock units that have something in common. Usually, the linkage involves common properties such as age, lithology and reservoir parameters such as porosity, permeability and reservoir pressure. In this chapter, we will also consider the correlation of clastic rock units with their sediment source areas (*provenance studies*) and the correlation of the isotopic properties of rock units with known global curves (or *secular* curves) developed for several isotopic systems throughout the Phanerozoic.

The objective of this chapter is to introduce the rationale behind correlation of rock properties and to outline the essential information a good correlation can provide for petroleum exploration and production. The bulk of the chapter comprises a series of case studies demonstrating the effectiveness (or otherwise) of inorganic geochemistry in helping to correlate. Inorganic geochemical techniques can assist in reservoir correlation on all scales, from exploration through to development and production. The first case history in this chapter describes the use of strontium isotope stratigraphy to improve the correlation of the Tertiary sediments in the Norwegian North Sea. The second describes how oxygen isotopes have been used in much the same way in the Gulf of Mexico, though only in the (admittedly thick) Plio-Pleistocene. The last two case histories are on a smaller scale. The first of these describes the use of strontium isotope analyses of rock *and* formation water to understand how the Chalk reservoir in the Ekofisk Field (Central North Sea) is compartmentalized. The final case history in this chapter — a study of the Gullfaks Field in the Northern North Sea — shows how Sm–Nd isotopes can provide qualitative information on the continuity of sandbodies deposited within the same sediment dispersal systems and supplement deterministic information available from logs and core.

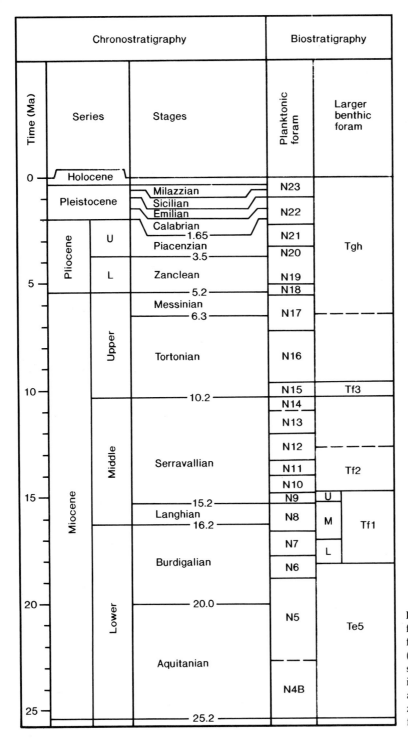

Fig. 8.1 Biostratigraphic schemes for planktonic and larger benthic foraminifera, Miocene to Holocene (from Haq *et al.*, 1987). Miocene shallow water carbonate platforms in Indonesia tend not to contain abundant planktonic forams. The zonation scheme based on benthic forams has a poorer resolution.

8.2 Stratigraphic correlation

Stratigraphic correlation is the linking together of rock units of similar age. In the oil industry, the most commonly applied stratigraphic techniques are seismic stratigraphy (Vail *et al.*, 1977) and, where rock samples are available, biostratigraphy. Magnetostratigraphy (which requires core samples) and isotope stratigraphy (also requiring rock samples) are less commonly used.

Seismic stratigraphy is arguably one of the most exciting developments in petroleum geoscience in the last two decades (Vail *et al.*, 1977; Berg & Woolverton, 1985; Wilgus *et al.*, 1988). It is a methodology which can be used to interpret seismic in the absence of well data and relies on recognizing key geometries at seismic reflector terminations, relying on the important (and not entirely proven) hypothesis that most seismic reflectors correspond to stratigraphic time-lines. Where rock (usually well) samples are available, biostratigraphy has been and will continue to be the standard method of stratigraphic correlation. Industrial biostratigraphy is almost entirely based on microfossil zonation schemes because it is not usual to drill through and core an ammonite. Furthermore, many microfossil species, particularly pelagic forms, evolved relatively rapidly, permitting fine stratigraphic subdivision in many parts of the Mesozoic and Tertiary. Micropalaeontology can also provide information about depositional environments (Hedgpeth, 1957).

Biostratigraphy suffers from several drawbacks, both intrinsic and practical. The main intrinsic difficulties arise when rock successions have no age-diagnostic fossils and where fossil assemblages are strongly facies- or province-dependent. An example of poor stratigraphic resolution resulting from the absence of age-diagnostic fossils and facies-dependence is the benthic foraminiferal scheme for the Miocene (Fig. 8.1). If, for example, we were to drill wells on an Early Miocene carbonate platform in Indonesia, where the bulk of the microfauna would comprise benthic foraminifera, our resolving power would only be about 3–4 Ma on average, and our biostratigraphic correlation would be correspondingly imprecise. A practical difficulty with biostratigraphy, particularly in the petroleum industry, is the absence of a uniform system of criteria to recognize series or subseries boundaries (Goll & Skarbo, 1990). This would not matter if a single analyst performed all the microfossil identifications but if, as is more usual, a whole range of

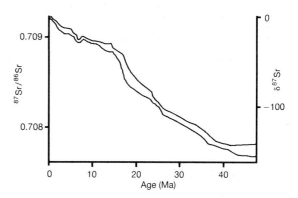

Fig. 8.2 Strontium isotope composition of seawater over the last 40 Ma.

independent microfossil experts is used, results may not always be easily comparable.

Isotope stratigraphy can provide a tool for stratigraphic correlation when the intrinsic and practical problems associated with conventional biostratigraphic correlation are severe. Taking the example of the Early Miocene carbonate platform in Indonesia, if we were to sample diagenetically unaltered carbonate fossils of any type and determine their $^{87}Sr/^{86}Sr$ isotopic ratio, we could potentially date most of the samples to ±1 Ma using the Sr isotopic evolution curve for Tertiary seawater (Fig. 8.2). This would permit good cross-platform correlation. If the carbonate platform happened to be of Late Miocene age, however, the flattening of the isotope evolution curve and the 'u' shape at approximately 7 Ma would render this method less appropriate because its precision would be poor (see also Fig. 5.21). There would also be little value in using the Sr isotope method on samples from which no other information is available because the Sr isotope curve for the whole of the Phanerozoic contains several rises and falls of the ratio (Section 5.5.5; Burke *et al.*, 1982). An $^{87}Sr/^{86}Sr$ measurement of 0.708000, gives eight possible ages ranging from Ordovician through to Tertiary! A further problem with strontium isotope stratigraphy (and indeed any isotope stratigraphic method) is the impact of diagenesis on the depositional isotopic signature. Dissolution, neomorphism and cement precipitation may have a profound effect on the original isotopic signature, and petrographic screening of sample material prior to analysis is an essential prerequisite.

Oxygen, carbon and sulphur isotopes can also be used for stratigraphic purposes. All require sample material of diagenetically unaltered marine origin,

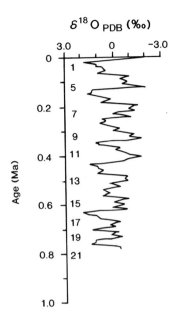

$\delta^{18}O$ PDB (‰)

Fig. 8.3 Oxygen isotopic composition of seawater is recorded by $\delta^{18}O$ of marine microfossils over the last 0.8 Ma. Negative isotopic compositions correspond to interglacials, positive isotopic compositions correspond to glaciations. Numbers 1 to 21 refer to isotopic stages.

either as fossils or as inorganic precipitates from seawater. Oxygen isotope stratigraphy has been very successfully applied in sediments deposited in the glacial–interglacial rhythms of the last 4 Ma or so. One hundred and fifty two oxygen isotopic stages have been recognized globally, giving an average resolution of 26 000 years (Trainor et al., 1988). As well as providing very high resolution, the method can be used to relate sediment flux to episodes of glaciation or deglaciation, providing an insight into the controls on sediment transport. By sampling and analysing both pelagic and benthic microfossils, clues to ancient bottom water circulation and seawater stratification can be obtained. Two major disadvantages with oxygen isotope stratigraphy are the nature of the record itself and the impact of diagenesis on isotope compositions. Because the record of seawater oxygen isotope compositions fluctuates between small negative and small positive values, the composition at any one time is not unique (Fig. 8.3). Unless a continuous sample record is available, oxygen isotope stratigraphy cannot be used alone as a stratigraphic tool and requires a coarse calibration with biostratigraphy or strontium isotopes. Diagenetic alteration will also

render the oxygen isotopic method virtually useless where complete or even partial alteration of the sediments has taken place. Oxygen isotope stratigraphy is usually restricted to deeper-water Pliocene to Recent deposits, largely because the effects of diagenesis obscure the depositional record before this time and because the amplitude of glacio-eustatic sea-level fluctuations − and hence the range of oxygen isotopic compositions − was less extreme prior to Plio-Pleistocene glaciations.

Carbon isotopes can also be used as a stratigraphic tool, although they are rather more difficult to interpret than oxygen isotopes (e.g. Leggett, 1985). The carbon isotopic composition of seawater recorded by calcareous tests of marine organisms results from the interplay of two main processes: fixation of ^{13}C-rich carbon in carbonate rocks and of ^{12}C-rich carbon in marine organic sediments produces a cyclicity rather like that displayed by oxygen isotopes. The record includes features such as the terminal Miocene event related to the desiccation of the Mediterranean and carbon isotopic excursions related to oceanic anoxic events (Scholle & Arthur, 1980). These can be used for a coarse stratigraphic correlation. As with oxygen isotopes, however, the carbon isotopic composition of sediments may be greatly affected by diagenetic processes and any carbon isotopic analysis for stratigraphic purposes must take account of dissolution, neomorphism and precipitation of later cements.

The marine sulphate sulphur and oxygen isotope secular curves are less well constrained than seawater oxygen and carbon curves, though they do go back further. They can only be used for very coarse correlation of marine evaporites (Section 4.5; Claypool et al., 1980).

One of the more obvious ways of using isotopes to correlate stratigraphically is by obtaining absolute ages of sedimentation by using radiogenic isotopes. Unfortunately, radiometric clocks are not generally reset at sediment deposition; mineral ages will tend to reflect either provenance or diagenesis and whole rock ages will generally be pretty meaningless. One mineral which can be used to date sediments directly is glauconite which precipitates in the marine environment, and is found in both siliciclastic and carbonate rocks. Glauconite can be dated using both Rb–Sr and K–Ar systems (Section 5.5.6) but is susceptible to diagenetic alteration, particularly by freshwater flushing, so that care must be taken in interpreting glauconite ages if subaerial exposure of the sediment is believed to have occurred.

Recently, attempts have been made to date the deposition of Precambrian and Palaeozoic carbonates using the Pb–Pb technique (Moorbath *et al.*, 1987). This has apparently been quite successful on Precambrian stromatolites, although it is difficult to know whether the dates are genuinely depositional and of stratigraphic significance, or whether diagenesis is being recorded.

The appropriate techniques to apply for stratigraphic correlation in petroleum geology are critically dependent on the nature of the material available and the depositional environment and broad stratigraphic age of the sediments of interest. If we only have seismic lines available, clearly seismic stratigraphy is our only tool. If we have seismic and well data available, we can choose between dating the seismic reflectors biostratigraphically and/or using isotope chronostratigraphy. In areas where age-diagnostic fossils are present

and assemblages are unlikely to be affected by provinciality or be strongly facies-dependent, biostratigraphy should be used. Where these criteria are not fulfilled, then Sr isotope stratigraphy can be used for correlation. For high resolution stratigraphic correlation in the Plio-Pleistocene, oxygen isotopes coarsely calibrated by biostratigraphy can be used. Direct radiometric dating of sediments is highly dependent on the presence of unaltered glauconite. One final cautionary note: any attempt at stratigraphic correlation using isotopes ignores the effects of diagenetic alteration at its peril!

8.3 Lithological and reservoir property correlation

Correlation of lithological and reservoir properties is usually an appraisal and development problem, following on from the discovery of a petroleum

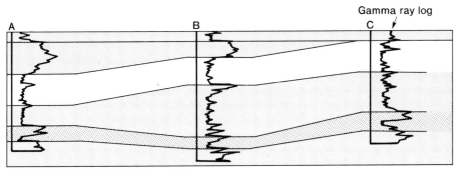

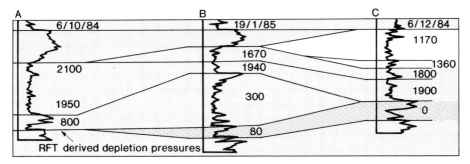

Fig. 8.4 Comparison of lithostratigraphic and pressure correlation in a reservoir from the Northern North Sea, UKCS. The hydrodynamic correlation is based on reservoir pressure data measured using the RFT (repeat formation tester) tool (Dake, 1982). The lithostratigraphic correlation is not justified by the pressure data. This does not mean that the lithostratigraphic scheme is incorrect, but there is no reason to correlate the interval as a continuous *reservoir zone*. The lithostratigraphic correlation of the lower stippled zone is supported by pressure data between wells B and C, but not between wells A and B.

accumulation. Typical correlation problems may involve predicting the distribution of the main reservoir sand, predicting the lateral distribution of horizontal barriers to flow such as mudstone beds or carbonate-cemented layers, or determining whether faulting is compartmentalizing an oil or gas field. All will exert a major influence on the continuity of petroleum filled intervals and on the behaviour of the underlying aquifer.

The nature of the problem to be tackled largely determines the appropriate correlation method. Although it is preferable to tackle almost any correlation problem within a stratigraphic framework, if a reservoir is significantly compartmentalized by faults or if lithologies and facies within a stratigraphic interval are highly variable or difficult to date, a stratigraphic correlation will probably be unable to provide adequate resolution and may be of limited use. If the correlation problem involves linking parts of a reservoir with related physical properties (for example, for input into reservoir simulation models), engineering data, particularly pressure and well performance can be used. An example of how pressure data from several wells may be used to link up units with a common pressure regime is shown in

Fig. 8.4 (p. 191). Note how the pressure and lithostratigraphic correlations do not correspond exactly. Well performance data may be used to correlate intervals based on the effect that production at one well has on neighbouring wells. Taking a hypothetical example, if production from lithological interval A in well X causes a pressure drop in lithological interval B in well Y, it is reasonable to assume that there is reservoir connectivity between the two wells, but between different lithological intervals. This correlative tool is known as interference testing, and provides vital dynamic data for reservoir modelling. Dynamic data may be supplemented by static geological correlations using wireline logs and core, as well as three-dimensional seismic data obtained using a very fine line spacing of only a few tens of metres. Finally, to incorporate a further element of geological heterogeneity present in most reservoirs, non-deterministic or *stochastic* data may be incorporated into a reservoir model. Stochastic models help describe unpredictable and uncorrelatable heterogeneities within a reservoir. Stochastic data may be obtained from outcrop or from better-constrained subsurface analogues. The interplay of deterministic and stochastic data in

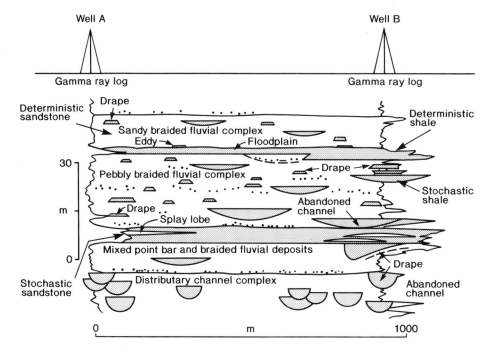

Fig. 8.5 Examples of reservoir heterogeneity in sandstones and shales deposited in a fluvial environment. Deterministic shales and sandstones can be correlated on a lithostratigraphic basis between wells A and B. Stochastic shales and sandstones cannot be correlated between the wells, and their distribution must be projected from statistical data.

building a model of reservoir heterogeneity and in understanding reservoir correlations is shown in Fig. 8.5.

8.4 Stratigraphic correlation in exploration: Tertiary of offshore Norway

8.4.1 Introduction

This example is based on the work of Rundberg and Smalley (1989). The objective of the work was to improve the correlation of Tertiary sedimentary rocks from the North Sea of offshore Norway (Fig. 8.6). The Tertiary units themselves are not of any interest as potential reservoir targets; rather the correlation was required to improve the understanding of their depositional history so that the burial and thermal history of petroleum-prone rocks at deeper stratigraphic levels could be refined. The decision to use strontium isotope stratigraphy rather than conventional biostratigraphy was driven by the frequent absence of key biostratigraphic markers in the studied material and the existence of sparse and in some cases contradictory well bio-

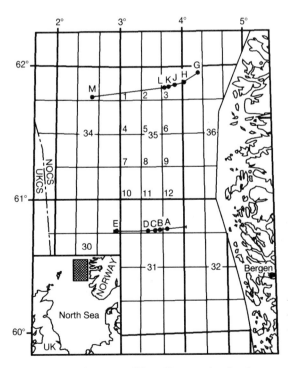

Fig. 8.6 Location map, offshore Norway, showing the two east–west transects through 11 wells (from Rundberg & Smalley, 1989).

stratigraphy (and, according to a reviewer of the manuscript, the fact that Craig Smalley likes using strontium isotopes).

8.4.2 Geological background

The Tertiary of the North Sea in the study area comprises marine clastic sediments deposited as a post-rift megasequence (*sensu* Hubbard *et al.*, 1985) over earlier syn-rift Jurassic and early post-rift Cretaceous sediments, within the overall extensional setting of the North Sea Basin. After the early Oligocene, basin-margin tectonic events resulted in uplift of Fennoscandia, the generation of significant seismic sequence boundaries and the corresponding shedding of large quantities of eroded material into the basin. These events in the basin history are of particular interest to the present study because of their influence on the burial and thermal history of the underlying Mesozoic reservoir and source intervals.

8.4.3 Approach

The data for the study were obtained from 11 wells across two east–west transects (Fig. 8.6). The seismic stratigraphy of the Tertiary units had already been unravelled by Rundberg (unpublished data) and the key boundaries identified for dating purposes. Lithological and palaeo-environmental information was obtained from the wells, chiefly from drill cuttings, and added to the seismic stratigraphic interpretation. Composite sections using all these data from both transects are illustrated in Fig. 8.7.

Following seismic mapping and lithological and environmental interpretation of existing data, samples for strontium isotope analysis were selected from seven of the wells at the sample points indicated on Fig. 8.8. Samples were obtained from well cuttings, which were washed and sieved to recover the 125–250 μm and > 250 μm size fractions. The analyses were carried out on individual hand-picked fragments of carbonate bivalve debris which were carefully examined for evidence of diagenetic alteration. One sample of phosphate (a fish tooth) was also used, along with three samples of glauconite from well E which were dated by the Rb–Sr method[*] (Smalley & Rundberg, 1990).

[*] Bivalve, fish tooth and glauconite samples were prepared by ion-exchange techniques and analysed by mass-spectrometer. Following analysis, the $^{87}Sr/^{86}Sr$ results from the bivalve and fish tooth samples were normalized to Holocene marine carbonate (HMC) standard.

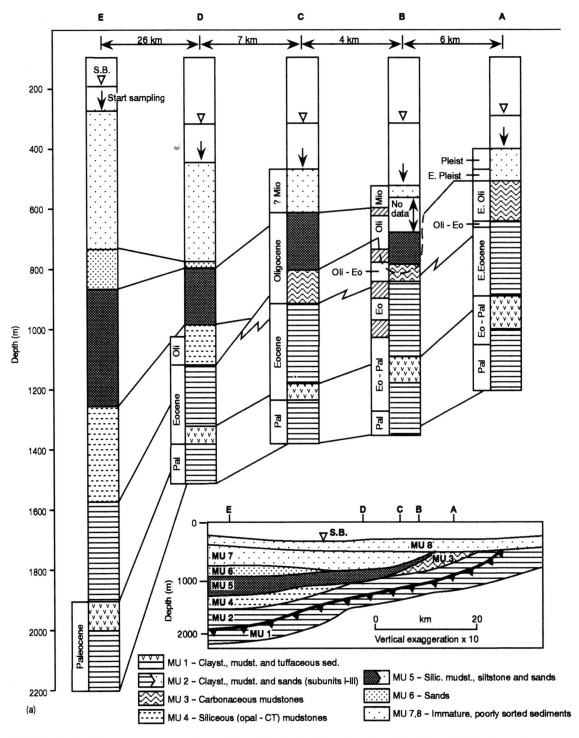

Fig. 8.7 Geological sections interpreted from wells and seismic data (from Rundberg & Smalley, 1989). (a) Transect through wells A–E. (b) Transect through wells G–M.

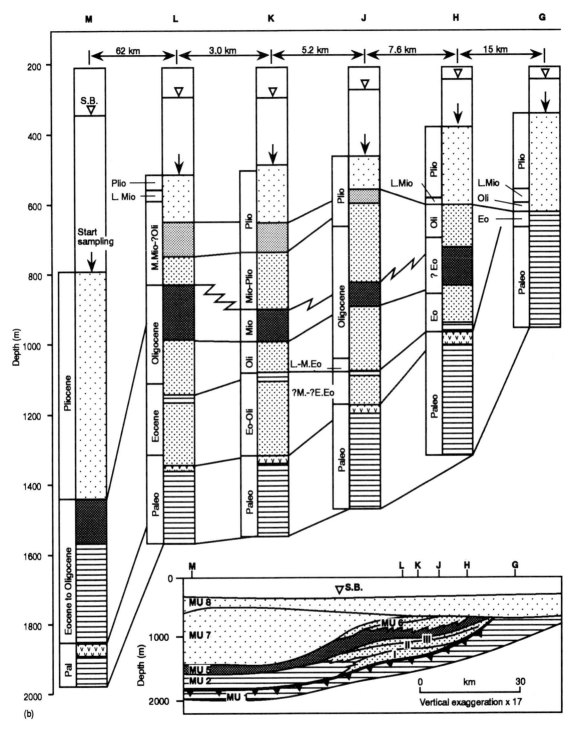

Fig. 8.7 *Continued.*

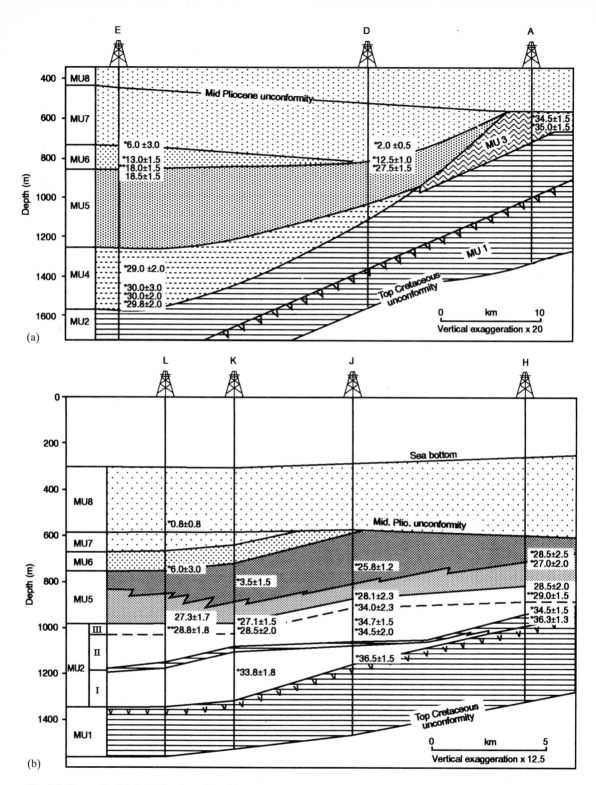

Fig. 8.8 Composite lithological and stratigraphic sections, with ages derived from Sr isotope analyses (from Rundberg & Smalley, 1989). Lithologies are based on the key on Fig. 8.7a. (a) Transect through wells A, D and E. (b) Transect through wells H, J, K and L.

The strontium isotopic ages for the bivalves and fish tooth were determined by plotting the 2σ limits of uncertainty for each $^{87}Sr/^{86}Sr$ analysis and reading off the corresponding dates from the upper and lower limits of the seawater curve (Fig. 8.2). The age of the sample is expressed as the median value ± half the possible age range. The ages determined by this method range in uncertainty from ± 0.5 to 3 Ma; this compares favourably with the resolution attainable by biostratigraphy. Rb–Sr glauconites ages are calculated from isochrons (Fig. 8.9) and are highly precise (± 0.6 to 0.2 Ma).

8.4.4 Strontium isotope ages

The Sr isotope ages for samples taken across each transect are shown in Fig. 8.8. In every well, the mean ages young up-section (except for two samples in wells K and H) and broadly confirm the correlation based on seismic stratigraphy. Figure 8.10 shows chronostratigraphies for both transects and highlights the major hiatuses evidenced from the isotopic information. The age constraints placed on sequence boundaries allowed a relationship between the Tertiary chronostratigraphy and lithostratigraphy to be established for the eight units mapped. MU 1 and MU 2 were deposited during the Palaeocene, Eocene and earliest Oligocene. The lithostratigraphic information from the wells indicates that these sediments were deposited in deep water (during times of high global sea-levels according to the eustatic sea-level curves of Haq *et al.*, 1987). In a landward direction, the coarser-grained, shallower-water sediments of MU 2 sub-units I and II are interpreted to have been deposited during basin margin uplift between 38 and 33 Ma ago (Fig. 8.8). From the Sr isotopic ages, it is evident that sub-unit II of the northern transect is the lateral equivalent of MU 3 in the southern transect. At about 30 Ma, the isotope ages suggest a major hiatus between sub-units II and III. This corresponds to a sudden change from deep to shallow marine facies, evidence of truncation below the seismic reflector corresponding to this junction, and a major sea-level fall on the Haq *et al.* (1987) curve. All indicate the development of a major sequence boundary at this time.

MU 5 is an upward-shallowing succession deposited between 29 and 24 Ma ago in the Late Oligocene. Although eustatic sea-level was rising at this time, passive-margin uplift and increased continental erosion gave rise to shallow water conditions which persisted for the remainder of the basin's history. A major hiatus and sequence

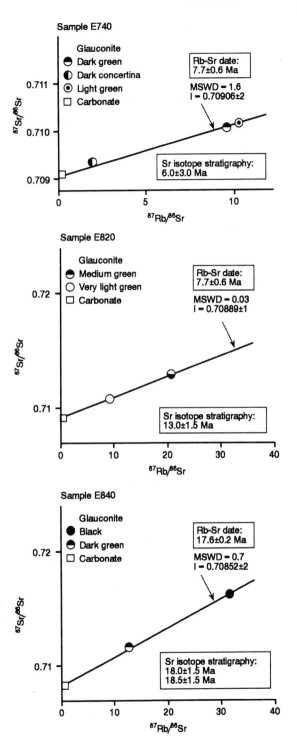

Fig. 8.9 Rb–Sr isochron ages of glauconites from well E (from Rundberg & Smalley, 1989). See Section 5.5.3 for an explanation of how the isochron ages are derived.

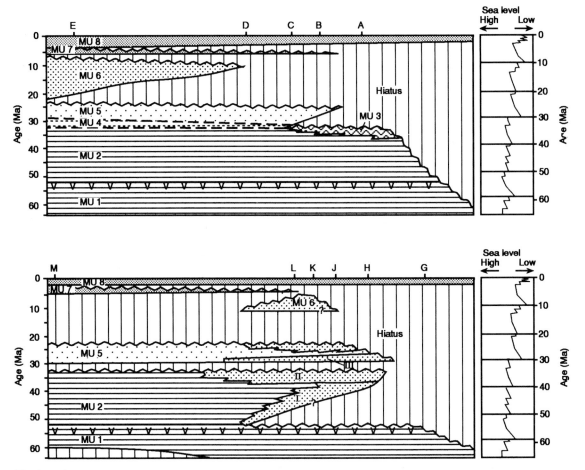

Fig. 8.10 Chronostratigraphic sections based on Sr isotope ages, seismic and well data. (from Rundberg & Smalley, 1989). Lithologies are based on the key on Fig. 8.7a.

boundary was developed at the top of MU 5 prior to deposition of MU 6, a widely distributed sand deposited between 18 and 7.7 Ma ago. In well E, Rb–Sr dates from glauconites complement the Sr isotope ages. Where MU 6 is absent, the hiatus above MU 5 is of much longer duration, with Pliocene sediments of MU 7 and MU 8 resting directly on MU 5.

Although few Sr isotope age data are available for MU 7 and MU 8, the evidence suggests that these units were deposited rapidly during the last 5 Ma. The important consequence of this is that some areas of the basin must have experienced rapid increases in burial depth to account for the significant thicknesses of Plio-Pleistocene sediments, especially across the northern transect (Fig. 8.8). Although it

cannot be dated with any accuracy due to sparse Sr isotope data, the regionally important Mid-Pliocene unconformity probably formed about 2 Ma ago. From the lithological evidence, MU 7 and 8 are sedimentologically immature and probably relate to rapid erosion of the hinterland during basin margin uplift.

8.4.5 Conclusions

From a combination of seismic stratigraphy, litho-logical information and Sr isotopic ages, the chrono-stratigraphy and correlation of the Tertiary of the Norwegian North Sea have been much improved. Sr isotopes assisted in both chronostratigraphic correla-tion and in the identification of major sequence boundaries in the area. These have been tied in to the regional tectonic framework and, where appropriate, to the eustatic sea-level curves of Haq *et al.* (1987). This improved understanding of the depositional sequences in this part of the North Sea has helped to constrain the burial and thermal history of underlying Mesozoic units of significant hydrocarbon potential.

8.5 Stratigraphic correlation in exploration: Plio-Pleistocene of the Gulf of Mexico

8.5.1 Introduction

This case study is based on the work of Trainor and Williams (1990) and the isotope stratigraphy group at the University of South Carolina, USA. The objectives were to:

1 establish a high resolution geochronology for Plio-Pleistocene sediments in the outer continental shelf-slope of the Gulf of Mexico (GOM) using oxygen isotopes;

2 identify unconformities for stratigraphic correlation; and

3 determine the relationship (if any) between phases of sand input and sea-level falls during glacial episodes.

Unlike the Plio-Pleistocene of offshore Norway discussed in the previous study, this portion of the stratigraphy in the GOM is an active exploration target, containing sandy reservoirs. Any correlation tool which can improve stratigraphic correlation and identify sand-prone portions of the stratigraphy is of major consequence to the exploration effort.

8.5.2 Geological background

The GOM was formed in the Mid to Late Jurassic by the breakup of Pangaea and the rifting of North America away from a joined Northern Africa–South America. Thick salt was deposited during the early stages of rifting, followed by the development of extensive carbonate platforms in the Cretaceous around a deep basin floored by oceanic crust. Major, siliciclastic deposition in the GOM began with the late Cretaceous–Palaeogene uplift and erosion of the North American Cordillera. Along the northern margin of the Gulf, siliciclastics were able to build out past the carbonate platform margins, creating a progradational shelf-slope system which differentially loaded the underlying salt, causing

pillowing and diapirism. Salt withdrawal has caused the formation of intrasalt basins which have acted as sediment traps from the earliest Tertiary, through the Plio-Pleistocene to the present day. Plio-Pleistocene depositional systems, of particular interest to this study, comprised shelf and deepwater fan complexes which can be linked back to sediment supply from the palaeo-Mississippi River system. Switching of the system has had a profound effect on the distribution of sediments in the GOM. The effects of glacio-eustasy have also influenced sediment supply, with the formation of deeply incised canyons in the shelf margin and slope which acted as feeders for deepwater fan sand deposition.

8.5.3 Approach

The data for this study were obtained from the OCS well G 1267 No. A-1, in the South Timbalier Block 198 (the well name is shortened to 'ST 198' hereafter), towards the south-east of the hydrocarbon exploration area (Fig. 8.11). The well was drilled in 1962 to a total depth of about 16 000 feet, with drilling cuttings obtained every 30 feet from 3000 feet downwards. Resistivity and self potential (SP) logs were run in the well to identify hydrocarbon-bearing sands (Fig. 8.12).

Samples of foraminifera for stable isotope analysis were obtained every 30 feet from sandstone, siltstone and mudstone drilling cuttings. Two types from different environments (shallow-dwelling planktonic *Globigerinoides* and bottom-dwelling

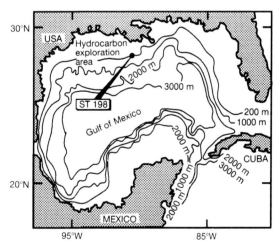

Fig. 8.11 Location of well ST 198 in the Gulf of Mexico (from Trainor & Williams, 1990).

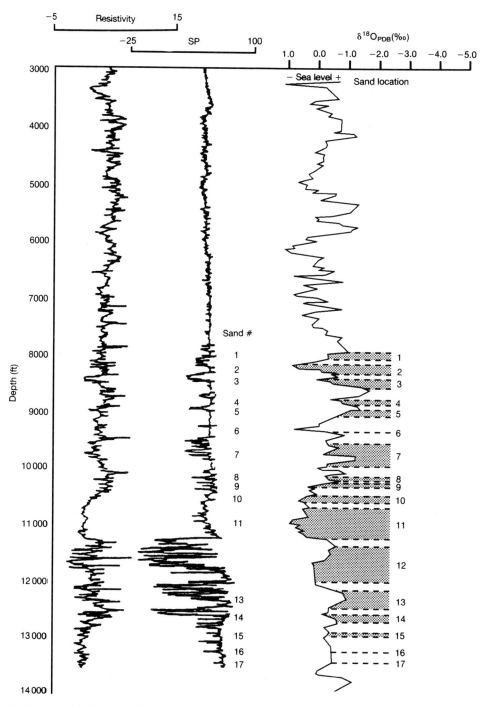

Fig. 8.12 Resistivity, SP (self potential) and sand location logs from well ST 198 (from Trainor & Williams, 1990). Foraminifera oxygen isotope ratios are also plotted against depth to identify possible correlations with sand abundance. Sands are numbered from 1 to 17.

Uvigerina spp.) were selected for analysis to search for differences that might exist between the surface and bottom-water isotopic records. Detailed biostratigraphy was performed on ST 198 to provide a coarse calibration for the oxygen isotope analyses and identified a major unconformity at 7320 ft. This was the only hiatus discernible from biostratigraphic analysis. In order to identify further unconformities not resolved biostratigraphically, the oxygen isotope data were plotted against age using the ages of isotope stage boundaries proposed by Trainor *et al.*, (1988) (Fig. 8.13). They were also plotted against depth to see whether oxygen isotope ratios could be related to the occurrence of sands (Fig. 8.12).

8.5.4 Oxygen isotope stratigraphy

Figure 8.13 shows two major hiatuses in the isotope record which divide the well into three broad sediment packages, T1, T2 and T3. The youngest, T1, begins at approximately 0.90 Ma, corresponding to stage 26/27, and continues until 0.84 Ma, corresponding to (glacial) stage 24. T2 extends from the beginning of stage 48 at 1.41 Ma to the beginning of stage 60 at 1.76 Ma. During this interval, negative $\delta^{18}O$ values for both planktonic and bottom-dwelling foraminifera suggest that glacial meltwater discharge affected both the surface and bottom waters. Interval T3 begins at the 104/105 isotope stage boundary and ends at glacial stage 132, approximately 3.92 Ma. The oxygen isotope stratigraphy provides better chronostratigraphic resolution than did biostratigraphy and identifies two major unconformities. Burial histories constructed using the two dating schemes are shown in Fig. 8.14. The unconformities can be tied to seismic lines at the well and extrapolated as sequence boundaries.

Positive seawater oxygen isotopic ratios reflect low sea-levels during glacial times when the lighter ^{16}O isotope is preferentially concentrated in the ice-sheet. Negative oxygen isotopic compositions reflect times of higher sea-levels when the ice sheets melt. From Fig. 8.12, it is evident that much of the sand

Fig. 8.13 Oxygen isotope compositions of foraminifera from well ST 198 (from Trainor & Williams, 1990). Isotope values on the left were derived from benthic forams, those on the right from planktonic forams. Numbers from 24 to 131 refer to oxygen isotope stages. The average timespan occupied by each is about 30 000 years.

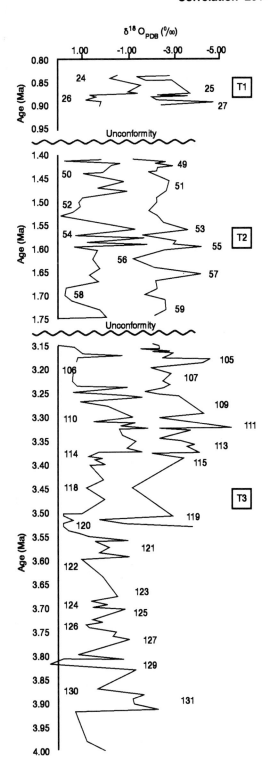

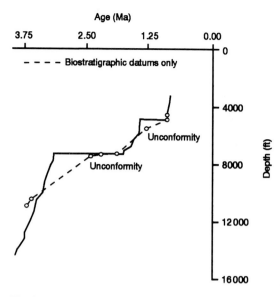

Fig. 8.14 Burial history for well ST 198 (from Trainor & Williams, 1990). The solid line represents the age–depth plot determined using oxygen isotopes showing two unconformities. The broken line is derived from biostratigraphy. The unconformities actually represent more geological time than has been recorded by sediment deposition.

was deposited during sea-level lowstands, and, to a lesser extent, during early transgressions as would be predicted by clastic sequence stratigraphy (Posamentier & Vail, 1988).

8.5.5 Conclusions

The main objectives of this work were to establish a high resolution stratigraphy for Plio-Pleistocene sediments of well ST 198 in the Gulf of Mexico, to identify unconformities for stratigraphic correlation in the area and to relate the sea-level changes (via the oxygen isotope curve) to periods of major sand input to the basin. By any standards, the study was pretty successful. Oxygen isotope stratigraphy of selected foraminiferal species satisfied most of these objectives. Two major unconformities were identified which could be tied to seismic for stratigraphic correlation away from wells, and much of the sand recorded in the well was deposited during times of sea-level lowstand or early transgression as predicted by clastic sequence stratigraphy.

8.6 Reservoir connectivity: Ekofisk Field, Cretaceous of offshore Norway

8.6.1 Introduction

The Ekofisk Field is the largest of several petroleum accumulations in the Cretaceous and earliest Tertiary (Danian) Chalks of the North Sea, of which most are in the Norwegian and Danish sectors (Fig. 8.15). The Chalk reservoirs are complex; reservoir units are mainly redeposited and fractured chalk, while marine hardgrounds (Kennedy & Garrison, 1975) form laterally extensive permeability barriers, which may compartmentalize reservoirs and give rise to different reservoir fluid chemistries above and below.

The objective of this study, from work by Smalley et al. (1992), was to demonstrate how strontium isotope analyses of fluid and rock samples could be used to determine the lateral extent and effectiveness of a cemented layer in the Ekofisk Field. The study also demonstrates how strontium isotope analyses of reservoir fluids can indicate the presence and origin of variable water salinities. This has implications for determining water (S_w) and oil (S_o) saturations.

8.6.2 Geological background

The Ekofisk Field is situated in the southern part of the Norwegian Sector of the North Sea (Figs 1.1 & 8.15). The Ekofisk structure was domed upwards by movement of underlying Permian Salt; the northwest flank of the field is faulted, but otherwise the structure is relatively simple. The stratigraphy of the field is shown in Fig. 8.16. Oil is reservoired in the Danian Ekofisk Formation and the Turonian to Maastrichtian Tor Formation. Both are composed largely of chalk debris flows, with individual flows reaching up to 10 m in thickness. The debris flows are interpreted to have been initiated by synsedimentary tectonism, probably related to salt diapirism and/or associated faulting. Below the Tor Formation lie the water-bearing Hod Formation and the Plenus Marl.

Major diagenesis of the Ekofisk chalks began in the marine environment, with extensive hardground formation at several levels throughout the field. The most significant is the 'Ekofisk Tight Zone', straddling the Cretaceous–Tertiary boundary, an autochthonous chalk which was heavily cemented and bored during exposure on the sea-floor (Fig.

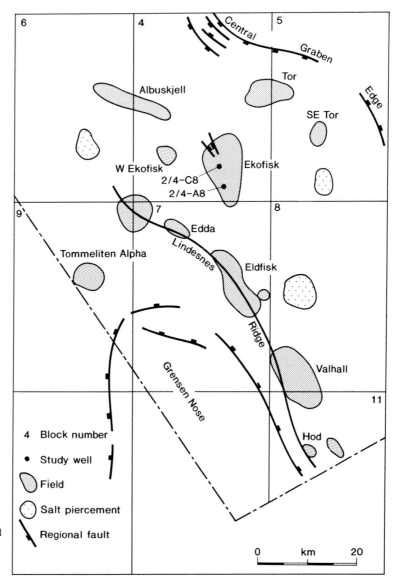

Fig. 8.15 Location of Ekofisk Field and wells 2/4-A8 and 2/4-C8 in the Norwegian North Sea (from Smalley *et al.*, 1992).

8.16). The Tight Zone has very low permeability, and production data indicate that it is a barrier to vertical fluid movement across parts of the field. Later diagenesis involved extensive calcite cementation, as well as fracturing. The fracturing of the chalk has resulted in permeabilities of up to 150 mD, even though permeability in the coccolith-dominated and calcite-cemented chalk matrix may be almost two orders of magnitude lower. Although some of the fractures have been infilled by later calcite, open fracture networks remain the most productive

intervals in the Ekofisk Field and other Chalk fields in the North Sea.

8.6.3 Approach

Two wells were sampled for this study, 2/4-A8 and 2/4-C8 (Fig. 8.15). Well A8 is situated on the southern flank of the Ekofisk structure, separated by a few kilometres from crestal well C8. Twenty-one samples of chalk matrix were collected from cores. For each sample, a few grams of rock were removed

Stage	Group	Formation	Zonation	Name/lithology
Danian	Montrose	Maureen eq.	MA	Maureen formation equivalent
	Chalk	Ekofisk	EY	Autochthonous shaly chalk, autochthonous and turbiditic chalk
			EX	Chalk debris flows with plastic slump structures
			EB	Tommeliten Tight Zone, chalk debris flows, slumps
			EC	Reworked Danian zone, chalk debris flows with plastic slumps, massive allochthonous chalk
			ED	Reworked Maastrichtian zone, massive and slumped allochthonous chalk
			EE	Ekofisk Tight Zone, cemented impermeable massive autochthonous chalk, "hardground"
Maas-trichtian to Turonian		Tor	TZ	Slumped chalk debris flows
			TY	Chalk debris flows with plastic and brittle slump structures
			TX	Chalk debris flows
			TB	Chalk debris flows
			TW	Chalk debris flows and massive allochthonous chalk
		Hod	HA–HF	Chalk
Ceno-manian		Plenus marl	PM	Marl

Fig. 8.16 Stratigraphy, lithologies and reservoir zonation of the Ekofisk Field (from Smalley *et al.*, 1992).

from the core closest to its centre and crushed. Strontium was recovered from the evaporated formation waters (lining the original pores of the crushed core as salts) by leaching in ultrapure water. Several of these samples were also analysed for carbon and oxygen isotope ratios. Porosity determinations were made on all the samples analysed.

8.6.4 Isotopic analyses of chalk and residual salts

Oxygen isotope analyses of the chalk samples range from −3.3 to −5.7‰ (Fig. 8.17). This range is too great to be accounted for solely by the variation of primary marine carbonate and is interpreted to represent a combination of primary carbonate and diagenetic cement, with the cements contributing lighter oxygen. Degree of cementation can be

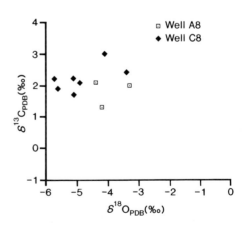

Fig. 8.17 $\delta^{13}C$ plotted against $\delta^{18}O$ for chalk samples, wells A8 and C8 (from Smalley *et al.*, 1992).

estimated from porosity as those samples with lower porosity contain more cement. The relationship between $\delta^{18}O$ and porosity (Fig. 8.18) allows the oxygen isotopic composition of the cements to be calculated. Assuming that the cements were precipitated from marine waters of Cretaceous to Palaeocene age (with a composition of about $-1‰$ SMOW), the approximate temperature of burial cement precipitation in the chalks ranges from about 90 to 160°C, suggesting that the cementation is a relatively high temperature phenomenon.

The carbon isotopic compositions of the samples are somewhat less variable than the oxygen isotope compositions, ranging from 1.3 to 3.0‰ PDB. These values fall within the range expected for unaltered marine chalks and show that no external carbon source has been involved in cementation.

Strontium isotope analyses of the Ekofisk Formation fall very close to what would be expected of unaltered chalks (Fig. 8.19), again suggesting that the strontium in the cements was derived from sedimentary chalk. In the Tor Formation, however, strontium isotope ratios are very different showing a much greater range of compositions and, in both wells, a steady increase in $^{87}Sr/^{86}Sr$ with depth. This probably represents a source of relatively radiogenic Sr that contributed a greater proportion of the Sr in the deeper cements.

In summary, the chalks of the Ekofisk Field comprise a mixture of primary calcite and calcite cement precipitated at relatively high temperatures during burial. In the Ekofisk Formation the cement was derived almost entirely from remobilized depositional calcite, whereas in the Tor Formation it was derived from a mixture of internally derived calcite and carbonate and strontium from elsewhere. The separate diagenetic evolution of the Ekofisk and the Tor formations suggests that the Ekofisk Tight Zone which separates them was acting as a barrier during diagenesis. Can strontium isotope analyses of the porewater salts confirm whether this barrier is still effective today, during oil production?

Strontium isotopic analyses of residual salts vary systematically with depth both within and between wells, correlating strongly with the strontium isotopic values measured in the chalk samples. $^{87}Sr/^{86}Sr$ nonetheless increases more rapidly with depth in the water than in the chalk (Fig. 8.19). The most plausible explanation for this difference is that porewater compositions are partially buffered by the enclosing chalk. In the Ekofisk Formation, the difference between chalk and porewater ratios is small

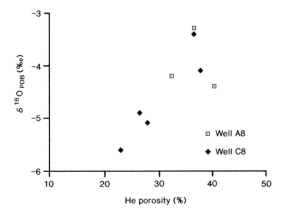

Fig. 8.18 $\delta^{18}O$ plotted against porosity for chalk samples, wells A8 and C8 (from Smalley *et al.*, 1992). $\delta^{18}O$ and porosity are clearly correlated.

and fairly constant, showing no depth-related trend. The small non-carbonate contribution of radiogenic Sr has either been derived locally, from silicate minerals within the chalk, or externally derived Sr has had time to homogenize throughout the formation. In the Tor Formation, the downward increase in $^{87}Sr/^{86}Sr$ does not correlate with the amount of silicate minerals. The most likely explanation is that the porewaters have been influenced by a flux of Sr with $^{87}Sr/^{86}Sr > 0.710$ from beneath. Irrespective of our lack of a full understanding of the source of the radiogenic Sr, the strontium isotopes clearly indicate that there is still a significant difference between the compositions of porewaters in the Ekofisk and Tor formations; the Ekofisk Tight Zone barrier is still effective today. This is corroborated by seismic data which indicate that the Tight Zone is continuous between the two study wells, and by pressure data which indicate significant pressure differences above and below the Tight Zone in some parts of the field.

The porewater strontium isotopic compositions can also be used as a tool for determining formation water salinities. Waters produced from the Ekofisk Formation and upper part of the Tor Formation have Sr isotopic ratios similar to those determined indirectly by residual salt analysis, and have total salinities of 50 000–85 000 mg.l^{-1}. The deeper part of the Tor Formation is characterized by more radiogenic Sr isotopic compositions and higher formation water total salinities, ranging from 80 000–140 000 mg.l^{-1}. It thus seems possible that the increase in $^{87}Sr/^{86}Sr$ with depth could be accompanied by a progressive increase in formation water

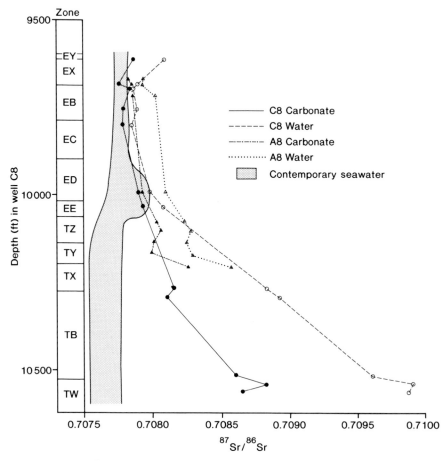

Fig. 8.19 Depth plotted against $^{87}Sr/^{86}Sr$ for chalk and residual salts, wells A8 and C8 (from Smalley *et al.*, 1992). TW to TZ, Tor Formation. EE to EY, Ekofisk Formation.

salinity, perhaps related to underlying Zechstein evaporites. The implications of any salinity gradient could be considerable as estimates of oil-in-place would be affected*.

8.6.5 Conclusions

The use of strontium isotopes on chalk and porewater salts in the Ekofisk Field demonstrated the long-lived effectiveness of the Ekofisk Tight Zone as a barrier between the Ekofisk and Tor formations,

*Oil saturation (S_o) is calculated by determining the water saturation (S_w), which is in turn estimated from the resistivity of the formation water, usually assuming a single formation water resistivity value. Resistivity is controlled largely by salinity, so if salinity variations do exist in the reservoir, this could have a significant impact on estimates of oil reserves calculations.

from the time of chalk burial diagenesis to the present day. The Tight Zone effectively divides the field into two across much of its extent. This conclusion is independently confirmed by seismic and pressure data. Isotopic analyses of porewater salts have also suggested that a salinity gradient may exist in the Tor Formation, controlled by diffusion of material from underlying Zechstein evaporites. This salinity gradient may impact upon oil reserve calculations for Ekofisk.

8.7 Reservoir correlation: Gullfaks Field, Triassic–Jurassic of offshore Norway

8.7.1 Introduction

The Gullfaks Field is one of several accumulations in the Northern North Sea in which the majority of

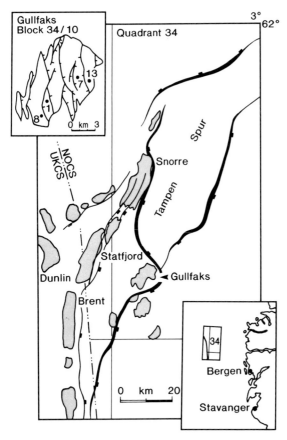

Fig. 8.20 Location map of Gullfaks, Snorre, Statfjord, Brent and Dunlin fields, Northern North Sea (from Mearns, 1989). Inset shows the location of studied wells in Gullfaks and fault compartmentalization of the field.

the oil is reservoired in Middle Jurassic Brent Group sandstones. Other fields near to Gullfaks and the UKCS–NOCS border include the giant Brent and Statfjord fields, Snorre (NOCS) and Dunlin (UKCS; Figs 1.1 & 8.20). In Gullfaks and Statfjord, significant quantities of petroleum are also contained within Lower Jurassic Cook Formation sandstones and uppermost Triassic (Rhaetian) to Pliensbachian Statfjord Formation sandstones.

Correlation of these reservoir sandstones across the large structural closures (many tens to a few hundred square kilometres) of these fields is often difficult, particularly as intrafield faulting has compartmentalized many of the structures and because the reservoir sandstone units (apart from the Cook Formation) are dominantly non-marine or marginal marine with few biostratigraphically useful taxa.

This case study, from the work of Mearns (1989), illustrates the application of neodymium isotope stratigraphy as a tool for:

1 correlating sedimentary units with their sediment source areas (provenance studies);

2 correlating on an intrareservoir scale, particularly between structurally complex fault-blocks a few kilometres apart; and

3 correlating on an interfield scale, in this case between the Gullfaks and Snorre fields, some 20 km apart.

8.7.2 Geological background

The Gullfaks Field is situated in Norwegian waters on the Tampen Spur structure of the northern Viking Graben of the Northern North Sea (Fig. 8.20). This structurally complex field contains approximately 200 Mbbl of oil in Triassic and Jurassic reservoir sandstones at approximately 2 km burial depth. Only 9% of Gullfaks oil is contained in the oldest reservoir unit, the Statfjord Formation which consists of interbedded sandstones, siltstones and shales interpreted to represent alluvial flood-plain and braided stream environments (Fig. 8.21). The Cook Formation of the Dunlin Group contains only 6% of the oil, in marine sandstones deposited following a major transgression over the terrestrial Statfjord sediments. The precise environment of deposition of the Cook sandstones is not well constrained; interpretations range from turbidites through distal deltaic sands to tidal sand bars.

The majority of the oil in Gullfaks (85%) is contained in Aalenian to Bathonian Brent Group sandstones. The Brent Group has been further divided into five formations (Fig. 8.21; Graue *et al.*, 1987). The oldest Brent Formation, the Broom, is a poorly sorted, shallow marine conglomeratic sandstone, showing an overall thinning from west to east across the field, interpreted as evidence of syndepositional faulting and/or a westerly source area for the Broom sediments. Importantly, the Broom Formation is not believed to be genetically linked to the overlying four Brent formations, but to represent sediment sourced from a different provenance area. The Rannoch Formation consists of micaceous siltstones and fine sandstones, with some cross-stratification, and is interpreted to represent deposition at a wave-influenced delta front. The Etive Formation comprises low-angle laminated and cross-stratified medium- to coarse-grained sandstones, interpreted respectively as upper shoreface and as mouth bar

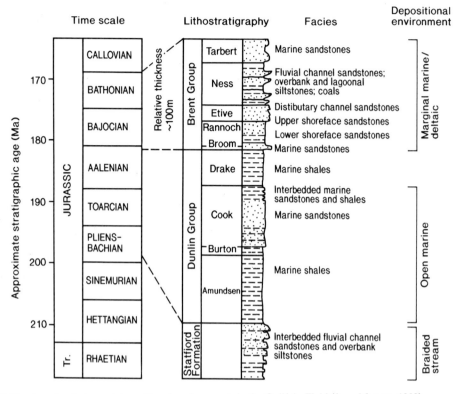

Fig. 8.21 Stratigraphy, facies and depositional environments in the Gullfaks Field (from Mearns, 1989).

and distributary channel deposits. The Ness Formation consists of shales, coals, siltstones and sandstones, deposited in a delta-plain environment. Fluvial channel, overbank, interdistributary bay and lagoonal subenvironments can be recognized. The youngest Tarbert Formation consists of marine sandstones, siltstones and shales and represents an overall transgressive phase, following northward deltaic progradation into the area.

8.7.3 Approach

Ninety-three samples of sandstones, siltstones and shales weighing between 100 and 300 g were obtained from four wells in different fault segments of the field, 34/10-1, 34/10-7, 34/10-8 and 34/10-13 (Fig. 8.20). The Statfjord Formation was sampled in well -13 only, whereas the Cook Formation was sampled in wells -1, -7 and -13, and the Brent Formation in wells -1 and -8. The samples were crushed and Sm and Nd isotope ratios analysed by standard techniques (Mearns, 1986). Analytical errors in the

$^{147}Sm/^{144}Nd$ ratio are estimated to be $\pm 0.25\%$, which computes through to provenance age as an error of ± 20 Ma.

8.7.4 Sm−Nd isotopic correlation

Figure 8.22 shows Nd concentrations, $^{147}Sm/^{144}Nd$, $^{143}Nd/^{144}Nd$ ratios and provenance ages plotted against depth for all the sampled wells. Most of the samples had Nd concentrations that ranged from 0 to 60 ppm, apparently independent of lithology and stratigraphy and therefore not of much use for correlation. Similarly, the $^{147}Sm/^{144}Nd$ ratio of the samples ranged from 0.10 to 0.12 and could not be used to discriminate lithology (although there was a general stratigraphic trend with lower mean and median $^{147}Sm/^{144}Nd$ ratios in Statfjord samples than in Dunlin and Brent Group units). Because of the general uniformity of $^{147}Sm/^{144}Nd$ ratios in most of the samples, the main control on provenance age is the $^{143}Nd/^{144}Nd$ ratio. The provenance age parameter was, however, preferred to the $^{143}Nd/^{144}Nd$

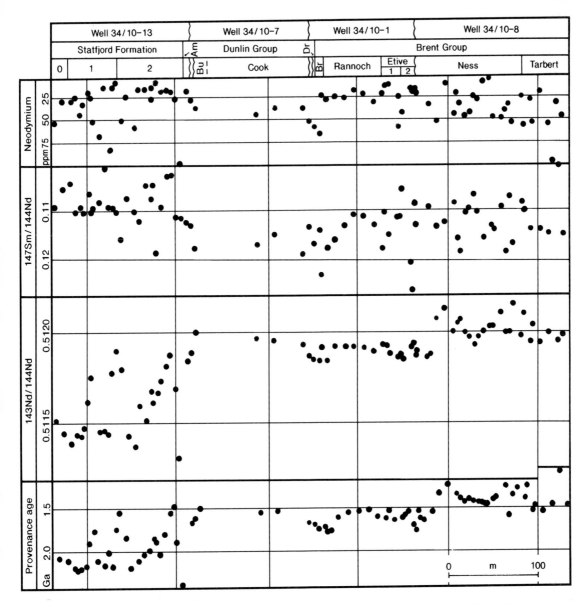

Fig. 8.22 Composite section from all four study wells in Gullfaks showing: (i) Nd concentrations; (ii) $^{147}Sm/^{144}Nd$; (iii) $^{143}Nd/^{144}Nd$; and (iv) provenance ages (from Mearns, 1989).

ratio as it is easier to understand and interpret than a dimensionless isotopic ratio.

Correlation of sediments with source areas
The general trend of provenance ages in Gullfaks is from oldest at the bottom of the section (in the Statfjord Formation) to youngest in the Brent Group. This trend is believed to reflect changes

in sediment source areas and thus to provide information about changes in provenance and palaeogeography with time. The majority of samples from the Statfjord Formation have provenance ages between 1850 and 2400 Ma, with a few samples as young as 1450 Ma. These results suggest a source area considerably older than most of those in Scotland or Norway, possibly the Archean shield to the north-west of Scotland. The range in provenance ages is believed to reflect a mixing of sediment types derived from the ancient Archean source, and a younger, more proximal source, possibly on the Shetland Platform.

The majority of samples from the Dunlin and Brent groups have provenance ages between 1350 and 1850 Ma which does not conclusively identify sediment provenance. However, some samples from the Ness and Tarbert formations have provenance ages of less than 1350 Ma, suggesting a source area younger than those in Scotland or Norway, possibly reflecting a sediment component from the Forties volcanic centre of the Central North Sea.

Correlation within the Gullfaks Field

On a field scale, the probability of time-equivalent strata yielding similar provenance ages is reasonably high, as the likelihood of several different sediment dispersal systems being involved in deposition at closely spaced localities is relatively small. This contention has been borne out by duplicate analyses of sediment samples collected from several hundred metres apart in modern Norwegian river systems which yield identical Sm–Nd provenance ages (Mearns, unpublished results). Sm–Nd model ages can therefore be used for well-to-well lithostratigraphic correlation (Fig. 8.23).

Figure 8.24 shows Ness Formation model ages and Nd isotope ratios plotted against depth for two Gullfaks wells about 2 km apart. The predefined lithostratigraphic units can apparently be correlated

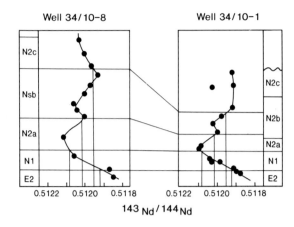

Fig. 8.24 Correlation of provenance ages in the Ness Formation (from Mearns, 1989).

using provenance ages. The provenance age patterns above the N2a zone are rather different, suggesting that the relatively uniform sediment deposition seen lower in the section has been replaced by more periodic, localized deposition with differential subsidence. Alternatively, these sediments may not be contemporaneous.

Correlation between the Gullfaks and Snorre fields

The correlation potential of sediment provenance ages between fields a few tens of kilometres apart is likely to be more limited than intrafield correlation because of the increased probability of different sediment dispersal systems actively depositing sediment at more widely spaced localities. The results from the Statfjord Formation of Gullfaks and Snorre

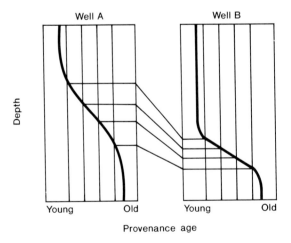

Fig. 8.23 How provenance ages can be used to correlate between wells (from Mearns, 1989).

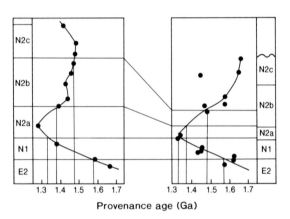

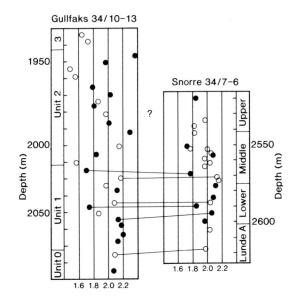

Fig. 8.25 Correlation of provenance ages in the Statfjord Formation between Gullfaks Field well 34/10-13 and Snorre Field well 34/7-5 (from Mearns, 1989).

indicate possible tie lines in the lower half of both successions (though the provenance ages for Gullfaks are slightly greater; Fig. 8.25). Although this cor-

relation is highly equivocal, it may represent deposition of sediments in this portion of the section by the same dispersal system. The upper section of both wells is, however, rather different, although the general trend upwards towards decreasing provenance ages appears to be consistent. The difficulty in correlating the upper section may reflect the onset of Jurassic rifting and the development of a tectonic control on sediment dispersal patterns.

8.7.5 Conclusions

Sm−Nd provenance age determinations have potential applications at several stages during exploration, field appraisal and development. Correlation of reservoir sediments with provenance areas may be used to map out the influence of different sediment dispersal systems on a regional scale and assist in the production of palaeogeographic and sand distribution maps at an early stage of exploration. At a more mature stage of exploration, comparison of interfield provenance ages can provide information about the likely continuity of sediment dispersal systems on an intermediate scale. On an intrafield scale, particularly where sediments are poor in biostratigraphically diagnostic forms and/or where the field is structurally complex, Sm−Nd provenance studies can assist in identifying sandbodies deposited by the same sediment dispersal system.

Chapter 9 **Petroleum Recovery**

9.1 Introduction

The petroleum in an oil or gas reservoir will never be completely recovered. A fraction will always remain in the reservoir for a number of reasons. For example, the reservoir may have insufficient energy of its own to expel the petroleum, capillary forces will prevent oil movement, the oil may be too viscous to move under reservoir conditions or some may be contained in blind pores which are not linked up to the overall reservoir porosity system (though this would beg the question of how petroleum might have entered in the first place). An ideal recovery mechanism would optimize the rate of petroleum production (and with it cash flow and the return on investment) and maximize the proportion of petroleum that will be eventually recovered. In practice, there tends to be a trade off between the two objectives: the recovery scheme that gets the oil out of the ground fastest (usually preferred by oil companies) frequently leaves a large proportion of it behind (usually not liked by governments). A consequence of improvements in recovery methods is that the recoverable reserves of fields tend to be occasionally revised upwards (which is good for share prices and cheaper than exploration). A 2% reserves growth in a 10 billion barrel oilfield, for example, could be worth hundreds of millions of pounds.

Petroleum recovery mechanisms are divided into three broad categories (Fig. 9.1).

1 *Primary recovery*, in which the inherent energy of the reservoir or adjacent aquifer provides the necessary drive for petroleum production. Primary recovery mechanisms include gas cap drive, solution gas drive, compaction drive and aquifer drive (Dake, 1978).

2 *Secondary recovery*, in which additional energy is introduced into the reservoir–aquifer system to accelerate production and increase ultimate recovery. Secondary recovery mechanisms include seawater and gas injection.

3 *Tertiary recovery*, more commonly referred to as *enhanced oil recovery* (EOR), which involves changing the physical and chemical properties of the

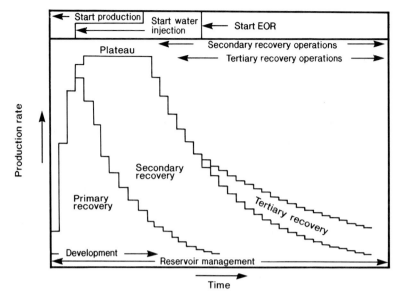

Fig. 9.1 Relationship between primary, secondary and tertiary recovery processes over the lifetime of an oilfield. Many (though far from all) reservoirs are initially left to produce without interference (primary recovery). Secondary recovery techniques may be employed a few years after the start of production in order to prolong peak plateau production rates. Tertiary recovery (EOR) methods are often used later on to retard the inevitable decline in production.

reservoir. EOR methods include the introduction of heat through fireflooding or steam-flooding to reduce the viscosity of heavy oils and improve their mobility (Latil, 1980) and the addition of chemicals such as surfactants and CO_2 to improve oil permeability.

Inorganic geochemistry can make contributions to the design of all three types of recovery processes. Its contribution to primary recovery is largely to early prediction of reservoir quality and as such has been covered in Chapter 6. In this chapter, we will review secondary and EOR mechanisms and the contribution which inorganic geochemistry can make in describing the reservoir prior to initiating a recovery programme and in monitoring its progress. Part of this monitoring process is the prediction of the nature and quantities of corrosive fluids. The main corrosive fluids affecting oil and gas fields are H_2S, known as *sour gas*, and CO_2. H_2S is corrosive to most materials used in petroleum production and is highly toxic. CO_2 is also highly corrosive. These gases may be natural reservoir fluids or may be generated as an unwelcome by-product of the petroleum recovery process itself. It is important to know which. Inorganic geochemistry can assist in identifying the origin of the fluid, likely quantities to be produced and reservoir intervals which are particularly prone to corrosive fluid production.

9.2 Secondary recovery

The two main secondary recovery mechanisms are water injection and gas injection. These will normally augment the natural aquifer or gas cap drive of the reservoir. The main objectives of both these recovery mechanisms are to maintain the pressure of the reservoir by replacing produced oil with injected water or gas (*voidage replacement*) and to improve the reservoir sweep*, as the injected fluid contacts portions of unproduced oil.

The key factors in assessing the suitability of a reservoir for secondary recovery are a knowledge of the connectivity of the reservoir and of any deterioration of reservoir quality across the field. The latter is especially important for water injection as the aquifers of many reservoirs commonly have poorer reservoir quality than the petroleum-bearing portions. In the Magnus Field, for example, the presence of abundant quartz cement and fibrous illite in the aquifer renders permeabilities so low that water could not be injected into the aquifer (Heaviside *et al.*, 1983). Water was in fact injected directly into the oil zone, which was necessary but which involved a loss of oil reserves. The contribu-

---

* Reservoir sweep refers to the proportion of unproduced oil which is contacted by injected seawater and pushed towards the production well. A high sweep efficiency means that the waterflood front has contacted a high proportion of the oil.

tion of inorganic geochemistry to aquifer permeability problems is in describing and predicting the diagenesis of the aquifer and its effect on reservoir quality (see also Chapter 6).

Once a secondary recovery scheme such as seawater injection has been initiated, monitoring the progress of the injected water is necessary to ensure that breakthrough of water into the production wells does not occur unexpectedly. During oil production, reservoir formation water is usually produced along with petroleum. As a water injection scheme proceeds, the produced water may comprise both the natural formation water and the injected water. The greater the quantity of injected water produced, the closer the water front is to the production well. There are several ways of monitoring the progress of injected water: production log analysis*, analysis of ionic species from produced fluids and tracer studies. The main cation used to detect seawater incursion is sulphate, normally present in much greater concentrations in seawater (approximately 2650 ppm) than in formation waters (usually a few to a few tens of parts per million). One of the difficulties in using sulphate concentrations to monitor breakthrough is that barium or strontium, often present in considerable amounts in formation waters, may remove sulphate from the injected seawater through the precipitation of barite or celestite in the reservoir. As well as causing formation damage (porosity and permeability reduction), the lowered concentrations of sulphate recorded from produced fluids will give rise to underestimation of the proportion of injection water in the petroleum–water mixture produced. Sulphate may also be removed from seawater by ion-exchange reactions in the formation. Tracers can overcome the problem of reaction in the reservoir. These include chemicals such as iodides, nitrates and alcohols and radioactive substances such as tritiated water (water with added ^3H) and solutions or complexes of radioactive isotopes of carbon, sodium, nickel and iodine. The injected seawater is normally spiked with a tracer at the injection well; if recovered at the production well, injected water has broken through. There are some obvious problems with tracers. They are expensive to buy and toxic or radioactive substances should not in general be tipped over the side of the rig so must be carefully disposed of (Smalley et al., 1988).

* Production logs include a range of tools which monitor the flow rate and saturation of hydrocarbons during field production.

A further limitation common to all tracer methods is that they yield no information at all until they are first detected in production wells which may be months or even years after injection.

The difficulties associated with tracer methods and with reactions of natural ionic species in the reservoir can be overcome to an extent by using natural isotopic tracers present both in the reservoir formation waters and in injected seawater. These include the oxygen and hydrogen isotopic compositions of the waters themselves and their strontium isotopic compositions and strontium concentrations. Several important criteria need to be fulfilled before natural isotopic tracers can be used, the most obvious of which is that injection and formation waters must be isotopically distinct. The first case study that we will look at in this chapter – from the Forties Field – illustrates how oxygen and hydrogen isotopes may be used to monitor produced waters and detect and quantify seawater breakthrough. A further prerequisite for the use of strontium isotopic tracers is that the strontium *concentrations* of injected and formation waters need to be similar; otherwise, the isotopic signature of the water with high Sr concentration will swamp that of the other, no matter how distinct their isotopic compositions are (Smalley et al., 1988).

9.3 Enhanced oil recovery

An EOR project will usually be initiated at some stage during secondary recovery as part of a phased recovery programme (Fig. 9.1). Some oilfields will even use EOR techniques from the start of production; many western Canadian heavy oilfields, for example, can only be produced by reducing the oil viscosity by *in situ* combustion or steamflooding (Outtrim & Evans, 1977).

The nine main EOR processes are listed in Table 9.1. In this chapter, we will deal mainly with thermal processes which alter the physical properties of the rock as well as that of the petroleum: *in situ* combustion and steamflooding. Before initiating any of these recovery processes, it is necessary to know the field's remaining oil-in-place and its distribution and to have a good geological description of the reservoir including connectivity, pore systems and mineralogies (especially clays) surrounding the pores (Ebanks, 1987). Connectivity may already be quite well known if the EOR process forms part of a phased recovery programme. However, if production involves EOR at the start, only static reservoir

Table 9.1 Enhanced oil recovery processes and their applicability. The processes in italics are of particular interest to this chapter. Depths, lithologies, oil gravities and permeabilities refer to the optimum reservoir properties for the recovery process. The processes marked with an asterisk are miscible displacement processes, in which the injected fluid mixes with the oil to form a mobile mixture which can then be displaced by secondary recovery techniques such as water or gas injection.

Process	Reservoir depth (m)	Lithology	Oil gravity (°API)	Average permeability (mD)
Steam injection	60–1200	S or CR	10–25	>100
Fireflooding	>160	S	<25	>100
Polymer injection (viscosity control agent)	<3000	S	>15	>20
Surfactant flooding	<3000	S	>25	>20
Alkaline flooding	<3000	S	>35	>20
Carbon dioxide*	>1000	S or CR	>30	>5
LPG/enriched gas*	>1000	S or CR	>30	<50
Nitrogen injection*	>3000	S or CR	>35	<10
High pressure lean gas injection	>2000	S or CR	>40	<50

S, siliciclastic reservoirs
CR, carbonate reservoirs

descriptors will be available, such as core/log information and pressure. These data can potentially be supplemented by strontium isotope analyses of porewater salts, which may provide information on reservoir connectivity (Section 8.6). A description of the pore-scale heterogeneity of the reservoir is also essential before initiating an EOR programme. Information on pore types and origin can be obtained from transmitted light microscopic observation of thin-sections and SEM (Chapter 2). Core analytical techniques (which conventionally include porosity, permeability and grain density measurements), combined with PIA, can provide information on the size distribution and shape of pores and on how they influence reservoir quality. Petrography can also provide information on the minerals that line pore systems. Clay mineral types and distributions are of particular importance as they can exhibit both a detrimental and enhancing effect on EOR processes. Abundant swelling clays such as smectites will expand and block pore throats during steam-flooding. By contrast, clay minerals may actually promote *in situ* combustion by promoting the deposition of small quantities of carbon ('coke') ahead of the fire front (Neasham, 1977). Clay mineral characterization and quantification can be achieved by TEM, XRD and TG-EWA.

The last two case histories in this chapter describe the effects of steam injection and fireflooding on the mineralogies and reservoir properties of heavy oil reservoirs in Alberta, Canada.

9.4 Production of corrosive fluids

Corrosive fluid production during petroleum recovery is hazardous to both production facilities and health. The principal corrosive fluids produced during petroleum recovery are H_2S and CO_2, both of which may form a proportion of the *in situ* reservoir fluids, and/or be generated as part of the production process.

H_2S can originate in three main ways. Firstly, it may be generated as a product of source rock maturation and be transported to the reservoir with the migrating petroleum. Marine carbonate source rocks are particularly prone to H_2S generation, owing to kerogen precursors with a high sulphide content. Secondly, sour gas may be generated by *in situ* reservoir reactions, particularly at elevated temperatures. For example, originally sulphur-rich oils reservoired in deeply buried Jurassic Smackover carbonates of the Gulf of Mexico have been thermally cracked to pyrobitumen, CH_4 and H_2S (Sassen, 1988). H_2S may also be generated by inorganic–organic reactions in reservoirs. In sulphate-bearing reservoirs such as the Permian Khuff Formation of Abu Dhabi, the reaction of anhydrite with CH_4 in the reservoir at high temperatures has produced H_2S and calcite. Finally, H_2S may be generated during petroleum production. It may be produced by the reduction of sulphate contained in seawater when injected into the reservoir, or it may originate by reactions within the reservoir formation fluids during production. The case study on the Bridport Sands reservoir of the Wytch Farm Oilfield outlines the problems of understanding the production of sour gas during petroleum recovery.

CO_2 is a common reservoir fluid in gas fields. The giant Natuna D-Alpha Gas Field of offshore Indonesia contains up to 100 tcf of CO_2, forming in excess of 70% of the total gas (Rudolph & Lehmann, 1989). This CO_2 was certainly derived from outside the reservoir, either by thermal decarboxylation in the source rock, or from an igneous source. CO_2 can be produced in a reservoir by reaction of carbonate with organic acids generated from organic matter in source rocks just prior to petroleum expulsion (though this process may be of minor importance). Hutcheon *et al.* (1980) have shown that CO_2 can potentially be produced by reaction of carbonates with clay minerals at high reservoir temperatures. CO_2 can also be generated during petroleum recovery. The main sources of carbon in most reservoirs are the hydrocarbons themselves and inorganic carbonate grains and cements. The case study on enhanced recovery of heavy oil from Cretaceous sandstones in western Canada shows how inorganic geochemistry can fingerprint the origin of CO_2.

9.5 Secondary recovery: Forties Field, offshore UK

9.5.1 Introduction

Forties is one of the largest oilfields in the North Sea, with an original oil-in-place of 4340 Mbbl, and original reserves estimated at 2480 Mbbl (Figs 1.1 & 6.47). The field was discovered in 1970 and started production in 1975 by a natural aquifer drive mechanism with seawater injection support. It currently supports five production platforms (Fig. 9.2), with 103 wells – 81 producers and 22 injectors – producing at 165 000 stock tank barrels of oil per day. The objective of this study was to investigate whether oxygen and hydrogen isotopes could be used to determine the proportions of formation water and injected seawater appearing in production wells in the field.

9.5.2 Geological background

The setting of the Forties Field is described in Section 6.7.2. The medium-grained subarkosic sandstones in the reservoir are generally poorly cemented, with limited amounts of kaolinite and quartz cement. Carbonate nodules and layers are also present, with the latter sometimes providing

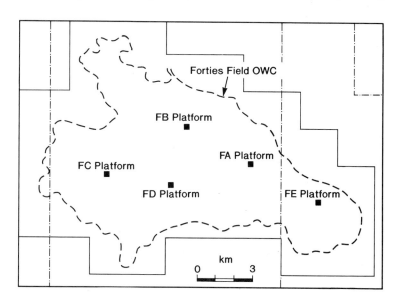

Fig. 9.2 Location of production platforms and oil–water contact (OWC) of the Forties Field, UK North Sea.

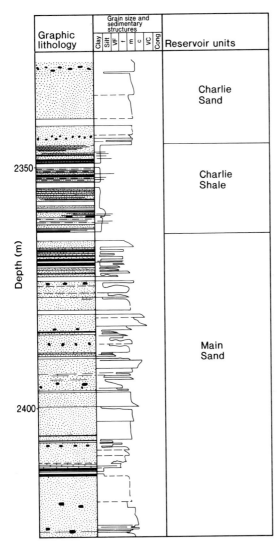

Fig. 9.3 Representative summary log through a Forties well showing Main Sand, Charlie Shale and Charlie Sand.

Table 9.2 Isotopic compositions and sulphate concentrations of Forties Field waters.

Well	δD (‰)	$\delta^{18}O$ (‰)	SO_4^{2-} (ppm)
Wells producing from Main Sand			
FA12	−18.4	−0.73	13
FA12	−22.0	−0.85	8
FA34	−13.8	−1.58	23
FA34	−23.5	−0.41	5
FB32	−20.6	−0.05	42
FB32	−23.1	+0.08	22
FD43	−25.5	+0.43	5
FD44	−24.5	+0.28	1
Wells producing from Charlie Sand			
FC13	−23.1	+1.66	7
FC32	−21.5	+0.81	9
FC44	−20.9	+0.60	9
FC51	−23.0	+0.91	8
FC63	−23.7	+1.41	6
Wells indicating seawater breakthrough			
from sulphate concentrations			
FA52	−19.5	+0.34	175
FC22	−18.5	+0.17	100
FC23	−14.8	+1.22	935
FC31	−22.1	+0.46	225
FC41	−13.8	+0.53	1200
FC41	−11.0	+0.57	1200
FC41	−9.1	+0.79	990
FC41	−4.2	+0.39	1255
Seawater	0.0	0.0	2650

9.5.3 Approach

Samples of produced Forties formation water which had *apparently* not suffered seawater breakthrough (according to their sulphate concentrations) were obtained from 10 wells; five producers from the Main Sand and five from the Charlie Sand. Five wells which were believed to be cutting seawater were also sampled. The sampled waters were analysed for their ionic constituents and for oxygen and hydrogen isotopic compositions (Table 9.2).

9.5.4 Chemical and isotopic analyses of produced fluids

$\delta^{18}O$ and δD are plotted in Fig. 9.4 for (i) Main Sand waters; (ii) Charlie Sand waters and (iii) breakthough waters. The pre-breakthrough waters from both Main and Charlie Sand show a relatively large

vertical permeability barriers. Porosities in the sands are generally good, ranging from 20 to 30%, with permeabilities of several hundred millidarcies.

Two main sandstone bodies have been identified in Forties, the Main Sand and the overlying Charlie Sand, separated by an extensive shale barrier, the Charlie Shale (Fig. 9.3). The Main Sand comprises a series of broadly sheet-like high density turbidites with a net to gross of 75–85%. The overlying Charlie Sand is less sandy, with a net to gross of 65–80%, and is interpreted as a submarine channel system.

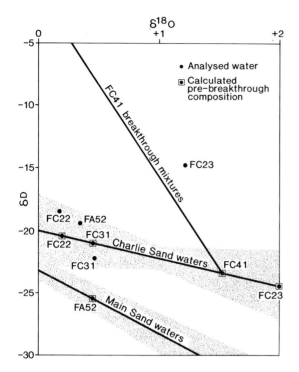

Fig. 9.4 $\delta^{18}O$ and δD analyses of Main and Charlie Sand formation waters, and seawater breakthrough mixtures. 95% confidence intervals are stippled either side of the best-fit lines.

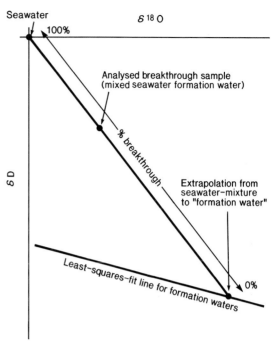

Fig. 9.5 Calculation of the degree of seawater breakthrough from oxygen and hydrogen compositions of injected seawater, formation waters and mixtures. The analysed breakthrough sample represents a mixture of 35% formation water and 65% seawater, calculated from the lever rule principle.

range in oxygen isotope compositions, clustering around 0‰, the value for seawater. Hydrogen isotopic compositions show proportionately less variation, but are distinctly different from the seawater hydrogen isotopic composition. This means that δD (rather than $\delta^{18}O$) is likely to be the most sensitive indicator of seawater breakthrough. An unexpected result of the isotopic analyses of the prebreakthrough waters was the separation of Main and Charlie Sand waters into two distinct populations, suggesting separate origins and/or evolutionary pathways for the Main and Charlie Sand waters. These isotopic differences also mean that for calculating the extent of seawater breakthrough it is essential to know whether production is from the Main or Charlie Sand.

Calculation of the extent of seawater breakthrough assumes a simple two-component mixture of seawater and formation water. The percentage breakthrough of seawater is calculated by applying a simple lever rule principle to both seawater–Main

Sand water and seawater–Charlie Sand water mixtures (Fig. 9.5). The main uncertainty associated with the calculation of breakthrough is the isotopic composition of uncontaminated formation water. Figure 9.4 shows the uncertainty range of Main and Charlie Sand water isotopic compositions. Using the simple two-component mixture, the extent of seawater breakthrough in the Main Sand from well FA52 is $23.4 \pm 3.0\%$ (Table 9.3). In the Charlie Sand, breakthrough in the study wells FC22 and FC23 is evident, but the most extensive seawater contamination is recorded from well FC41 which over a period of 11 months increased its production of seawater as a percentage of total water produced from approximately 43 to 85%.

Estimates of breakthrough using both ionic and isotopic methods are compared in Table 9.3. Sulphate concentrations tend to underestimate the extent of seawater breakthrough by comparison with the isotopic method, probably due to in-reservoir reaction of sulphate and ion exchange.

Table 9.3 Comparison of the extent of seawater breakthrough from stable isotope and sulphate concentrations. The percentage breakthrough for sulphate was calculated on the assumption that seawater contains 2650 ppm sulphate and uncontaminated formation water contains 10 ppm sulphate.

Well	Date sampled	Percentage breakthrough (isotopic)	Percentage breakthrough (sulphate)
Main Sand			
FA52	31-1-83	23.4 ± 3.0	6.5
Charlie Sand			
FC22	12-11-84	9.8 ± 4.4	3.4
FC23	12-11-84	39.6 ± 2.4	35.0
FC41	22-6-83	42.7 ± 3.3	45.1
FC41	24-6-83	55.1 ± 2.6	45.1
FC41	25-6-83	63.5 ± 2.1	37.1
FC41	6-7-83	85.3 ± 0.9	47.2

9.5.5 Conclusions

In the Forties Field, hydrogen and oxygen isotopic analyses of produced waters provide a robust method for determining the extent of seawater breakthrough during seawater injection. The sensitivity of the method is considerably greater than that provided by analysis of sulphate ion concentrations, but isotopically well-characterized pre-breakthrough formation waters are required for successful application of the method. The study also turned up an interesting additional result: that the Main and Charlie Sands contained water of different isotopic compositions and hence of probable differing origins or with different histories.

9.6 Secondary recovery and gas souring: Wytch Farm Oilfield, Dorset, UK

9.6.1 Introduction

During oil production from the Jurassic Bridport Sands, one of the two reservoirs of the Wytch Farm Oilfield in Dorset, spatially and temporally variable amounts of corrosive H_2S were produced from wells sampled over the period 1984–85. The origin of the H_2S was not known: it was unlikely to have been derived in significant quantities from Liassic mudstones, the petroleum source rock for the Wytch Farm Field (which would have generated relatively low sulphur crude) or to have formed in the reservoir by reaction of CH_4 with sulphate minerals as these are very rare in the Bridport Sands. The objective of this study was to distinguish between two plausible origins for the produced H_2S:

1 reaction of formation water with seawater which

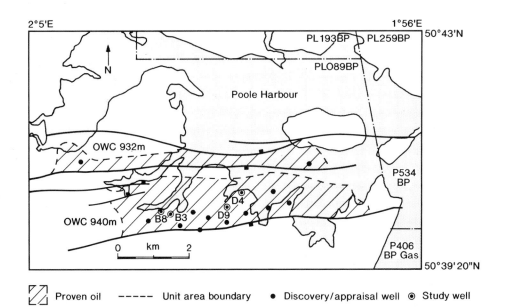

Fig. 9.6 Location of study wells in the Wytch Farm Oilfield.

was being injected into the Bridport Sands as a secondary recovery mechanism; and

2 generation of sulphide within the formation waters themselves, with no external sulphur source involved.

9.6.2 Geological background

The Wytch Farm Field, discovered in 1973–74, is the largest onshore oilfield in Western Europe with original reserves in excess of 280 Mbbl (Figs 1.1 & 9.6). The field comprises several reservoir intervals (Fig. 9.7). The main reservoir unit is the Sherwood Sandstone, a Triassic alluvial-braided stream deposit approximately 150 m thick, which contains 85% of the petroleum (Dranfield et al., 1987). The Bridport Sands, Toarcian shallow marine sandstones, contain up to 9% of the reserves. Less significant production is also obtained from a relatively thin Middle Jurassic limestone.

The Bridport Sands comprise bioturbated, moderately to well sorted fine-grained sandstones and siltstones, with few visible sedimentary structures. The sands are interpreted to have been deposited in a storm-dominated shallow marine environment, and glauconite and shell debris are common. Where the Bridport Sands outcrop along the Dorset coast, spectacular extensive layers of calcite-cemented fine sandstones can be seen, alternating with poorly cemented friable siltstones on a decimetre scale. Other diagenetic components are of relatively minor volumetric importance, but include kaolinite and quartz cement. The oil in the Bridport, Sherwood and the other minor reservoir units in the Wytch Farm Field was sourced from Liassic mudstones which are downthrown against the field's main bounding fault to the south (Fig. 9.6). These are known, from geochemical analysis, to have generated allow sulphur crude.

9.6.3 Approach

The approach used here was to characterize the chemistry of the fluids in the reservoir. Samples of produced liquids from the Bridport reservoir were obtained from four wells: B3, B8, D4 and D9 (Fig. 9.6) on three occasions between 1984 and 1985. Two samples of injected seawater were also obtained.

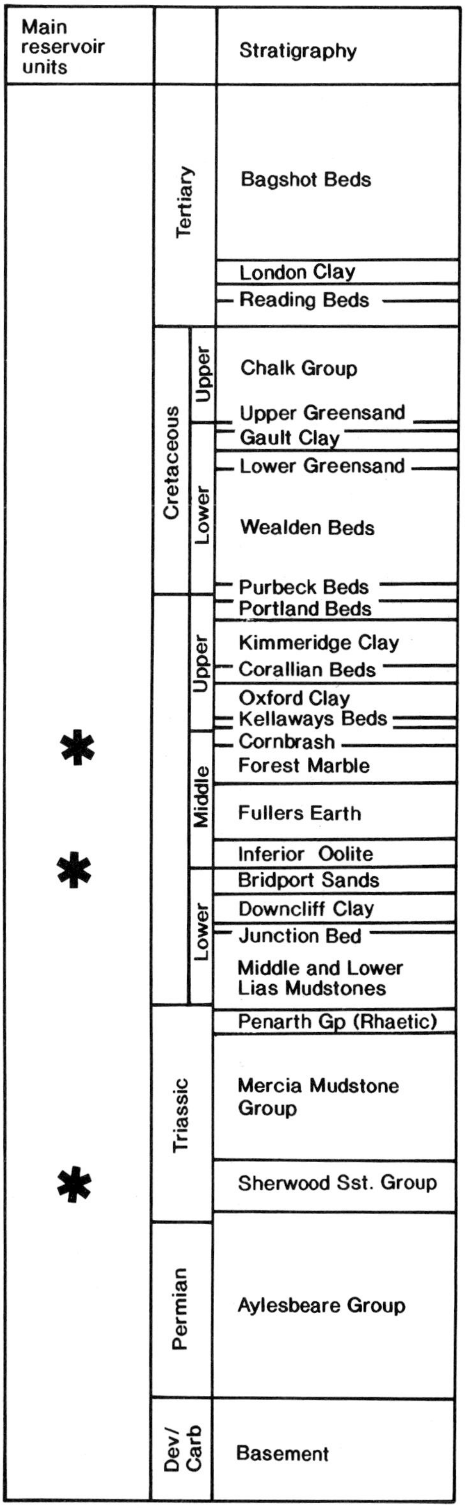

Fig. 9.7 Stratigraphy and main reservoir units of the Wytch Farm Oilfield.

Main reservoir units		Stratigraphy
	Tertiary	Bagshot Beds
		London Clay
		Reading Beds
	Cretaceous Upper	Chalk Group
		Upper Greensand
		Gault Clay
		Lower Greensand
	Cretaceous Lower	Wealden Beds
		Purbeck Beds
		Portland Beds
✱	Upper	Kimmeridge Clay
		Corallian Beds
		Oxford Clay
		Kellaways Beds
		Cornbrash
	Middle	Forest Marble
		Fullers Earth
		Inferior Oolite
✱		Bridport Sands
	Lower	Downcliff Clay
		Junction Bed
		Middle and Lower Lias Mudstones
		Penarth Gp (Rhaetic)
	Triassic	Mercia Mudstone Group
✱		Sherwood Sst. Group
	Permian	Aylesbeare Group
	Dev/ Carb	Basement

Well	Date sampled	δD (‰)	$\delta^{18}O$ (‰)	Na (mg.l^{-1})	Cl (mg.l^{-1})
B3	18-7-84	−34.4	−4.7	27910	48240
B3	7-8-84	−35.6	−5.4	27330	47300
B3	5-9-85	−35.4	−5.3	27020	46691
D4	18-7-84	−33.4	−5.3	25710	44330
D4	7-9-84	−34.1	−5.0	25550	43900
D4	5-9-85	−38.0	−5.2	25120	43261
B8	18-7-84	−33.6	−4.8	25550	44290
B8	5-9-85	−33.6	−5.2	25798	44911
D9	18-7-84	−23.7	−3.5	20160	34430
D9	7-9-84	−24.9	−3.2	19630	33600
D9	5-9-85	−21.3	−2.8	17140	29155
Injected water			−1.6	+0.2	10540	18740
Injected water			−2.3	0.0	10170	18550

Table 9.4 Stable isotopic compositions, Na and Cl concentrations of Bridport waters.

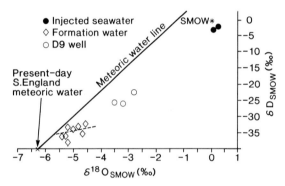

Fig. 9.8 $\delta^{18}O$ and δD compositions of injected seawater, present-day meteoric water and formation waters in the Bridport Sands reservoir.

Chemical analyses of water separated from the well-head liquids and seawater samples were performed, along with hydrogen and oxygen isotopic analyses. Sulphur isotopic analyses were also made on dissolved sulphate and gaseous H_2S.

9.6.4 Chemical and isotopic analyses of produced fluids

Chemical and isotopic analyses are listed in Table 9.4. The hydrogen and oxygen isotopic compositions of Bridport Sands formation waters fall between −33 and −38‰ and −4.5 and −5.4‰ respectively, quite distinct from seawater isotopic compositions and indicating a meteoric water source. The Bridport formation waters do not, however, plot on the present-day meteoric water line, but form a field to the right of the line (Fig. 9.8). This oxygen shift is a

commonly observed feature of formation waters and is attributed to oxygen isotope exchange between the water and isotopically heavier carbonate and silicate minerals (see Section 4.3.1 and Clayton et al., 1966).

There are three produced water isotope analyses, all from well D9, which do not fall within the formation water range described above. Both oxygen and hydrogen isotopes show heavier values, trending towards those of injected seawater (Fig. 9.8), suggesting breakthrough of seawater into well D9. However, when the Na and Cl concentrations of the D9 waters are plotted against their isotopic compositions (Fig. 9.9), they fall outside a simple seawater–formation water mixing trend, suggesting that a third component is contributing to their composition. This third component is interpreted to be recent surficial groundwaters, as the D9 data trend towards δD and $\delta^{18}O$ values of present-day southern England meteoric water. From the interpretation of these three components present in the produced waters of D9, the amount of water breakthrough in the well can be calculated using the lever rule principle (Fig. 9.9). The two end-member components in this case were assumed to be a mixture of 85% seawater and 15% entrained modern groundwater injected into the Bridport reservoir and a formation water of average isotopic and chemical composition. From this calculation, it is possible to demonstrate that seawater + modern groundwater in the produced waters of well D9 has increased from approximately 34% in 1984 to 52% in 1985.

Armed with this information, it is now possible to assess the affect of breakthrough on generation of

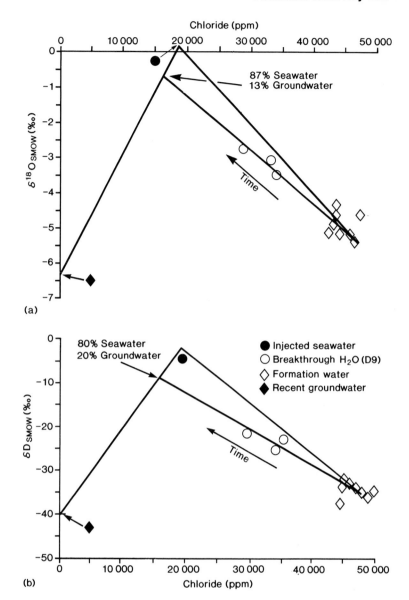

Fig. 9.9 Chloride content plotted against (a) $\delta^{18}O$ and (b) δD for injected seawater, meteoric water and well D9 mixtures.

H_2S. Table 9.5 lists the sulphur isotope ratios and sulphur species concentrations of produced fluids from the four wells. Unsurprisingly, the highest concentrations of sulphate are measured in the D9 well where seawater breakthrough has occurred. However, the concentration of H_2S measured in D9 is not as high as that in B3 or B8. It thus appears that the sulphur for the H_2S is not simply supplied by seawater. The heavy sulphate $\delta^{34}S$ values recorded in all the Bridport Sands formation waters are incompatible with either present-day or Jurassic seawater sulphate and are interpreted to have resulted from reduction of SO_4^{2-} to H_2S. (Note that the samples from D9 fall very close to the seawater sulphate isotope values reflecting the seawater breakthrough.) In a closed system, sulphate reduction enriches the H_2S in ^{32}S leaving residual sulphate enriched in ^{34}S. Excluding the D9 data, the correlation between $\delta^{34}S$ for the sulphate and sulphide in the formation waters is positive and quite strong, compatible with a closed system reduction model (Fig. 9.10). However, plots of sulphate against

Well	Date sampled	Water cut (%)	H_2S (ppm)	$\delta^{34}S$ H_2S (‰)	SO_4^{2-} (mg.l^{-1})	$\delta^{34}S$ SO_4^{2-} (‰)
B3	18-7-84	20	180	+13.5	55	+45.9
B3	7-8-84	21	180	+14.4	60	+51.5
B3	5-9-85	18	40	+0.2	45	+29.3
D4	18-7-84	8.5	11	—	180	+32.7
D4	7-9-84	11	12	+4.5	195	+33.8
D4	5-9-85	6	12	+6.8	165	+32.7
B8	18-7-84	16	35	+3.3	185	—
B8	5-9-85	40	70	+16.0	80	+38.2
D9	18-7-84	26	25	—	725	+24.9
D9	7-9-84	26	31	−4.6	740	+25.3
D9	5-9-85	30	40	—	985	+26.6
Seawater					2580	+21.3
Seawater					2250	+21.3

Table 9.5 Water cut, concentrations and sulphur isotopic composition of sulphur species in Bridport Formation waters.

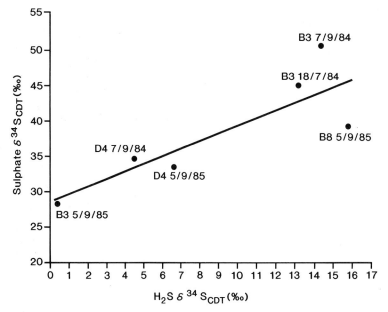

Fig. 9.10 $\delta^{34}S_{SO_4}$ vs. $\delta^{34}S_{H_2S}$ for Bridport Sands Formation waters, showing a positive correlation.

sulphide concentrations and of both against their respective isotopic compositions depart from expected closed system trends, suggesting that for pristine Bridport Sands formation waters there was differential loss of reduced sulphur and/or that the formation waters had different initial sulphate concentrations. Loss of isotopically light, early formed H_2S can be demonstrated by calculation of the mean isotopic composition of the total sulphur during a period of well production. In a closed system, the total sulphur isotopic composition should equal that of the initial sulphur (which would have been Jurassic seawater with $\delta^{34}S$ of approximately 16‰; Claypool et al., 1980). By calculating the proportions of sulphur species produced over one month and apportioning isotopic compositions accordingly, a mean total sulphur isotopic composition works out at 34 ± 4‰. Significant loss of reduced sulphur must indeed have occurred to drive the total sulphur composition towards such heavy values. Different initial sulphate concentrations in the formation waters are demonstrated by data from (for example) well B3. The sulphur chemistry of B3 water produced during September 1985 differs from that of fluids produced elsewhere in the Bridport reservoir (and from that of fluids produced at other times from well B3; Table 9.5).

The results indicate that although seawater breakthrough is occurring, it is not responsible for enhanced H$_2$S production. The Bridport reservoir fluids have a spatially and temporally variable sulphur chemistry and are losing H$_2$S, although it is not possible to quantify this loss. These conclusions indicate that prediction of H$_2$S production will be difficult or impossible. However, one final piece of evidence may be of predictive use. Production data from the wells indicate that there is a relationship between the H$_2$S content of produced fluids and water cut (Table 9.5). This relationship is interpreted to indicate greater H$_2$S concentrations in the water zone of the Bridport Sands, caused by the greater extent of the closed system sulphate reduction in the water leg. Produced water from well D4, for example, which has a low water cut and low H$_2$S concentration, is interpreted to have been derived solely from the oil leg, whereas the rapid rise in water cut and H$_2$S concentration in well B8 is interpreted to represent a major contribution from the water leg.

9.6.5 Conclusions

Oxygen and hydrogen isotopic data and chemical analyses of Bridport Sands reservoir fluids indicate that seawater breakthrough has occurred in one well, but breakthrough does not appear to be responsible for any increase in H$_2$S production. Analysis of the sulphur chemistry and sulphur isotopic composition of the produced fluids indicates that closed system sulphate reduction is responsible for H$_2$S production, chiefly in the water leg of the reservoir, although an unquantifiable amount of H$_2$S is being lost to the formation during sulphate reduction.

Although the complexities of variable sulphur chemistry in the Bridport Sands waters, along with unquantifiable H$_2$S loss, make quantitative prediction of H$_2$S concentrations somewhat fraught, the relationship of increasing H$_2$S with water cut from the water zone (rather than from seawater breakthrough) is a useful qualitative tool. Rising water production from any Bridport Sands well which is unrelated to seawater breakthrough may herald an increase in H$_2$S production.

9.7 Enhanced oil recovery – steam injection: Cold Lake Area Oil Sands, Alberta, Canada

9.7.1 Introduction

Large volumes of heavy oil (*bitumen*) are reservoired in Cretaceous sandstones at relatively shallow depths of a few hundred metres along the Alberta–Saskatchewan border in western Canada, outcropping as the famous Athabasca Tar Sands (Fig. 9.11). The Cold Lake Oil Sands in Alberta are

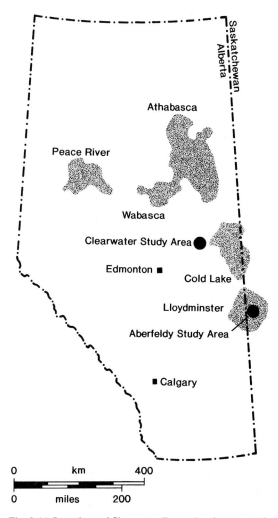

Fig. 9.11 Locations of Clearwater Formation (Section 9.7) and Aberfeldy Field (Section 9.8) study areas in western Canada (from Lefebvre & Hutcheon, 1986). Stippled areas show locations of heavy oil accumulations.

estimated to contain $75 \times 10^9\,m^3$ of bitumen* with an API gravity of 10° and a viscosity of 4000–10 000 centipoise at 37.8°C (Proctor *et al.*, 1984). The most viable recovery methods for such heavy oil are techniques which raise the temperature of the formation, promote thermal cracking and reduce bitumen viscosity. The high temperature technique normally used on Cold Lake sands is cyclic steam injection, in which 80% steam and 20% water is injected at high pressures (12 MPa) and temperatures (300–320°C) into the reservoir for periods of 40–90 days. The injected mixture is left to heat the reservoir for a further 40 days, following which steam, condensed water, oil and gas are produced from the injection wells at rates of approximately $100\,m^3$ per day (Shepherd, 1981). This process may be repeated for 10 or more cycles.

The Cretaceous reservoir sandstones of the Cold Lake area are buried to depths of 400–500 m, with ambient reservoir temperatures of about 15°C. During steam recovery, the physical conditions in the reservoir are changed dramatically, altering not only the properties of the bitumen, but also the properties of the rock. Porosities and permeabilities may be significantly reduced as new minerals are precipitated, or existing water-sensitive minerals such as swelling clays expand to fill porespaces. Minerals may also be dissolved during steam injection, enhancing porosities and permeabilities. Carbonate minerals, particularly calcite, may be susceptible to dissolution and may also contribute to CO_2 produced at the wellhead during bitumen recovery (Hutcheon *et al.*, 1990). An understanding of the pre-injection mineralogy of the rock matrix and its likely behaviour during steam injection are essential if oil recovery is to be maximized. The objective of the study described below (Hutcheon *et al.*, 1989, 1990) was to constrain the types and effects of rock–water reaction during steam injection using a combination of detailed petrography and petrophysics, along with isotopic analysis of carbonate minerals and produced CO_2.

9.7.2 Geological background

The Cold Lake deposit covers an area of approximately $9000\,km^2$ with most bitumen reservoired in the upper part of the Clearwater Formation, part of the Lower Cretaceous (Aptian–Albian) Mannville Group (Minken, 1974). The Clearwater Formation

---

* This 'oil' is difficult to get into barrels.

is up to 80 m thick in the Cold Lake area, over 50 m of which comprises sandstones deposited as deltaic distributary mouth bars and in distributary channels (Harrison *et al.*, 1981). The distributary mouth bar deposits contain the highest oil saturations and have the best porosities and permeabilities. A representative summary log is shown in Fig. 9.12. The sediments are believed to have been sourced from a distant, dominantly igneous terrain to the west and south-west and the compositional immaturity and relative coarseness of the sandstones suggest rapid deposition (Putnam & Pedskalny, 1983). Petrographic analysis of the Clearwater sandstones shows them to be fine- to medium-grained, angular to subangular and lithofeldspathic in composition. Diagenesis of the sandstones was complex, with early pyrite, quartz, potassium feldspar and albite overgrowths and calcite, chlorite, illite, siderite and smectite cements. Diagenesis also resulted in framework grain dissolution and precipitation of clinoptilolite and opal, and was followed by oil emplacement. The latest diagenetic effects in the sandstones are related to the influx of relatively fresh water causing kaolinite precipitation and oil fermentation and degradation. An additional product of oil degradation was an isotopically distinct post-bitumen calcite cement (Hutcheon *et al.*, 1989).

9.7.3 Approach

Sampled cores came from a steam injection project approximately 10 km west of Cold Lake (Fig. 9.11). The first core, from well T1, was cut before steam injection started. The second core, from well EX only 15 m away from T1, was cut after more than 2 years of continuous steam injection on a five-spot pattern (one central injection well and four surrounding production wells). Temperatures in this well had reached 260°C but were lower (200°C) by the time the core was cut. The Clearwater Formation interval selected for steam injection and sampled in both wells was a relatively clean sandstone, probably of distributary mouth bar or channel origin. Following core description of both wells, 100 thin-sections were examined in detail from both wells: 40 samples were examined by analytical SEM and a further 80 were analysed by XRD to determine bulk-rock and < 2 μm fraction mineralogies. The changes in mineralogy between the two wells were then compared to changes in porosity and permeability and to observed changes in oil production rate.

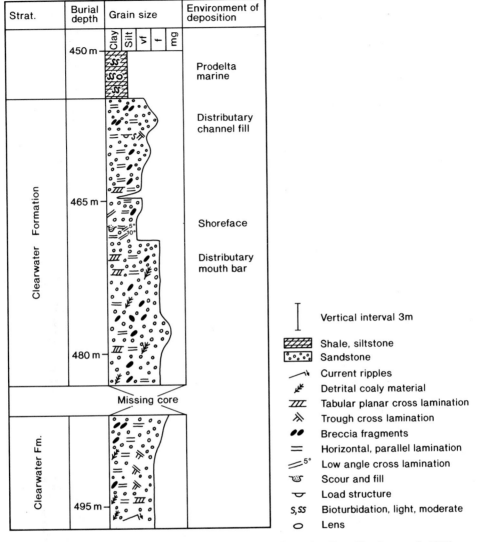

Fig. 9.12 Summary stratigraphy and sedimentology of the Clearwater Formation (from Hutcheon *et al.*, 1989).

Later work on another Cold Lake steam injection project in Clearwater sandstones was concerned more specifically with the origin of the CO_2 generated during steam injection (Hutcheon *et al.*, 1990). The approach taken here was to obtain produced water, oil and gas samples from three wells in their first cycle of steam injection and recovery. Water samples were filtered in the field, and total alkalinity, pH, sulphide and ammonia measured. Carbonate species in produced waters were preserved for isotopic analysis and CO_2 and CH_4 contents of produced gases were measured by

gas chromatography. The isotopic compositions of the separated gases were also measured, along with those of about 40 calcite samples extracted from Clearwater sandstones.

9.7.4 Petrographic and isotopic investigations of steam-induced reactions

Petrographic and petrophysical studies of wells T1 and EX showed a distinct change in rock mineralogy and reservoir quality in the steamed interval. The early diagenetic minerals were particularly susceptible to alteration, changing from an assemblage of illite, chlorite, zeolites and smectite to one dominated by smectite and analcime. A two-fold increase in the relative amount of smectite was detected by < 2 μm fraction XRD analysis; the new smectite was also coarser (4–10 μm). Analcime, a sodic zeolite, petrographically the last phase to form in the steamed interval, formed euhedral crystals 10–15 μm across. The framework grains were not substantially affected by steam injection with only a limited amount of pitting and etching of grain surfaces, although chemically less stable volcanogenic grains acted as nuclei for the precipitation of new minerals.

New mineral growth, especially that of smectite, led to decreased bulk porosity and permeability in the steamed interval and thus to decreased oil production rates. Thick smectite rims on framework grains increased the surface area of the porespace mineral assemblage, which led to an increase in the proportion of microporosity and reduced permeability. Smectite and analcime are believed to have formed by reaction of kaolinite and feldspar, mainly on the surface of lithic clasts, with sodic waters. Porosity and permeability reduction are also believed to have been partly caused by compaction of the rock framework during flushing of the oil.

As well as reducing reservoir quality, steam injection has also been associated with the production of CO_2 from the Clearwater Sandstones. CO_2 comprised about 40% of the produced gas (Hutcheon et al., 1990). Other produced gases included CH_4, water vapour and air entrained during steam injection. Isotopic analyses of the CO_2, dissolved carbonate and carbonate cements (predominantly calcite) in the reservoir indicated that nearly all the CO_2 was derived from dissolution of diagenetic calcite at temperatures between 70 and 220°C (Fig. 9.13). Alkalinity measurements of the produced waters also confirm that calcite was being dissolved (though the concentration of Ca^{2+} in the waters did not equal alkalinity, suggesting that Ca^{2+} was being consumed by reaction with clay minerals to form zeolites). Very little CO_2 was apparently derived from the pyrolysis of bitumen and this small quantity was contributed early on in the production cycle. When combined with data from other steam injection pilot studies (Cathles et al., 1987), it is evident that significant quantities of CO_2 production should be expected from formations with appropriate mineralogy: carbonates (calcite, dolomite and siderite) and silicates (kaolinite, smectite and illite).

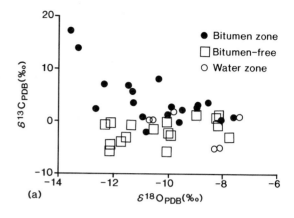

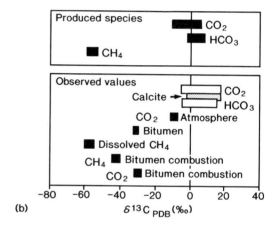

Fig. 9.13 Isotopic compositions of carbon-bearing substances from thermal recovery wells in the Cold Lake area (from Hutcheon et al., 1990). (a) Whole rock carbon and oxygen isotopic compositions of calcite in samples from Clearwater Formation sandstones. (b) Compositions of produced species from several steam injection pilots in the Clearwater Formation. Measured compositions are in black, and the range of CO_2 and HCO_3^- isotopic compositions (shown in white) is calculated from the observed range of calcite isotopic compositions (stippled). The ^{13}C-depleted CO_2 of the produced species may have been derived from several sources as illustrated (i.e. dissolved CH_4, bitumen), but the majority of the ^{13}C-enriched CO_2 must have been supplied from the breakdown of calcite.

9.7.5 Conclusions

Steam injection into Clearwater Formation sandstones has resulted in a change in rock mineralogy, with the precipitation of additional smectite and a new mineral, analcime. In general, this has resulted in a loss of porosity and reduced permeability, even in the relatively clean sandstones chosen for the steam injection pilot. Relationships between mineral growth, porosity reduction and decreased recovery rates suggest that if steam were injected into sandstones of initially lower porosity and/or higher clay content, recovery rates would be significantly reduced. CO_2 production during steam injection has apparently been caused by carbonate dissolution with Ca^{2+} being consumed by reactions with clay minerals.

9.8 Enhanced oil recovery — fireflooding: Lloydminster Area Oil Sands, Saskatchewan, Canada

9.8.1 Introduction

South of the Cold Lake Oil Sands lie the heavy oil fields of the Lloydminster area (Fig. 9.11). Like their Cold Lake counterparts, these fields contain heavy oil reservoired in Cretaceous sandstones, although the oil here can be produced in part by conventional primary and secondary recovery methods as it has densities of 16–24° API and viscosities of 400–1500 centipoise.

This case study involves monitoring a fireflooding pilot in the Aberfeldy Field, 15 km to the east of Lloydminster (Lefebvre & Hutcheon, 1986; Hutcheon & Lefebvre, 1988). The reservoir of interest is the Lower Cretaceous Sparky Interval of the Mannville Group, which has yielded between 5.2 and 8.3% of original oil-in-place during primary recovery and secondary recovery by water flooding. Enhanced oil recovery by fireflooding and steamflooding is expected to yield a recovery of 9.5% of oil-in-place (Adams, 1982). For the purposes of this case study, only the results of the fireflood will be described in detail.

9.8.2 Geological background

The Lower Cretaceous Sparky Interval is the most prolific oil producer in the Lloydminster area, containing up to 60% of the proven reserves and having the best reservoir properties (Orr *et al.*, 1977). The Sparky Interval is up to several tens of metres thick and is divided into four depositional units (Smith, 1984) with a basal bioturbated lagoonal mudstone, followed by three coarsening-upward units interpreted as shoreface deposits, which are in turn overlain by a bioturbated carbonaceous inter-

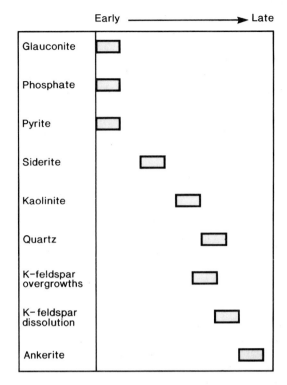

Fig. 9.14 Paragenetic history for Sparky Interval sandstones (from Hutcheon & Lebebvre, 1988).

distributary bay mudstone with crevasse splay sandstones and finally capped by carbonaceous mudstones and a coal. The overall environment of deposition is interpreted as a wave-dominated delta (van Hulten & Smith, 1984). The best reservoir properties are found in the sandstones capping the coarsening-upward units. Unlike the lithofeldspathic Clearwater sandstones of Cold Lake, these are sublitharenites with some glauconite, well sorted and mature. A variety of diagenetic minerals have been precipitated in the Sparky sandstones (Fig. 9.14) but except where carbonate cementation has been locally pervasive, porosities have remained high (up to 30%) and permeabilities may be as high as a few darcys (Adams, 1982).

9.8.3 Approach

Two cored wells were studied in detail from the fireflood area of the Aberfeldy Field by Lefebvre (1984), B7A and B7, some 50 m apart. Both wells were drilled in 1975, with only B7 being contacted by the burn front. As with the Cold Lake study described in Section 9.7, both wells were examined in detail using thin-section petrography, SEM and XRD (bulk and < 2 μm fraction) analysis of 32 samples. Porosity, permeability and residual water and oil saturation measurements were also made on core from both wells.

9.8.4 Petrographic investigation of mineralogical changes

The main mineralogical changes observed following fireflooding were illite formation, the thermal decomposition of diagenetic kaolinite (Lefebvre & Hutcheon, 1986) and the crystallization of potassium feldspar and haematite. Illite growth is generally restricted to the periphery of burned zones where coke is present. Decomposed kaolinite (as indicated by XRD) is still recognizable with SEM but is poorly formed; it is still the most volumetrically significant clay mineral present. Feldspar is most commonly observed to have grown in the intensely burned zones, forming a few small euhedral crystals a few micrometres across. The burned zones in the core show a distinct reddish coloration caused by the precipitation of small amounts of haematite (identified by XRD and SEM). Haematite is most abundant in association with shale and volcanic rock fragments, from which much of its iron was presumably derived (some of this iron was also from diagenetic siderite).

Changes in reservoir properties associated with the fireflooding were relatively minor. Porosity and permeability were slightly reduced in the burned zones, due to the formation of illite and haematite and the deposition of coke. However, these poroperm changes are of no significance to future recovery of any heavy oil in the burned

zones as oil saturation was reduced to zero in these intervals. It is important to note that the best developed burned zones actually lie beneath permeability barriers which have prevented overriding of injected air during fireflooding.

9.8.5 Conclusions

The main conclusion from this study is that fireflooding of the Sparky Interval has not resulted in volumetrically significant mineralogical changes. Consequently, the reservoir quality of the fireflooded interval has been little affected, apart from minor poroperm reductions in the burned zone in which oil saturation has been reduced to zero in any case. This contrasts markedly with the major mineralogical changes and associated deterioration of reservoir quality observed in the steamflooding pilot of the Clearwater sandstones. Lefebvre and Hutcheon (1986) attribute this different behaviour not to the different thermal recovery techniques, but to the depositional and diagenetic mineralogies of the Clearwater and Sparky sandstones. The relatively unstable mineralogies of the Clearwater sandstones − containing abundant feldspars and lithic fragments, carbonate cements and other diagenetic phases, including smectite − underwent volumetrically significant changes in mineralogy during steamflooding causing poroperm reduction. By contrast, the more stable non-reactive quartzose sands of the Sparky Interval underwent little diagenetic change and poroperm loss. There is obviously a need for a careful and thorough mineralogical and chemical description of any reservoir interval before thermal recovery techniques are initiated and irreversible damage done to the reservoir.

References

Adams, A.E., Mackenzie, W.S. & Guildford, C. (1984) *Atlas of Sedimentary Rocks under the Microscope.* Longman, London.

Adams, D.M. (1982) Experiences with waterflooding Lloydminster heavy oil reservoirs. *J. Petrol. Tech.,* **34,** 1643–1650.

Allen, P.A. & Allen, J.R. (1990) *Basin Analysis.* Blackwell Scientific Publications, Oxford.

Amaefule, J.O., Ajufo, A., Peterson, E. & Durst, K. (1987) Understanding formation damage processes: an essential ingredient for improved measurement and interpretation of relative permeability data. *Soc. Petrol. Eng.,* **16 232,** 25 pp.

Andrawes, F., Holzer, G., Roedder, E., Gibson, E.K. & Oro, J. (1984) Gas chromatographic analysis of volatiles in fluid inclusions. *J. Chromatography,* **302,** 181–193.

Aplin, A.C., Warren, E.A, Grant, S.M. & Robinson, A.G. (1993) Mechanisms of quartz cementation in North Sea reservoir sands: constraints from fluid compositions. In Horbury, A.S. & Robinson, A.G. (eds), *Diagenesis and Basin Development, AAPG Memoir* (in press).

Archer, J.S. & Wall, C.G. (1986) *Petroleum Engineering: Principles and Practice.* Graham and Trotman, London.

Aronson, J.L. & Lee, M. (1986) K/Ar systematics of bentonite and shale in a contact metamorphic zone, Cerillos, New Mexico. *Clays and Clay Minerals,* **34,** 483–487.

Awwiller, D.N. & Mack, L.E. (1991) Diagenetic modification of Sm–Nd model ages in Tertiary sandstones and shales, Texas Gulf Coast. *Geology,* **19,** 311–314.

Ayalon, A. & Longstaffe, F.J. (1988) Oxygen isotope studies of diagenesis and pore water evolution in the Western Canada sedimentary basin: evidence from the Upper Cretaceous Belly River Sandstone, Alberta. *J. Sed. Petrol.,* **58,** 489–505.

Barker, C.E. & Halley, R.B. (1988) Fluid inclusions in vadose cement with consistent vapour to liquid ratios, Pleistocene Miami Limestone, southeastern Florida. *Geochim. Cosmochim. Acta,* **52,** 1019–1025.

Barnaby, R.J. & Rimstidt, J.D. (1989) Redox conditions of calcite cementation interpreted from Mn and Fe contents of authigenic calcites. *Geol. Soc. Am. Bull.,* **101,** 795–804.

Barres, O., Burneau, A., Dubessy, J. & Pagel, M. (1987) Application of micro-FTIR spectroscopy to individual hydrocarbon fluid inclusion analysis. *Appl. Spectroscopy,* **41,** 1000–1008.

Bathurst, R.G.C. (1975) *Carbonate Sediments and their Diagenesis.* Developments in Sedimentology, 12. Elsevier, Amsterdam.

Berg, O.R. & Woolverton, D.G. (1985) Seismic stratigraphy II − an integrated approach to hydrocarbon exploration. *AAPG Memoir,* **39,** 276 pp.

Bethke, C. (1985) A numerical model of compaction-driven groundwater flow and heat transfer and its application to the paleohydrology of intracratonic sedimentary basins. *J. Geophys. Res.,* **90,** 6817–6828.

Bethke, C., Harrison. W.J., Upson, C. & Altaner, S.P. (1988) Supercomputer analysis of sedimentary basins. *Science,* **239,** 261–267.

Binns, P.R. & Bodnar, R.J. (1986) Decrepitation behaviour of fluid inclusions in quartz at one atmosphere confining pressure. *EOS,* **67,** 399.

Bird, M.I. & Chivas, A.R. (1988) Stable-isotope evidence for low-temperature kaolinitic weathering and post-formational hydrogen-isotope exchange in Permian kaolinites. *Chem. Geol. (Isotope Geosciences Section),* **72,** 249–265.

Bjørlykke, K. (1984) Secondary porosity: how important is it? *AAPG Memoir,* **37,** 277–286.

Bloch, S. (1991) Empirical prediction of porosity and permeability in sandstones. *AAPG Bull.,* **75,** 1145–1160.

Bodnar, R.J. & Bethke, P.M. (1984) Systematics of stretching of fluid inclusions I: fluorite and sphalerite at 1 atmosphere confining pressure. *Econ. Geol.,* **79,** 141–161.

Bodnar, R.J., Binns, P.R. & Hall, D.L. (1989) Synthetic fluid inclusions IV: quantitative evaluation of the decrepitation behaviour of fluid inclusions in quartz at one atmosphere confining pressure. *J. Metamorphic. Geol.,* **7,** 229–242.

Bray, C.J., Spooner, E.T.C., Hall, C.M., York, D., Bills, T.M. & Krueger, H.W. (1987) Laser probe ^{40}Ar/^{39}Ar and conventional K/Ar dating of illites associated with the McClean unconformity-related uranium deposits, north Saskatchewan, Canada. *Can. J. Earth Sci.,* **24,** 10–23.

Brereton, N.R. (1970) Corrections for interfering isotopes in the ^{40}Ar/^{39}Ar dating method. *Earth*

Planet. Sci. Lett., **8**, 427–433.

Brindley, G.W. (1981) X-ray diffraction (with ancillary techniques) of clay minerals. In Longstaffe, F.J. (ed.), *Clays and the Resource Geologist*, pp. 22–35. *Min. Soc. Canada Short Course Notes*.

Brindley, G.W. & Brown, C. (1980) *Crystal Structures of Clay Minerals and their X-ray Identification*. Mineralogical Society, London.

Brint, J.F., Hamilton, P.J., Haszeldine, R.S., Fallick, A.E. & Brown, S. (1991) Oxygen isotopic analysis of diagenetic quartz overgrowths from the Brent Sands: a comparison of two preparation methods. *J. Sed. Petrol.*, **61**, 527–533.

Brown, D.W., Floyd, A.J. & Sainsbury, M. (1988) *Organic Spectroscopy*. John Wiley and Sons, Chichester.

Burke, W.H., Denison, R.E., Hetherington, E.A., Koepnick, R.B., Nelson, H.F. & Otto, J.B. (1982) Variation of seawater $^{87}Sr/^{86}Sr$ throughout Phanerozoic time. *Geology*, **10**, 516–519.

Burley, S.D. (1984) Patterns of diagenesis in the Sherwood Sandstone Group (Triassic), United Kingdom. *Clay Minerals*, **19**, 403–440.

Burley, S.D. (1986) The development and destruction of porosity within Upper Jurassic reservoir sandstones of the Piper and Tartan Fields, Outer Moray Firth, North Sea. *Clay Minerals*, **21**, 649–694.

Burley, S.D., Kantorowicz, J.D. & Waugh, B. (1985) Clastic diagenesis. In Brenchley, P.J. & Williams, B.P.J. (eds), *Sedimentology: Recent Developments and Applied Aspects*, pp. 189–226. Geological Society Special Publication, 18. Blackwell Scientific Publications, Oxford.

Burley, S.D., Mullis, J. & Matter, A. (1989) Timing diagenesis in the Tartan Reservoir (UK North Sea): constraints from combined cathodoluminescence microscopy and fluid inclusion studies. *Mar. Petrol. Geol.*, **6**, 98–120.

Burruss, R.C. (1987) Diagenetic palaeotemperatures from aqueous fluid inclusions: re-equilibration of inclusions in carbonate cements by burial heating. *Min. Mag.*, **51**, 477–481.

Burruss, R.C. (1989) Palaeotemperatures from fluid inclusions: advances in theory and technique. In Naeser, N.D. (ed.), *Thermal History of Sedimentary Basins: Methods and Case Histories*, pp. 119–131. Springer-Verlag, Berlin.

Carothers, W.W., Adami, L.H. & Rosenbauer, R.J. (1988) Experimental oxygen isotope fractionation between siderite–water and phosphoric acid liberated CO_2–siderite. *Geochim. Cosmochim. Acta*, **52**, 2445–2450.

Cathles, L.M., Schoell, M. & Simon, R. (1987) CO_2 generation during steamflooding: a geologically based kinetic theory that includes carbon isotope effects and application to high temperature steamfloods. In *SPE Int. Symp. Oilfield Chem., Feb. 4–6, 1987, San Antonio*, pp. 255–270. *Soc. Petrol. Eng.*, **16267**.

Chaudhuri, S., Broedel, V. & Clauer, N. (1987) Strontium isotopic evolution of oil-field waters from carbonate reservoir rocks from Bindley Field, central Kansas, USA. *Geochim. Cosmochim. Acta*, **51**, 43–53.

Chaudhuri, S. & Cullers, R.L. (1979) The distribution of rare earth elements in deeply buried Gulf Coast sediments. *Chem. Geol.*, **24**, 327–338.

Chiba, H., Kusakabe, M., Hirano, S., Matsuo, S. & Somiya, S. (1981) Oxygen isotope fractionation factors between anhydrite and water from 100°C to 550°C. *Earth Planet. Sci. Lett.*, **53**, 55–62.

Chiba, H. & Sakai, H. (1981) Oxygen isotope exchange rate between dissolved sulfate and water at hydrothermal temperatures. *Geochim. Cosmochim. Acta*, **49**, 993–1000.

Choquette, P.W. & Pray, L.C. (1970) Geologic nomenclature and classification of porosity in sedimentary carbonates. *AAPG Bull.*, **54**, 207–250.

Claypool, G.E., Holser, W.T., Kaplan, I.R., Sakai, H. & Zak, I. (1980) The age curves of sulfur and oxygen isotopes in marine sulfate and their mutual interpretation. *Chem. Geol.*, **28**, 199–260.

Clayton, C.J. (1991) Effect of maturity on carbon isotope ratios of oils and condensates. *Org. Geochem.*, **17**, 887–899.

Clayton, R.N., Friedman, I., Graf, D.L., Mayeda, T.K., Meents, W.F. & Shimp, N.F. (1966) The origin of saline formation waters: I. Isotopic composition. *J. Geophys. Res.*, **71**, 3869–3882.

Clayton, R.N., Muffler, L.J.P. & White, D.E. (1968) Oxygen isotope study of calcite and silicates of the River Ranch No. 1 well, Salton Sea geothermal field, California. *Am. J. Sci.*, **266**, 968–979.

Coleman, M.L. & Raiswell, R. (1981) Carbon, oxygen and sulphur isotope variations in concretions from the Upper Lias of N.E. England. *Geochim. Cosmochim. Acta*, **45**, 329–340.

Collins, P.L.F. (1979) Gas hydrates in CO_2-bearing fluid inclusions and the use of freezing data for estimation of salinity. *Econ. Geol.*, **74**, 1435–1444.

Comings, B.D. & Cercone, K.R. (1986) Experimental contamination of fluid inclusions in calcite. *SEPM Annual Mid-year Meeting Abstracts*, **3**, 24.

Coustau, H., Gauthier, J., Kulbicki, G. & Winnock, E. (1970) Hydrocarbon distribution in the Aquitaine Basin of SW France. In *The Exploration for Petroleum in Europe and North Africa*, pp. 73–85. Institute of Petroleum.

Craig, H. (1957) Isotopic standards for carbon and oxygen and correction factors for mass-spectrometric analysis of carbon dioxide. *Geochim. Cosmochim. Acta*, **12**, 133–149.

Craig, H. (1961) Isotopic variation in meteoric waters. *Science*, **133**, 1702–1703.

Crowe, C.W. (1986) Precipitation of hydrated silica from spent hydrofluoric acid: how much of a problem is it? *J. Petrol. Tech.*, **38**, 1234–1240.

Curnelle, R., Dubois, P. & Sequin, J.C. (1982) The Mesozoic–Tertiary evolution of the Aquitaine Basin. *Phil. Trans. R. Soc. Lond.*, **A305**, 63–84.

Curtis, C.D., Petrowski, C. & Oertel, G. (1972) Stable carbon isotope ratios within carbonate concretions: a clue to place and time of formation. *Nature*, **235**, 98–100.

Dake, L.P. (1978) *Fundamentals of Reservoir Engineering*. Elsevier, Amsterdam.

Dake, L.P. (1982) Application of the repeat formation tester in vertical and horizontal pulse testing in the Middle Jurassic Brent Sands. Paper 270, *European Petroleum Conference*, London.

Dalrymple, G.B. & Lanphere, M.A. (1969) *Potassium–Argon Dating*. W.H. Freeman and Co., New York.

Dalrymple, G.B. & Lanphere, M.A. (1971) ^{40}Ar/^{39}Ar technique of K/Ar dating: a comparison with the conventional technique. *Earth Planet. Sci. Lett.*, **12**, 300–308.

Damon, P.E. (1970) A theory of 'real' K–Ar clocks. *Eclogae Geol. Helv.*, **63**, 69–76.

De'Ath, N.G. & Schuyleman, S.F. (1981) The geology of the Magnus oilfield. In Illing, L.V. & Hobson, G.D. (eds), *Petroleum Geology of the Continental Shelf of North-west Europe*, pp. 342–351. Heyden and Sons, London.

Deegan, C.E. & Scull, B.J. (1977) A proposed lithostratigraphic nomenclature for the Central and Northern North Sea. *Inst. Geol. Sci. Rep.*, **77/25**.

Deer, W.A., Howie, R.A. & Zussman, J. (1977) *An Introduction to the Rock Forming Minerals*. Longman, London.

Deines, P. (1980) The isotopic composition of reduced organic carbon. In Fritz, A.P. & Fontes, J.C. (eds), *Handbook of Environmental Isotope Geochemistry, Volume 1: the Terrestrial Environment*, pp. 329–406. Elsevier, Amsterdam.

Deines, P., Langmuir, D. & Harmon, R.S. (1974) Stable carbon isotope ratios and the existence of a gas phase in the evolution of carbonate ground water. *Geochim. Cosmochim. Acta*, **38**, 1147–1164.

Deloule, E., Albarede, F. & Sheppard, S.M.F. (1991) Hydrogen isotope heterogeneities in the mantle from ion probe analysis of amphiboles from ultramafic rocks. *Earth Planet. Sci. Lett.*, **105**, 543–559.

Demming, D., Nunn, J.A. & Evans, D.G. (1990) Thermal effects of compaction-driven groundwater flow from overthrust belts. *J. Geophys. Res.*, **95**, 6669–6683.

DePaolo, D.J. (1986) Detailed record of the Neogene Sr isotopic evolution of seawater from DSDP Site 590B. *Geology*, **14**, 103–106.

DePaolo, D.J. & Ingram, B.L. (1985) High resolution stratigraphy with strontium isotopes. *Science*, **227**, 938–941.

Dickson, J.A.D. (1966) Carbonate identification and genesis as revealed by staining. *J. Sed. Petrol.*, **36**, 491–505.

Dickson, J.A.D. (1983) Graphical modelling of crystal aggregates and its relevance to cement diagnosis. *Phil. Trans. R. Soc. London*, **A309**, 465–502.

Dickson, J.A.D. (1991) Disequilibrium carbon and oxygen isotope variations in natural calcite. *Nature*, **353**, 842–844.

Dickson, J.A.D., Smalley, P.C. & Kirkland, B.L. (1991) Carbon and oxygen isotopes in Pennsylvanian biogenic and abiogenic aragonite (Otero County, New Mexico): a laser microprobe study. *Geochim. Cosmochim. Acta*, **55**, 2607–2613.

Dixon, S.A., Summers, D.M. & Surdam, R.C. (1989) Diagenesis and preservation of porosity in Norphlet Formation (Upper Jurassic), southern Alabama. *AAPG Bull.*, **73**, 707–728.

Dorobek, S.L. (1987) Petrography, geochemistry and origin of burial diagenetic facies, Siluro-Devonian Heldeberg Group (Carbonate rocks), Central Appalachians. *AAPG Bull.*, **71**, 492–514.

Dranfield, P., Begg, S.H. & Carter, R.R. (1987) Wytch Farm Oilfield: reservoir characterization of the Triassic Sherwood Sandstone for input to reservoir simulation studies. In Brooks, J. & Glennie, K. (eds), *Petroleum Geology of North West Europe*, pp. 149–160. Graham and Trotman, Dordrecht.

Dravis, J.D. & Yurewicz, D.A. (1985) Enhanced carbonate petrography using fluorescence microscopy. *J. Sed. Petrol.*, **55**, 795–804.

Dubessy, J., Audeoud, D., Wilkins, R. & Kosztolanyi, C. (1982) The use of the Raman microprobe MOLE in the determination of the electrolytes dissolved in the aqueous phase of fluid inclusions. *Chem. Geol.*, **37**, 137–150.

Ebanks, W.J. (1987) Geology in enhanced oil recovery. In Tillman, R.W. & Weber, K.J. (eds), *Reservoir Sedimentology*, pp. 1–14. *SEPM Special Publication*, **40**.

Egeberg, P.K., Smalley, P.C. & Aagaard, P. (1990) Strontium isotope geochemistry of Leg 113 interstitial waters and carbonates. In Barker, P.F. *et al.* (eds), *Proc. ODP Sci. Results, 113*, pp. 147–157.

Ocean Drilling Program, College Station, TX.

Ehrlich, R., Crabtree, S.J., Kennedy, S.K. & Cannon, R.L. (1984) Petrographic image analysis I: analysis of reservoir pore complexes. *J. Sed. Petrol.*, **54**, 1365–1376.

Ehrlich, R. & Davies, D.K. (1989) Image analysis of pore geometry: relationships to reservoir engineering and modelling. *Soc. Petrol. Eng.*, **19054**, SPE Gas Technology Conference, Dallas, Texas, 1530.

Elderfield, H. & Gieskes, J.M. (1982) Sr isotopes in interstitial waters from Deep Sea Drilling Project cores. *Nature*, **300**, 493–497.

Eldridge, C.S., Compston, W., Williams, I., Both, R.A., Walshe, J.C. & Ohmoto, H. (1988) Sulfur isotope variability in sediment-hosted massive sulfide deposits as determined using the ion microprobe SHRIMP: I. An example from the Rammelsberg orebody. *Econ. Geol.*, **83**, 443–449.

Emery, D. & Dickson, J.A.D. (1989) A syndepositional meteoric phreatic lens in the Middle Jurassic Lincolnshire Limestone, England, UK. *Sed. Geol.*, **65**, 273–284.

Emery, D., Dickson, J.A.D. & Smalley, P.C. (1987) The strontium isotopic composition and origin of burial cements in the Lincolnshire Limestone (Bajocian) of central Lincolnshire, England. *Sedimentology*, **34**, 795–806.

Emery, D., Hudson, J.D., Marshall, J.D. & Dickson, J.A.D. (1988) The origin of late spar cements in the Lincolnshire Limestone, Jurassic of central England. *J. Geol. Soc. Lond.*, **145**, 621–633.

Emery, D. & Marshall, J.D. (1989) Zoned calcite cements: has analysis outpaced interpretation? *Sed. Geol.*, **65**, 205–210.

Emery, D. & Myers, K.J. (1990) Ancient subaerial exposure and freshwater leaching in sandstones. *Geology*, **18**, 1178–1181.

Erlich, R.N., Barrett, S.F. & Guo Bai Ju (1990) Seismic and geological characteristics of drowning events on carbonate platforms. *AAPG Bull.*, **74**, 1523–1537.

Eslinger, E. & Pevear, D. (1988) Clay minerals for petroleum geologists and engineers. *SEPM Short Course*, **22**, 411 pp.

Eslinger, E.V. & Savin, S.M. (1973) Oxygen isotope geothermometry of the burial metamorphic rocks of the Precambrian Belt Supergroup, Glacier National Park, Montana. *Geol. Soc. Am. Bull.*, **84**, 2549–2560.

Etminan, H. & Hoffman, C.F. (1989) Biomarkers in fluid inclusions: a new tool in constraining source regimes and its implications for the genesis of Mississippi Valley-type deposits. *Geology*, **17**, 19–22.

Fabre, D. & Couty, R. (1986) Investigations in the density effects in the Raman spectrum of methane up to 3000 bars. *C. R. Acad. Sci.*, **303**, 1305–1308.

Fairchild, I.J. (1983) Chemical controls of cathodoluminescence of natural dolomites and calcites: new data and review. *Sedimentology*, **30**, 579–583.

Faure, G. (1986) *Principles of Isotope Geology*, 2nd edn. John Wiley and Sons, Chichester.

Fisher, R.S. & Land, L.S. (1986) Diagenetic history of Eocene Wilcox sandstones, south-central Texas. *Geochim. Cosmochim. Acta*, **50**, 551–561.

Folk, R.L. (1974) *Petrology of Sedimentary Rocks*. Hemphill, Austin, TX.

Foster, D.G. & Robinson, A.G. (1993) Geological history of the Flemish Pass Basin, offshore Newfoundland. *AAPG Journal* (in press).

Frank, J.R., Carpenter, A.B. & Oglesby, T.W. (1982) Cathodoluminescence and composition of calcite cement in the Taum Sauk Limestone (Upper Cambrian), southeast Missouri. *J. Sed. Petrol.*, **52**, 631–638.

Friedman, I. & O'Neil, J.R. (1977) Compilation of stable isotope fractionation factors of geochemical interest. In Fleischer, M. (ed.), *Data of Geochemistry*, 6th edn. *USGS Prof. Paper*, **440-KK**, 12 pp.

Froehlich, P.N., Klinkhammer, G.P., Bender, M.L., Luedtke, N.A., Heath, G.R., Cullen, D., Dauphin, P., Hammond, D., Hartman, B. & Maynard, V. (1979) Early oxidation of organic matter in pelagic sediments of the eastern equatorial Atlantic: suboxic diagenesis. *Geochim. Cosmochim. Acta*, **43**, 1075–1090.

Galehouse, J.S. (1971a) Point counting. In Carver, R.E. (ed.), *Procedures in Sedimentary Petrology*, pp. 385–407. Wiley-Interscience, New York.

Galehouse, J.S. (1971b) Sedimentation analysis. In Carver, R.E. (ed.), *Procedures in Sedimentary Petrology*, pp. 65–94. Wiley-Interscience, New York.

Games, L.M., Hayes, J.M. & Gunsalis, R.P. (1978) Methane producing bacteria: natural fractionation of the stable carbon isotopes. *Geochim. Cosmochim. Acta*, **42**, 1295–1297.

Garven, G. (1989) A hydrogeologic model for the formation of the giant oil sands deposits of the Western Canada Sedimentary Basin. *Am. J. Sci.*, **289**, 105–166.

Girard, J.P., Aronson, J.L. & Savin, S.M. (1988) Separation, K/Ar dating and $^{18}O/^{16}O$ ratio measurements of diagenetic K-feldspar overgrowths: an example from the Lower Cretaceous arkoses of the Angola Margin. *Geochim. Cosmochim. Acta*, **52**, 2207–2214.

Glasmann, J.R. (1987) Argon diffusion in illite during diagenesis: how good is the K/Ar clock? In *Clay*

Minerals Society, 24th Annual Meeting, Socorro, New Mexico, p. 60.

Gluyas, J.G. & Coleman, M.L. (1992) Material flux and porosity changes during sandstone diagenesis. *Nature*, **356**, 52–54.

Gluyas, J.G. Grant, S.M. & Robinson, A.G. (1993) Geochemical evidence for a temporal control on sandstone cementation. In Horbury, A.S. & Robinson A.G. (eds), *Diagenesis and Basin Development, AAPG Memoir* (in press).

Goldhaber, M.B. & Kaplan, I.R. (1974) The sulfur cycle. In Goldberg, E.D. (ed.), *The Sea*, Vol. 5, pp. 569–665. John Wiley and Sons, Chichester.

Goldstein, R.H. (1986) Reequilibration of fluid inclusions in low-temperature calcium-carbonate cement. *Geology*, **14**, 792–795.

Goldstein, S.J. & Jacobsen, S.B. (1988) Nd and Sm systematics of river water suspended material: implications for crustal evolution. *Earth Planet. Sci. Lett.*, **87**, 249–265.

Goll, R.M. & Skarbo, O. (1990) High resolution dating of Cenozoic sediments from northern North Sea using ^{87}Sr/^{86}Sr stratigraphy: discussion. *AAPG Bull.*, **74**, 1283–1286.

Grant, S.M. & Oxtoby, N.H. (1992) The timing of quartz cementation in Mesozoic sandstones from Haltenbanken, offshore mid-Norway: fluid inclusion evidence. *J. Geol. Soc. Lond.*, **149**, 479–482.

Graue, E., Helland-Hansen, W., Johnson, J., Lomo, L., Nottvedt, A., Ronning, K., Ryseth, A. & Steel, R. (1987) Advance and retreat of the Brent delta system, Norwegian North Sea. In Brooks, J. & Glennie, K.W. (eds), *Petroleum Geology of North West Europe*, pp. 915–937. Graham and Trotman, Dordrecht.

Griffin, O.G. (1954) A new international standard for the quantitative X-ray analysis of shales and mine dust. *Res. Rep. S. Af. Mines Res. Est.*, **101**, 1–25.

Grover, G.H. & Read, J.F. (1983) Paleoaquifer and deep burial related cements defined by regional cathodoluminescence patterns, Middle Ordovician carbonates, Virginia. *AAPG Bull.*, **67**, 1275–1303.

Guilhaumou, N., Zanier-Szydlowski, N. & Enguehard, F. (1990a) Micro FTIR analysis of hydrocarbon bearing fluid inclusions: qualitative and quantitative aspects. In *Third Biennial Pan-American Conference on Research on Fluid Inclusions (PACROFI) Abstracts*, p. 39. University of Toronto.

Guilhaumou, N., Szydlowski, N. & Pradier, B. (1990b) Characterisation of hydrocarbon fluid inclusions by infra-red and fluorescence microspectrometry. *Min. Mag.*, **54**, 311–324.

Hagemann, H.W. & Hollerbach, A. (1986) The fluorescence behaviour of crude oils with respect to their thermal maturation and degradation. *Org. Geochem.*, **10**, 473–480.

Hagemann, R., Nief, G. & Roth, E. (1970) Absolute isotopic scale for deuterium analysis of natural waters. Absolute D/H ratio for SMOW. *Tellus*, **22**, 712–715.

Hamilton, P.J., Giles, M.R. & Ainsworth, P. (1992) K–Ar dating of illites in Brent Group reservoirs: a regional perspective (in press).

Hamilton, P.J., Kelley, S. & Fallick, A.E. (1989) K–Ar dating of illite in hydrocarbon reservoirs. *Clay Minerals*, **24**, 215–231.

Hanor, J.S. (1980) Dissolved methane in sedimentary brines: potential effect on the *PVT* properties of fluid inclusions. *Econ. Geol.*, **75**, 603–617.

Haq, B.U., Hardenbol, J. & Vail, P.R. (1987) Chronology of fluctuating sea levels since the Triassic. *Science*, **235**, 1156–1167.

Hardy, R.G. & Tucker, M.E. (1988) X-ray powder diffraction of sediments. In Tucker, M.E. (ed.), *Techniques in Sedimentology*, pp. 191–228. Blackwell Scientific Publications, Oxford.

Harrison, A.G. & Thode, H.G. (1957) The kinetic isotope effect in the chemical reduction of sulphate. *Transactions of the Faraday Society*, **53**, 1648–1651.

Harrison, D.B., Glaister, R.P & Nelson, H.W. (1981) Reservoir description of the Clearwater oil sand, Cold Lake, Alberta. In Meyer, R.F. & Steele, C.T. (eds), *The Future of Heavy Crude Oils and Tar Sands*, pp. 264–280. McGraw-Hill, London.

Harrison, T.M. & McDougall, I. (1981) Excess ^{40}Ar in metamorphic rocks from Broken Hill, New South Wales: implications for ^{40}Ar/^{39}Ar age spectra and the thermal history of the region. *Earth Planet. Sci. Lett.*, **55**, 123–149.

Harrison, W.J. & Summa, L.L. (1991) Paleohydrology of the Gulf of Mexico basin. *Am. J. Sci.*, **291**, 109–176.

Hart, S.R. (1964) The petrology and isotopic mineral age relations of a contact zone in the Front Range, Colorado. *J. Geol.*, **72**, 493–525.

Harwood, G. (1988) Microscopical techniques: II. Principles of sedimentary petrography. In Tucker, M.E. (ed.), *Techniques in Sedimentology*, pp. 108–173. Blackwell Scientific Publications, Oxford.

Hearn, P.P. (1987) A quantitative technique for determining the mass fractions of authigenic and detrital K-feldspar in mineral separates. *Scanning Microscopy*, **1**, 1039–1043.

Hearn, P.P. & Sutter, J.F. (1985) Authigenic potassium feldspar in Cambrian carbonates: evidence of Alleghanian brine migration. *Science*, **228**, 1529–1531.

Hearn, P.P., Sutter, J.F. & Belkin, H.E. (1987) Evidence for late-Paleozoic brine migration in Cambrian carbonate rocks of the central and southern Appalachians: implications for Mississippi Valley-type sulfide mineralisation. *Geochim. Cosmochim. Acta*, **51**, 1323–1334.

Heaviside, J., Langley, G.O. & Pallatt, N. (1983) Permeability characteristics of Magnus Reservoir Rock. In *8th European Formation Evaluation Symposium, March, 1983*, pp. 1–29. London.

Hedgpeth, J.W. (1957) *Treatise on Marine Ecology and Palaeoecology. Geol. Soc. Amer. Mem.*, **67**, 1269 pp.

Hess, J., Bender, M.L. & Schilling, J.G. (1985) Evolution of the ratio strontium-87 to strontium-86 in seawater from Cretaceous to Present. *Science*, **231**, 979–984.

Hewitt, C.H. (1963) Analytical techniques for recognizing water-sensitive reservoir rocks. *J. Petrol. Tech.*, **15**, 813–818.

Hiscott, R.N., Wilson, R.C.L., Gradstein, F.M., Pujalte, V., Garcia-Mondejar, J., Boudreau, R.R. & Wishart, H.A. (1990) Comparative stratigraphy and subsidence history of Mesozoic rift basins of North Atlantic. *AAPG Bull.*, **74**, 60–76.

Hoefs, J. (1987) *Stable Isotope Geochemistry*, 3rd edn. Springer-Verlag, Berlin.

Hogg, A.J.C. (1989) *Petrographic and isotopic constraints on the diagenesis and reservoir properties of the Brent Group sandstones, Alwyn South, Northern UK North Sea*. PhD Thesis, University of Aberdeen.

Horbury, A.D. & Adams, A.E. (1989) Meteoric phreatic diagenesis in cyclic late Dinantian carbonates, northwest England. *Sed. Geol.*, **65**, 319–344.

Horsfield, B. & McLimans, R.K. (1984) Geothermometry and geochemistry of aqueous and oil-bearing fluid inclusions from Fateh Field, Dubai. *Org. Geochem.*, **6**, 733–740.

Houghton, H.F. (1980) Refined techniques for staining plagioclase and alkali feldspars in thin section. *J. Sed. Petrol.*, **50**, 529–631.

Houseknecht, D.W. (1988) Intergranular pressure solution in four quartzose sandstones. *J. Sed. Petrol.*, **58**, 228–246.

Hubbard, R.J., Pape, J. & Roberts, D.G. (1985) Depositional sequence mapping as a technique to establish tectonic and stratigraphic framework and evaluate hydrocarbon potential on a passive continental margin. In Berg, O.R. & Woolverton, D.G. (eds), *Seismic Stratigraphy II − an Integrated Approach to Hydrocarbon Exploration*, pp. 79–92. *AAPG Memoir*, **39**.

Hudson, J.D. (1977) Stable isotopes and limestone lithification. *J. Geol. Soc. Lond.*, **133**, 637–660.

Huggett, J.M. (1984) Controls on mineral authigenesis in Coal Measures sandstones of the East Midlands, UK. *Clay Minerals*, **19**, 342–357.

Huggett, J.M. (1986) An SEM study of phyllosilicate diagenesis in sandstones and mudstones in the Westphalian Coal Measures using back-scattered electron microscopy. *Clay Minerals*, **21**, 603–616.

Hutcheon, I., Abercrombie, H.J., Putnam, P., Gardner, R. & Krouse, H.R. (1989) Diagenesis and sedimentology of the Clearwater Lake Formation at Tucker Lake. *Bull. Can. Petrol. Geol.*, **37**, 83–97.

Hutcheon, I., Abercrombie, H.J. & Krouse, H.R. (1990) Inorganic origin of carbon dioxide during low temperature thermal recovery of bitumen: chemical and isotopic evidence. *Geochim. Cosmochim. Acta*, **54**, 165–171.

Hutcheon, I. & Lefebvre, R. (1988) Sedimentology, diagenesis and thermal effects on petrophysical properties in the Aberfeldy Field, Saskatchewan. *Bull. Can. Petrol. Geol.*, **36**, 70–85.

Hutcheon, I., Oldershaw, A. & Ghent, E.P. (1980) Diagenesis of Cretaceous sandstones of the Koutenay Formation at Elk Valley (southeastern British Columbia) and Mt. Allen (southwestern Alberta). *Geochim. Cosmochim. Acta*, **44**, 1425–1435.

Ireland, B.J., Curtis, C.D. & Whiteman, J.A. (1983) Compositional variations within some glauconites and illite and implications for their stability and origins. *Sedimentology*, **30**, 769–786.

Irwin, H., Curtis, C.D. & Coleman, M. (1977) Isotopic evidence for source of diagenetic carbonates formed during burial of organic-rich sediments. *Nature*, **269**, 209–213.

Jackson, M.L. (1979) *Soil Chemical Analysis − Advanced Course*, 2nd edn. (Published by author).

Jacobsen, S.B. & Wasserburg, G.J. (1980) Sm−Nd isotopic evolution of chondrites. *Earth Planet. Sci. Lett.*, **50**, 139–155.

Jahn, B.M. (1988) Pb−Pb dating of young marbles from Taiwan. *Nature*, **332**, 429–432.

Jensenius, J. (1987) High temperature diagenesis in shallow Chalk reservoir, Skjold oil field, Danish North Sea: evidence from fluid inclusions and oxygen isotopes. *AAPG Bull.*, **71**, 1378–1386.

Jensenius, J. & Burruss, R.C. (1990) Hydrocarbon−water interactions during brine migration: evidence from hydrocarbon inclusions in calcite cements from Danish North Sea oil fields. *Geochim. Cosmochim. Acta*, **54**, 705–713.

Jensenius, J., Buchardt, B., Jørgensen, N.O. & Pedersen, S. (1988) Carbon and oxygen isotopic studies of the chalk reservoir in the Skjold oilfield, Danish North Sea: implications for diagenesis. *Chem. Geol. (Isotope Geoscience Section)*, **73**, 97–107.

Jørgensen, N.O. (1987) Oxygen and carbon isotope compositions of Upper Cretaceous chalk from the Danish sub-basin and the North Sea Central Graben. *Sedimentology*, **34**, 559–570.

Jourdan, A., Thomas, M., Brevart, O., Robson, P., Sommer, F. & Sullivan, M. (1987) Diagenesis as the control of the Brent Sandstone reservoir properties in the Greater Alwyn area (East Shetland Basin). In Brooks, J. & Glennie, K. (eds), *Petroleum Geology of North West Europe*, pp. 951–961. Graham and Trotman, Dordrecht.

Kantorowicz, J.D. (1985) The petrology and diagenesis of Middle Jurassic clastic sediments, Ravenscar Group, Yorkshire. *Sedimentology*, **32**, 833–853.

Kaplan, I.R. (1983) Stable isotopes of sulfur, nitrogen and deuterium in Recent marine environments. In *Stable Isotopes in Sedimentary Geology*, pp. 2.1–2.108. *SEPM Short Course*, **10**.

Kastner, M. (1971) Authigenic feldspars in carbonate rocks. *Am. Miner.*, **56**, 1403–1442.

Kearsley, A. & Wright, V.P. (1988) Geological applications of scanning cathodoluminescence microscopy. *Microscopy & Analysis*, **September**.

Kennedy, W.J. & Garrison, R.E. (1975) Morphology and genesis of nodular chalks and hardgrounds in the Upper Cretaceous of southern England. *Sedimentology*, **22**, 311–386.

Kerr, P.F. (1959) *Optical Mineralogy*. McGraw-Hill, London.

Khilar, K.C. & Fogler, H.S. (1984) The existence of a critical salt concentration for particle release. *J. Colloid Interface Sci.*, **101**, 214–224.

Kita, I., Taguchi, S. & Matsubaya, O. (1985) Oxygen isotope fractionation between amorphous silica and water at 34–93°C. *Nature*, **314**, 63–64.

Klug, H.P. & Alexander, L.E. (1974) *X-ray Diffraction Procedures for Polycrystalline and Amorphous Materials*. John Wiley and Sons, New York.

Knauth, L.P. & Beeunas, M.A. (1986) Isotope geochemistry of fluid inclusions in Permian halite with implications for the isotopic history of ocean water and the origin of saline formation waters. *Geochim. Cosmochim. Acta*, **50**, 419–433.

Koepnick, R.B., Burke, W.H., Denison, R.E., Hetherington, E.A., Otto, J.B. & Waite, L.E. (1985) Construction of the seawater $^{87}Sr/^{86}Sr$ for the Cenozoic and Cretaceous: supporting data. *Chem. Geol. (Isotope Geoscience Section)*, **58**, 55–81.

Kralik, M. (1984) Effects of cation exchange treatments and acid leaching on the Rb–Sr system of illite from Fithian, Illinois. *Geochim. Cosmochim. Acta*, **48**, 527–533.

Krouse, H.R., Ritchie, R.G.S. & Roche, R.S. (1987) Sulphur isotope composition of H_2S evolved during the non-isothermal pyrolysis of suphur-containing materials. *J. Anal. Appl. Phys.*, **12**, 19–29.

Krouse, H.R. & Tabatabai, M.A. (1986) Stable sulfur isotopes. In *Sulfur in Agriculture*, Agronomy Monograph, 27, pp. 169–205. Soil Science Society of America, Madison, WI.

Krouse, H.R., Vian, C.A., Eliuk, L.S., Ueda, A. & Halas, S. (1988) Chemical and isotopic evidence for thermochemical sulphate reduction by light hydrocarbon gases in deep carbonate reservoirs. *Nature*, **333**, 415–419.

Lambert, S.J. & Epstein, S. (1980) Stable isotope investigations of an active geothermal system in Valles Caldera, Jemez Mountains, New Mexico. *J. Volcanological Geothermal Res.*, **8**, 111–129.

Land, L.S. (1983) The application of stable isotopes to studies of the origin of dolomite and problems of diagenesis of clastic sediments. In *Stable Isotopes in Sedimentary Geology*, pp. 4.1–4.22. *SEPM Short Course*, **10**.

Land, L.S. & Dutton, S.P. (1978) Cementation of a Pennsylvanian deltaic sandstone: isotopic data. *J. Sed. Petrol.*, **48**, 1167–1176.

Laskowski, T.E., Fluegeman, R.H. & Grant, N.K. (1988) Rb–Sr glauconite systematics and the uplift of the Cincinnati Arch. *Geology*, **8**, 368–370.

Latil, M. (1980) *Enhanced Oil Recovery*. Editions Technip, Paris.

Lee, M.C. (1984) *Diagenesis of the Permian Rotliegendes sandstone, North Sea: K/Ar, O^{18}/O^{16} and petrologic evidence*. PhD thesis, Case Western University, Cleveland, OH.

Lee, M.C., Aronson, J.L. & Savin, S.M. (1985) K/Ar dating of time of gas emplacement in Rotliegendes sandstone, Netherlands: *AAPG Bull.*, **69**, 1381–1385.

Lee, M.C. & Savin, S.M. (1985) Isolation of diagenetic overgrowths on quartz sand grains for oxygen isotopic analysis. *Geochim. Cosmochim. Acta*, **49**, 497–501.

Lee, M.R. & Harwood, G.M. (1989) Dolomite calcitization and cement zonation related to uplift of the Raisbury Formation (Zechstein carbonate), northeast England. *Sed. Geol.*, **65**, 285–306.

Lefebvre, R. (1984) A study of pre and post recovery cores from Aberfeldy thermal recovery pilots, Lloydminster area, Saskatchewan. Unpublished MSc thesis, University of Calgary.

Lefebvre, R. & Hutcheon, I. (1986) Mineral reactions in quartzose rocks during thermal recovery of heavy oil, Saskatchewan, Canada. *Appl. Geochem.*, **1**, 395–405.

Leggett, J.K. (1985) Deep-sea pelagic sediments. In Brenchley, P.J. & Williams, B.P.J. (eds), *Sedimentology: Recent Developments and Applied*

Aspects, pp. 95–122. Geological Society Special Publication, 18. Blackwell Scientific Publications, Oxford.

Leroy, J. (1979) Contribution à l'étalonnage de la pression interne des inclusions fluides lors de leur décrépitation. *Bull. Mineral.*, **102**, 584–593.

Liewig, N., Clauer, N. & Sommer, F. (1987) Rb–Sr and K–Ar dating of clay diagenesis in Jurassic sandstone oil reservoir, North Sea. *AAPG Bull.*, **71**, 1467–1474.

Lloyd, R.M. (1968) Oxygen isotope behaviour in the sulfate-water system. *J. Geophys. Res.*, **73**, 6099–6110.

Long, J.V. & Agrell, S.O. (1965) The cathodoluminescence of minerals in thin section. *Min. Mag.*, **34**, 318–326.

Longstaffe, F.J. (1986) Oxygen isotope studies of diagenesis in the Basal Belly River Sandstone, Pembina-I pool, Alberta. *J. Sed. Petrol.*, **56**, 77–88.

Longstaffe, F.J. & Ayalon, A. (1990) Hydrogen-isotope geochemistry of diagenetic clay minerals from Cretaceous sandstones, Alberta, Canada: evidence for exchange. *Appl. Geochem.*, **5**, 657–668.

Lovell, J.P.B. (1990) Cenozoic. In Glennie, K.W. (ed.), *Introduction to the Petroleum Geology of the North Sea*, 3rd edn, pp. 273–293. Blackwell Scientific Publications, Oxford.

Lundegard, P.D., Land, L.S. & Galloway, W.E. (1984) Problems of secondary porosity: Frio Formation (Oligocene), Texas Gulf Coast. *Geology*, **12**, 399–402.

McBride, E.F. (1989) Quartz cement in sandstones: a review. *Earth Sci. Rev.*, **26**, 69–112.

McBride, E.F., Land, L.S. & Mack, L.E. (1987) Diagenesis of eolian and fluvial feldspathic sandstones, Norphlet Formation (Upper Jurassic), Rankin County, Mississippi, and Mobile County, Alabama. *AAPG Bull.*, **71**, 1019–1034.

McCreesh, C.A., Ehrlich, R. & Crabtree, S.J. (1991) Petrography and reservoir physics II: relating thin section porosity to capillary pressure, the association between pore types and throat size. *AAPG Bull.*, **75**, 1563–1578.

MacDonald, A.J. & Spooner, E.T.C. (1981) Calibration of a Linkham TH600 programmable heating-cooling stage for microthermometric examination of fluid inclusions. *Econ. Geol.*, **74**, 1248–1258.

MacGowan, D. (1989) Prediction of permeability from a combination of mercury injection and pore image analysis data. In *Proceedings of Joint IMA/SPE European Conference on the Mathematics of Oil Recovery, 25–27 July, 1989, Cambridge, UK*.

McHardy, W.J. & Birney, A.C. (1987) Scanning electron microscopy. In Wilson, M.J. (ed.), *A Hand-book of Determinative Methods in Clay Mineralogy*, pp. 173–208. Blackie, Glasgow.

McHardy, W.J., Wilson, M.J. & Tait, J.M. (1982) Electron microscope and X-ray diffraction studies of filamentous illitic clay from sandstones of the Magnus Field. *Clay Minerals*, **17**, 23–29.

Mackenzie, A.S., Brassell, S.C., Eglinton, G. & Maxwell, J.R. (1982) Chemical fossils: the geological fate of steroids. *Science*, **217**, 491–504.

Mackenzie, A.S. & McKenzie, D. (1983) Isomerisation and aromatisation of hydrocarbons in sedimentary basins formed by extension. *Geol. Mag.*, **120**, 417–528.

McLimans, R.K. (1987) The application of fluid inclusions to migration of oil and diagenesis in petroleum reservoirs. *Appl. Geochem.*, **2**, 585–603.

McLimans, R.K. & Videtich, P.E. (1987) Reservoir diagenesis and oil migration: Middle Jurassic Great Oolite Limestone, Wealden Basin, Southern England. In Brooks, J. & Glennie, K. (eds), *Petroleum Geology of North West Europe*, pp. 119–128. Graham and Trotman, Dordrecht.

McLimans, R.K. & Videtich, P.E. (1989) Diagenesis and burial history of Great Oolite limestone, southern England. *AAPG Bull.*, **73**, 1195–1205.

Malm, O.A. (1985) Statfjordformasjonen, mineralogi av sandstein og leirstein i 34/10-13 (Mineralogy of sandstones and mudstones in the Statfjord Formation, well 34/10-13). *Report, Statoil*, 28 pp.

Mancini, E.A., Mink, R.M., Bearden, B.L. & Wilkerson, R.P. (1985) Norphlet Formation (Upper Jurassic) of southwestern and offshore Alabama: environments of deposition and petroleum geology. *AAPG Bull.*, **69**, 881–898.

Manum, S.B., Boulter, M.C., Gunnarsdottir, H., Rangnes, K. & Scholze, A. (1989) Eocene to Miocene palynology of the Norwegian Sea (ODP Leg 104). In Eldholm, O., Thiede, J., Taylor, E. *et al.* (eds), *Proc. ODP Sci. Results, 104*, pp. 611–662. Ocean Drilling Program, College Station, TX.

Marshall, D.J. (1988) *Cathodoluminescence of Geological Materials*. Unwin Hyman.

Mason, R.A. (1987) Ion microprobe analysis of trace elements in calcite with an application to the cathodoluminescence zonation of limestone cements from the Carboniferous of South Wales. *Chem. Geol.*, **64**, 209–264.

Mason, R.A. & Mariano, A.N. (1989) Cathodoluminescence and chemistry of synthetic calcites. In Abstract volume. *Theoretical Aspects and Practical Applications of Cathodoluminescence, British Sedimentological Research Group Workshop Meeting, Manchester, April 1989*, 14.

Matsuhisa, Y., Goldsmith, J.R. & Clayton, R.N.

(1979) Oxygen isotopic fractionation in the system quartz–albite–anorthite–water. *Geochim. Cosmochim. Acta*, **43**, 1131–1140.

Matter, A. & Ramseyer, K. (1985) Cathodoluminescence petrography as a tool for provenance studies of sandstones. In Zuffa, G.G. (ed.), *The Provenance of Arenites*, Proc. Cetraro, Cosenza, 1984, NATO ASI Ser C148, pp. 191–211. Reidel, Dordrecht.

Mearns, E.W. (1986) Sm–Nd ages for Norwegian garnet peridotite. *Lithos*, **19**, 269–278.

Mearns, E.W. (1988) A samarium–neodymium survey of modern river sediments from northern Britain. *Chem. Geol. (Isotope Geoscience Section)*, **73**, 1–13.

Mearns, E.W. (1989) Neodymium isotope stratigraphy of Gullfaks oilfield. In Collinson, J.D. (ed.), *Correlation in Hydrocarbon Exploration*, pp. 201–215. Graham and Trotman, Dordrecht.

Mernagh, T.P. & Wilde, A.R. (1989) The use of the laser Raman microprobe for the determination of salinity in fluid inclusions. *Geochim. Cosmochim. Acta*, **53**, 765–771.

Meshri, I. (1990) An overview of chemical models and their relationship to porosity prediction in the subsurface. In Meshri, I. & Ortoleva, P.J. (eds), *Prediction of Reservoir Quality through Chemical Modelling*, pp. 45–54. *AAPG Memoir*, **49**.

Meshri, I & Ortoleva, P.J. (eds) (1990) *Prediction of Reservoir Quality through Chemical Modelling. AAPG Memoir*, **49**.

Meshri, I. & Walker, J.M. (1990) A study of water–rock interaction and simulation of diagenesis in the Upper Almond Sandstones of the Red Desert and Washakie Basins, Wyoming. In Meshri, I. & Ortoleva, P.J. (eds), *Prediction of Reservoir Quality through Chemical Modelling*, pp. 55–70. *AAPG Memoir*, **49**.

Meyers, W.J. (1974) Carbonate cement stratigraphy of the Lake Valley Formation (Mississippian), Sacramento Mts., New Mexico. *J. Sed. Petrol.*, **44**, 837–861.

Meyers, W.J. (1978) Carbonate cements: their regional distribution and interpretation in Mississippian limestones of southwestern New Mexico. *Sedimentology*, **25**, 371–399.

Miller, J. (1988) Cathodoluminescence microscopy. In Tucker, M.E. (ed.), *Techniques in Sedimentology*, pp. 174–190. Blackwell Scientific Publications, Oxford.

Milliken, K.L., Land, L.S. & Loucks, R.G. (1981) History of burial diagenesis determined from isotopic geochemistry, Frio Formation, Brazoria County, Texas. *AAPG Bull.*, **65**, 1397–1413.

Milne, A.A. (1928) *The House at Pooh Corner*. Methuen, London.

Minken, D.F. (1974) The Cold Lake oil sands: geology and a reservoir estimate. In Hills, L.V. (ed.), pp. 84–99. *Canadian Society of Petroleum Geologists, Memoir*, **3**.

Mitchener, B.C., Lawrence, D.A., Partington, M.A., Bowman, M.B.J. & Gluyas, J.G. (1992) Brent Group: sequence stratigraphy and regional implications. In Morton, A.C., Haszeldine, R.S., Giles, M.R. & Brown, S. (eds), *Geology of the Breat Group*, Geological Society Special Publication, 61, pp. 45–80. Blackwell Scientific Publications, Oxford.

Moorbath, S., Taylor, P.N., Orpen, J.L., Treloar, P. & Wilson, J.F. (1987) First direct radiometric dating of Archaean stromatolitic limestone. *Nature*, **326**, 865–867.

Moore, C.H. (1989) *Carbonate Diagenesis and Porosity*, Developments in Sedimentology, 46. Elsevier, Amsterdam.

Morton, J.P. (1983) *Rb–Sr dating of clay diagenesis*. PhD thesis, University of Texas at Austin.

Morton, J.P. & Long, L.E. (1984) Rb–Sr ages of glauconite recrystallization: dating times of regional emergence above sea level. *J. Sed. Petrol.*, **54**, 495–506.

Moser, M.R., Rankin, A.H. & Milledge, H.J. (1990) The application of FTIR microspectroscopy to the characterisation of hydrocarbon-bearing fluid inclusions from the eastern part of the Derbyshire orefield. In *Third Biennial Pan-American Conference on Research on Fluid Inclusions (PACROFI) Abstracts*, p. 61. University of Toronto.

Mossman, J.R., Aplin, A.C., Curtis, C.D. & Coleman, M.L. (1991) Geochemistry of inorganic and organic sulphur in organic-rich sediments from the Peru Margin. *Geochim. Cosmochim. Acta*, **55**, 3581–3595.

Nadeau, P. & Tait, J.M. (1987) Transmission electron microscopy. In Wilson, M.J. (ed.), *A Handbook of Determinative Methods in Clay Mineralogy*, pp. 209–247. Blackie, Glasgow.

Narr, W. & Burruss, R.C. (1984) Origin of reservoir fractures in Little Knife Field, North Dakota. *AAPG Bull.*, **68**, 1087–1100.

Neasham, J.W. (1977) The morphology of dispersed clay in sandstone reservoirs and its effect on sandstone shaliness, pore space and fluid flow properties. In *52nd Annual Fall Technical Conference and Exhibition of SPE and AIME, Denver. Soc. Petrol. Eng.*, **6858**, 8 pp.

Nelson, B.K. & DePaolo, D.J. (1988) Comparison of isotopic and petrographic provenance indicators in sediments from Tertiary sedimentary basins of New Mexico. *J. Sed. Petrol.*, **58**, 348–357.

Odin, G.S. (1982) *Numerical Dating in Stratigraphy*. Wiley, New York.

Odin, G.S. & Matter, A. (1981) De glauconiarum origine. *Sedimentology*, **28**, 611–641.

O'Grady, M.R., Bodnar, R.J., Hellgeth, J.W., Conroy, C.M., Taylor, L.T. & Knight, C.L. (1989) Fourier-transform infrared (FTIR) microspectrometry of individual petroleum fluid inclusions in geological samples. In Russell, P.E. (ed.), *Microbeam Analysis – 1989*, pp. 579–582. San Francisco Press, San Francisco.

Ohmoto, H. & Lasaga, A.C. (1982) Kinetics of reactions between aqueous sulfates and sulfides in hydrothermal systems. *Geochim. Cosmochim. Acta*, **46**, 1727–1745.

Ohmoto, H. & Rye, R.O. (1979) Isotopes of sulfur and carbon. In Barnes, H.C. (ed.), *Geochemistry of Hydrothermal Ore Deposits*, pp. 509–567. John Wiley and Sons, New York.

Ohr, M., Halliday, A.N. & Peacor, D.R. (1991) Sr and Nd isotopic evidence for punctuated clay diagenesis, Texas Gulf Coast. *Earth Planet. Sci. Lett.*, **105**, 110–126.

O'Neil, J.R. (1986) Theoretical and experimental aspects of isotopic fractionation. In Valley, J.W., Taylor, H.P. Jr & O'Neil, J.R. (eds), *Stable Isotopes in High Temperature Processes*, pp. 1–40. Reviews in Mineralogy, 16. Mineral Society of America, Washington, DC.

O'Neil, J.R. (1987) Preservation of H, C and O isotopic ratios in the low temperature environment. In Kyser, T.K. (ed.), *Stable Isotope Geochemistry of Low Temperature Fluids*, pp. 85–128. Min. Assoc. of Canada Short Course, **13**.

O'Neil, J.R. & Kharaka, Y.F. (1976) Hydrogen and oxygen isotope exchange reactions between clay minerals and water. *Geochim. Cosmochim. Acta*, **40**, 241–246.

O'Neil, J.R. & Taylor, H.P. Jr (1967) The oxygen isotope and cation exchange chemistry of feldspars. *Am. Mineral.*, **52**, 1414–1437.

Orr, W.L. (1977) Geologic and geochemical controls on the distribution of hydrogen sulfide in natural gas. In Campos, R. & Goni, J. (eds), *Advances in Organic Geochemistry 1975, Proceedings 7th International Meeting of Organic Geochemistry, Madrid*, pp. 891–899. Empresa Nac. Adaro Invest. Min., Madrid.

Orr, R.D., Johnston, J.R. & Manko, E.M. (1977) Lower Cretaceous geology and heavy oil potential of the Lloydminster area. *Bull. Can. Petrol. Geol.*, **25**, 1187–1221.

Osborne, M. & Haszeldine, R.S. (1993) Fluid inclusions in diagenetic quartz record oilfield burial temperatures, not precipitation temperatures. *Mar. Petrol. Geol.* (in press).

Outtrim, C.P. & Evans, G.P. (1977) Alberta's oil sands reserves and their evaluation. In Redford, D.A. & Winestock, A.G. (eds), *Heavy Oil Symposium, 28th Annual Technological Meeting of the Petroleum Society of the Canadian Institute of Mining and Metallurgy*, pp. 36–66.

Pagel, M., Walgenwitz, F. & Dubessy, J. (1986) Fluid inclusions in oil and gas bearing sedimentary formations. In Burrus, J. (ed.), *Thermal Modelling in Sedimentary Basins*, pp. 565–583. Editions Technip, Paris.

Palmer, D.A. & Drummond, S.E. (1986) Thermal decarboxylation of acetate. Part I: the kinetics and mechanism of reaction in aqueous solution. *Geochim. Cosmochim. Acta*, **50**, 813–823.

Pankhurst, R.J. & O'Nions, R.K. (1973) Determination of Rb/Sr and $^{87}Rb/^{86}Sr$ ratios of some standard rocks and evaluation of X-ray fluorescence spectrometry in Rb–Sr geochemistry. *Chem. Geol.*, **12**, 127–136.

Parkhurst, D.L., Thorstenstenson, D.C. & Plummer, L.N. (1980) PHREEQE – a computer program for geochemical calculations. *US Geol. Surv. Water Resources Investigation*, **80–96**.

Parsley, A.J. (1990) North Sea hydrocarbon plays. In Glennie, K.W. (ed.), *Introduction to the Petroleum Geology of the North Sea*, 3rd edn, pp. 362–388. Blackwell Scientific Publications, Oxford.

Pasteris, J.D., Wopenka, B. & Seitz, J.C. (1988) Practical aspects of quantitative laser Raman microprobe spectroscopy for the study of fluid inclusions. *Geochim. Cosmochim. Acta*, **52**, 979–988.

Pettijohn, F.J. (1975) *Sedimentary Rocks*, 3rd edn. Harper and Row, London.

Pettijohn, F.J., Potter, P.E. & Siever, R. (1973) *Sand and Sandstone*. Springer-Verlag, Heidelberg.

Phakey, P.P., Curtis, C.D. & Oertel, G. (1972) Transmission electron microscopy of fine-grained phyllosilicates in ultra-thin rock sections. *Clays and Clay Minerals*, **20**, 193–197.

Popp, B.N., Anderson, T.F. & Sandberg, P.A. (1986a) Textural, elemental and isotopic variations among constituents in middle Devonian Limestones, North America. *J. Sed. Petrol.*, **56**, 715–727.

Popp, B.N., Podosek, F.A., Brannon, J.C., Anderson, T.F. & Pier, J. (1986b) $^{87}Sr/^{86}Sr$ ratios in Permo-Carboniferous sea water from the analysis of well-preserved brachiopod shells. *Geochim. Cosmochim. Acta*, **50**, 1321–1328.

Posamentier, H.W. & Vail, P.R. (1988) Eustatic controls on clastic deposition II – sequence and systems tract models. In Wilgus, C.K., Hastings, B.S.,

Kendall, C.G. St C., Posamentier, H.W., Ross, C.A. & Van Wagoner, J.C. (eds), *Sea-level Changes: an Integrated Approach*, pp. 125–154. *SEPM Special Publication*, **42**.

Prezbindowski, D.R. & Larese, R.E. (1987) Experimental stretching of fluid inclusions in calcite – implications for diagenetic studies. *Geology*, **15**, 333–336.

Prezbindowski, D.R. & Tapp, J.B. (1991) Dynamics of fluid inclusion alteration in sedimentary rocks: a review and discussion, *Org. Geochem.*, **17**, 131–142.

Primmer, T.J. & Thornley, D. (1991) Thermogravimetric-evolved water analysis of sandstones. In *Am. Clay Miner. Soc. Abstract. 26th Meeting of Clay Min. Soc., Houston, 1991*, p. 150.

Proctor, R.M., Taylor, G.C. & Wade, J.A. (1984) Oil and natural gas resources of Canada. *Geological Survey of Canada Paper*, **83–31**.

Putnam, P.E. & Pedskalny, M.A. (1983) Provenance of Clearwater Formation reservoir sandstones, Cold Lake, Alberta, with comments on feldspar composition. *Bull. Can. Petrol. Geol.*, **31**, 148–160.

Pye, K. & Krinsley, D.H. (1984) Petrographic examination of sedimentary rocks in the SEM using backscattered electron detectors. *J. Sed. Petrol.*, **54**, 877–888.

Quigley, T.M. & Mackenzie, A.S. (1988) The temperatures of oil and gas formation in the sub-surface. *Nature*, **333**, 549–552.

Rainey, S.C.R. (1987) *Sedimentology, diagenesis and geochemistry of the Magnus Sandstone Member, Northern North Sea*. PhD Thesis, University of Edinburgh.

Raiswell, R. (1988) Chemical model for the origin of minor limestone–shale cycles by anaerobic methane oxidation. *Geology*, **16**, 641–644.

Ramseyer, K., Baumann, J., Matter, A. & Mullis, J. (1988) Cathodoluminescence colours in α-quartz. *Min. Mag.*, **52**, 669–677.

Rankin, A.H., Hodge, B.L. & Moser, M. (1990) Unusual oil-bearing inclusions in fluorite from Baluchistan, Pakistan. *Min. Mag.*, **54**, 335–342.

Reeder, R.J. (1986) Zoning types and their origins in sedimentary carbonate minerals. In Rodriguez-Clemente, R. & Tardy, Y. (eds), *Proceedings of the International Meeting on Geochemistry of the Earth Surface and Processes of Mineral Formation, Granada, Spain*, pp. 743–752. CSIC, Madrid.

Reeder, R.J., Fagioli, R.O. & Meyers, W.J. (1990) Oscillatory zoning of Mn in solution-grown calcite crystals. *Earth Sci. Rev.*, **29**, 39–46.

Reeder, R.J. & Grams, J.C. (1987) Sector zoning in calcite cement crystals; implications for trace element distributions in carbonates. *Geochim.*

Cosmochim. Acta, **51**, 187–194.

Reuter, A. & Dallmeyer, R.G. (1987) $^{40}Ar/^{39}Ar$ dating of cleavage formation in tuffs during anchizonal metamorphism. *Contrib. Mineral. Petrol.*, **97**, 352–360.

Robinson, A.G., Coleman, M.L. & Gluyas, J.G. (1992a) The age and cause of illite cement growth, Village Fields area, Southern North Sea: evidence from K–Ar ages and $^{18}O/^{16}O$ ratios. *AAPG Bull.*, **77**, 68–80.

Robinson, A.G. & Gluyas, J.G. (1992a) Duration of quartz cementation in sandstones, North Sea and Haltenbanken basins. *Mar. Petrol. Geol.*, **9**, 324–327.

Robinson, A.G. & Gluyas, J.G. (1992b) Model calculations of loss of porosity in sandstones as a result of compaction and quartz cementation. *Mar. Petrol. Geol.*, **9**, 319–323.

Robinson, A.G., Grant, S.M. & Oxtoby, N.H. (1992b) Geological evidence against natural non-elastic deformation of fluid inclusions in diagenetic quartz cement. *Mar. Petrol. Geol.*, **9**, 568–572.

Roedder, E. (1984) Fluid inclusions. *Min. Soc. America Reviews in Mineralogy*, **12**.

Roedder, E. (1990) Fluid inclusion analysis – prologue and epilogue. *Geochim. Cosmochim. Acta*, **54**, 495–507.

Rudolph, K.W. & Lehmann, P.J. (1989) Platform evolution and sequence stratigraphy of the Natuna Platform, South China Sea. In Crevello, P.D., Wilson, J.L., Sarg, J.F. & Read, J.F. (eds), *Controls on Carbonate Platform and Basin Development*, pp. 353–361. *SEPM Special Publication*, **44**.

Rundberg, Y. & Smalley, P.C. (1989) High resolution dating of Cenozoic sediments from northern North Sea using $^{87}Sr/^{86}Sr$ stratigraphy. *AAPG Bull.*, **73**, 298–308.

Rush, P.F. & Chafetz, H.S. (1990) Fabric-retentive, non-luminescent brachiopods as indicators of original $\delta^{13}C$ and $\delta^{18}O$ compositions: a test. *J. Sed Petrol.*, **60**, 968–981.

Russel, J.D., Birnie, A. & Fraser, A.R. (1984) High gradient magnetic separation (HGMS) in soil clay mineral studies. *Clay Minerals*, **19**, 771–778.

Sassen, R. (1988) Geochemical and carbon isotope studies of crude oil destruction, bitumen precipitation, and sulfate reduction in the deep Smackover Formation. *Org. Geochem.*, **12**, 351–361.

Savin, S.M. & Lee, M.C. (1988) Isotopic studies of phyllosilicates. In Bailey, S.W. (ed.), *Hydrous Phyllosilicates (exclusive of micas)*, pp. 189–223. *Reviews in Mineralogy*, **19**.

Scherer, M. (1987) Parameters influencing porosity in sandstones: a model for sandstone porosity prediction. *AAPG Bull.*, **71**, 485–491.

Schmidt, V. & McDonald, D.A. (1979) Texture and recognition of secondary porosity in sandstones. In Scholle, P.A. & Schluger, P.R. (eds), *Aspects of Diagenesis*, pp. 209–225. *Spec. Publ. Soc. Econ. Paleont. Miner.*, **26**.

Schmidt, V., McDonald, D.A. & Platt, R.L. (1977) Pore geometry and reservoir aspects of secondary porosity in sandstones. *Bull. Can. Petrol. Geol.*, **25**, 271–290.

Schmoker, J.W. & Gautier, D.L. (1988) Sandstone porosity as a function of thermal maturity. *Geology*, **16**, 1007–1010.

Scholle, P.A. (1978) A color illustrated guide to carbonate rock constituents, textures, cements and porosities. *AAPG Mem.*, **27**, 241 pp.

Scholle, P.A. (1979) A color illustrated guide to constituents, textures, cements and porosities of sandstones and associated rocks. *AAPG Mem.*, **28**, 201 pp.

Sellwood, B.W. (ed.) (1989) Zoned carbonate cements: techniques, applications and implications. *Sed. Geol.*, **65**, 205–355.

Sellwood, B.W., Shepherd, T.J., Evans, M.R. & James, B. (1989) Origin of late cements in oolitic reservoir facies: a fluid inclusion and isotopic study (mid-Jurassic, southern England). *Sed. Geol.*, **61**, 223–237.

Shackleton, N.J. & Opdyke, N.D. (1973) Oxygen isotope and paleomagnetic stratigraphy of equatorial Pacific core V28-238: oxygen isotope temperatures and ice volumes on a 105 year and 106 year scale. *Quaternary Res.*, **3**, 39–55.

Shafiqullah, M. & Damon, P.E. (1974) Evaluation of K–Ar isochron methods. *Geochim. Cosmochim. Acta*, **38**, 1341–1358.

Shepherd, D.W. (1981) Steam stimulation recovery of Cold Lake bitumen. In Meyer, R.F. & Steele, C.T. (eds), *The Future of Heavy Crude Oils and Tar Sands*, pp. 349–360. McGraw-Hill, London.

Shepherd, T., Rankin, A.H. & Alderton, D.H.M. (1985) *A Practical Guide to Fluid Inclusion Studies*. Blackie, Glasgow.

Sheppard, S.M.F. (1986) Characterization and isotopic variations in natural waters. In Valley, J.W., Taylor, H.P. Jr & O'Neil, J.R. (eds), *Stable Isotopes in High Temperature Processes*, pp. 165–183. *Min. Soc. America, Reviews in Mineralogy*, **16**.

Sippel, R.F. (1968) Sandstone petrology, evidence from luminescence petrography. *J. Sed. Petrol.*, **38**, 530–554.

Smalley, P.C., Lønøy, A. & Råheim, A. (1992) Spatial $^{87}Sr/^{86}Sr$ variations in formation water and calcite from the Ekofisk Chalk oilfield: implications for reservoir connectivity and fluid composition. *Appl. Geochem.*, **7**, 341–350.

Smalley, P.C., Qvale, G. & Qvale, H. (1989) Some ages from Leg 104 Site 642 obtained by Rb–Sr glauconite dating and Sr isotope stratigraphy. In Eldholm, O., Thiede, E. *et al.* (eds), *Proc. ODP Sci. Results, 104*, pp. 249–253. Ocean Drilling Program, College Station, TX.

Smalley, P.C., Råheim, A., Dickson, J.A.D. & Emery, D. (1988) $^{87}Sr/^{86}Sr$ in waters from the Lincolnshire Limestone aquifer, England, and the potential of natural strontium isotopes as a tracer for a secondary recovery seawater injection process in oilfields. *Appl. Geochem.*, **3**, 591–600.

Smalley, P.C. & Rundberg, Y. (1991) High resolution dating of Cenozoic sediments from northern North Sea using $^{87}Sr/^{86}Sr$ stratigraphy: reply. *AAPG Bull.*, **74**, 1287–1290.

Smith, P.E. & Farquhar, R.M. (1989) Direct dating of Phanerozoic sediments by the $^{283}U-^{206}Pb$ method. *Nature*, **241**, 518–521.

Smith, S.R. (1984) The Lower Cretaceous Sparky Formation, Aberfeldy steamflood pilot, Saskatchewan: a wave-dominated delta? In Stott, D.F. & Glass, D.J. (eds), *The Mesozoic of Middle North America*, pp. 413–429. *Can. Soc. Petrol. Geol. Memoir*, **9**.

Starkey, H.C. Blackman, P.D. & Hauff, P.L. (1984) The routine mineralogical analysis of clay-bearing samples. *USGS Bull.*, **1563**.

Steiger, R.H. & Jager, E. (1977) Subcommission on geochronology: convention on the use of decay constants in geo- and cosmochronology. *Earth Planet. Sci. Lett.*, **36**, 359–362.

Sterner, S.M. & Bodnar, R.J. (1989) Synthetic fluid inclusions − VII: re-equilibration of fluid inclusions in quartz during laboratory-simulated metamorphic burial and uplift. *J. Metamorphic. Geol.*, **7**, 243–260.

Stow, D.A.V. & Miller, J. (1984) Mineralogy, petrology and diagenesis of sediments at Site 530, southeast Angola Basin. In *Initial Reports of the Deep Sea Drilling Project*, Vol. 75, pp. 857–873. US Government Printing Office, Washington, DC.

Stueber, A.M., Pushkar, P. & Hetherington, E.A. (1984) A strontium isotopic study of Smackover brines and associated solids, southern Arkansas. *Geochim. Cosmochim. Acta*, **48**, 1637–1649.

Suchecki, R.K. & Land, L.S. (1983) Isotopic geochemistry of burial-metamorphosed volcanogenic sediments, Great Valley sequence, northern California. *Geochim. Cosmochim. Acta*, **47**, 1487–1499.

Sudo, T., Shimoda, S., Yotsumoto, H. & Aita, S. (1981) *Electron Micrographs of Clay Minerals*. Elsevier, Amsterdam.

Syers, J.K., Chapman, S.L., Jackson, M.L., Rex, R.W. & Clayton, R.N. (1968) Quartz isolation from rocks, sediments and soils for determination of oxygen isotope composition. *Geochim. Cosmochim. Acta*, **32**, 1022–1025.

Tankard, A.J. & Welsink, H.J. (1987) Extensional tectonics and stratigraphy of Hibernia oilfield, Grand Banks, Newfoundland. *AAPG Bull.*, **71**, 1210–1232.

Taylor, H.P. (1974) The application of oxygen and hydrogen isotope studies to problems of hydrothermal alteration and ore deposition. *Econ. Geol.*, **69**, 843–883.

Taylor, S.R. & McClennan, S.M. (1981) The rare earth element evidence in Precambrian sedimentary rocks: implications for crustal evolution. In Kroner, A. (ed.), *Precambrian Plate Tectonics*, pp. 527–548. Elsevier, Amsterdam.

Tellier, K.E., Hlucki, M.M. & Walker, J.R. (1988) Application of high gradient magnetic separation (HGMS) to structural and compositional studies of clay mineral mixtures. *J. Sed. Petrol.*, **58**, 761–763.

Thode, H.G., Munster, J. & Dunford, H.B. (1961) Sulphur isotope geochemistry. *Geochim. Cosmochim. Acta*, **25**, 150–174.

Towe, K.M. (1974) Quantitative clay petrology: the trees but not the forest? *Clays and Clay Minerals*, **22**, 375–378.

Trainor, D.M. & Williams, D.F. (1990) Quantitative analysis and correlation of oxygen isotope records from planktonic and benthic foraminifera and well log records from OCS well G 1267 no. A-1 South Timbalier Block 198, northcentral Gulf of Mexico. In *GCSSEPM Foundation Eleventh Annual Research Conference Program and Abstracts*, pp. 363–377. SEPM, Houston.

Trainor, D.M., Williams, D.F. & Lerche, I. (1988) Refinement and spectral analysis of Plio-Pleistocene oxygen and carbon isotopic records. *Gulf Coast Geol. Soc. Trans.*, **38**, 435–442.

Trewin, N.H. (1988) The SEM in sedimentology. In Tucker, M.E. (ed.), *Techniques in Sedimentology*, pp. 229–273. Blackwell Scientific Publications, Oxford.

Tucker, M.E. (1981) *Sedimentary Petrology, an Introduction*. Blackwell Scientific Publications, Oxford.

Tucker, M.E. (1988) *Techniques in Sedimentology*. Blackwell Scientific Publications, Oxford.

Turner, G. & Cadogan, P.H. (1974) Possible effects of ^{39}Ar recoil in ^{40}Ar–^{39}Ar dating. In *Proceedings of the Fifth Lunar Science Conference, Geochim. Cosmochim. Acta, Supplement 5*, **2**, 1601–1615.

Turner, N.L. & Ping Zhong Hu (1990) The Lower Miocene Liuhua carbonate reservoir, Pearl River Mouth Basin, offshore China (abstract). *AAPG Bull.*, **74**, 781.

Turner, N.L. & Ping Zhong Hu (1991) The Lower Miocene Liuhua carbonate reservoir, PRMB, offshore China. *Offshore Technology Conference Paper*, **6511**, 113–123.

Ulrich, M.R. & Bodnar, R.J. (1988) Systematics of stretching of fluid inclusions II: barite at 1 atm. confining pressure. *Econ. Geol.*, **83**, 1037–1046.

Urey, H.C. (1947) The thermodynamic properties of isotopic substance. *J. Chem. Soc. Lond.*, 562–581.

Vail, P.R., Todd, R.G. & Sangree, J.B. (1977) Chronostratigraphic significance of seismic reflections. In Payton, C.E. (ed.), *Seismic Stratigraphy – Applications to Hydrocarbon Exploration*, pp. 99–116. *AAPG Memoir*, **26**.

Van der Plas, L. & Tobi, A.C. (1965) A chart for judging the reliability of point counting results. *Am. J. Sci.*, **263**, 87–90.

van Hulten, F.F.N. & Smith, S.R. (1984) The Lower Cretaceous Sparky Formation, Lloydminster area: stratigraphy and palaeoenvironment. In Stott, D.F. & Glass, D.J. (eds), *The Mesozoic of Middle North America*, pp. 431–440. *Can. Soc. Petrol. Geol. Memoir*, **9**.

Veizer, J., Fritz, P. & Jones, B. (1986) Geochemistry of brachiopods: oxygen and carbon isotopic records of Paleozoic oceans. *Geochim. Cosmochim. Acta*, **50**, 1679–1696.

Visser, W. (1982) Maximum diagenetic temperature in a petroleum source-rock from Venezuela by fluid inclusion geothermometry. *Chem. Geol.*, **37**, 95–101.

Vry, J., Brown, P.E. & Beauchaine, J. (1987) Application of micro-FTIR spectroscopy to the study of fluid inclusions. *EOS*, **44**, 1538.

Walderhaug, O. (1990) A fluid inclusion study of quartz cemented sandstones from offshore mid-Norway – possible evidence for continued quartz cementation during oil emplacement. *J. Sed. Petrol.*, **60**, 203–210.

Walgenwitz, F., Pagel, M., Meyer, A., Maluski, H. & Monie, P. (1990) Thermo-chronological approach to reservoir diagenesis in the offshore Angola basin: a fluid inclusion, ^{40}Ar–^{39}Ar and K–Ar investigation. *AAPG Bull.*, **74**, 547–563.

Walker, G. (1985) Mineralogical applications of luminescence techniques. In Berry, F.J. & Vaughan, D.J. (eds), *Chemical Bonding and Spectroscopy in Mineral Chemistry*, pp. 103–140. Chapman and Hall, London.

Walker, G., Abumere, O.E. & Kamaluddin, B. (1989) Luminescence spectroscopy of rock-forming carbonates. *Miner. Mag. Spec. Issue*, **April**.

Walmsley, P.J. (1975) The Forties Field. In Woodland, A.W. (ed.), *Petroleum and the Continental Shelf of North-West Europe, Vol. 1: Geology*, pp. 477–485. Elsevier Applied Science.

Warren, E.A. (1987) The application of a solution–mineral equilibrium model to the diagenesis of Carboniferous sandstones, Bothamsall oilfield, East Midlands, England. In Marshall, J. (ed.), *Diagenesis of Sedimentary Sequences*, pp. 55–69. Geological Society of London Special Publication, 36

Warren, E.A. & Curtis, C.D. (1989) The chemical composition of authigenic illite within two sandstones as analysed by ATEM. *Clay Minerals*, **24**, 137–156.

Waugh, B. (1978) Authigenic K-feldspar in British Permo-Triassic sandstones. *J. Geol. Soc. Lond.*, **135**, 51–56.

Weaver, O.D. & VanDamme, A. (1988) Renewed interest seen in Aquitaine Basin. *Oil & Gas J.*, **86**, 55–58.

Weir, A.H., Ormerod, E.C. & El-Mansey, M.I. (1975) Clay mineralogy of sediments of the western Nile Delta. *Clay Minerals*, **10**, 369–386.

Whiticar, M.J. & Faber, E. (1986) Methane oxidation in sediment and water column environments – isotope evidence. *Org. Geochem.*, **10**, 759–768.

Whittle, C.K. (1985) *Analytical transmission microscopy of authigenic chlorites*. PhD Thesis, University of Sheffield.

Wilgus, C.K., Hastings, B.S., Kendall, C.G. St C., Posamentier, H.W., Ross, C.A. & Van Wagoner, J.C. (eds) (1988) *Sea-level Changes: an Integrated Approach. SEPM Special Publication*, **42**.

Wopenka, B. & Pasteris, J.D. (1986) Limitations to quantitative analysis of fluid inclusions in geological samples by laser Raman microprobe spectroscopy. *Appl. Spectroscopy*, **40**, 144–151.

Wopenka, B., Pasteris, J.D. & Freeman, J.J. (1990) Analysis of individual fluid inclusions by Fourier transform infrared and Raman microspectroscopy. *Geochim. Cosmochim. Acta*, **54**, 519–533.

Yeh, H.W. (1980) D/H ratios and late-stage dehydration of shales during burial. *Geochim. Cosmochim. Acta*, **44**, 341–352.

Yeh, H.W. & Savin, S.M. (1977) Mechanism of burial metamorphism of argillaceous sediments: 3. O-isotope evidence. *Geol. Soc. Am. Bull.*, **88**, 1321–1330.

York, D. (1969) Least squares fitting of a straight line with correlated errors. *Earth Planet. Sci. Lett.*, **5**, 320–324.

York, D., Hall, C.M., Yanase, Y., Hanes, J.A. & Kenyon, W.J. (1981) ^{40}Ar–^{39}Ar dating of terrestrial minerals with a continuous laser. *Geophys. Res. Lett.*, **8**, 1136–1138.

Zeitler, P.R. & Fitzgerald, J.D. (1986) Saddle shaped ^{40}Ar–^{39}Ar age spectra from young microstructurally complex potassium feldspars. *Geochim. Cosmochim. Acta*, **50**, 1185–1199.

Zinkernagel, U. (1978) Cathodoluminescence of quartz and its application to sandstone petrography. *Contrib. Sediment*, **8**, 69 pp.

Index